Introduction to Neuro-Fuzzy Systems

Advances in Soft Computing

Editor-in-chief
Prof. Janusz Kacprzyk
Systems Research Institute
Polish Academy of Sciences
ul. Newelska 6
01-447 Warsaw, Poland
E-mail: kacprzyk@ibspan.waw.pl

Esko Turunen
Mathematics Behind Fuzzy Logic
1999. ISBN 3-7908-1221-8

Robert Fullér

Introduction to Neuro-Fuzzy Systems

**With 185 Figures
and 11 Tables**

Springer-Verlag Berlin Heidelberg GmbH

Prof. Robert Fullér
Department of Operations Research
Eötvös Loránd University
Rakoczi ut 5
1088 Budapest
Hungary
Email: rfuller@cs.elte.hu
and
Institute of Advanced Management Systems Research
Åbo Akademi University
Lemminkäisenkatu 14B
2520 Turku
Finland
Email: rfuller@ra.abo.fi

ISBN 978-3-7908-1256-5

Cataloging-in-Publication Data applied for
Die Deutsche Bibliothek – CIP-Einheitsaufnahme
Fullér, Robert: Introduction to neuro-fuzzy systems / Robert Fullér. –
Springer-Verlag Berlin Heidelberg 2000
  (Advances in soft computing)
  ISBN 978-3-7908-1256-5   ISBN 978-3-7908-1852-9 (eBook)
  DOI 10.1007/978-3-7908-1852-9

© Springer-Verlag Berlin Heidelberg 2000
Originally published by Physica-Verlag Heidelberg in 2000

Softcover Design: Erich Kirchner, Heidelberg

SPIN 10747248   88/2202-5 4 3 2 1 0 – Printed on acid-free paper

Preface

Fuzzy sets were introduced by Zadeh (1965) as a means of representing and manipulating data that was not precise, but rather fuzzy. Fuzzy logic provides an inference morphology that enables approximate human reasoning capabilities to be applied to knowledge-based systems. The theory of fuzzy logic provides a mathematical strength to capture the uncertainties associated with human cognitive processes, such as thinking and reasoning. The conventional approaches to knowledge representation lack the means for representating the meaning of fuzzy concepts. As a consequence, the approaches based on first order logic and classical probablity theory do not provide an appropriate conceptual framework for dealing with the representation of commonsense knowledge, since such knowledge is by its nature both lexically imprecise and noncategorical.

The developement of fuzzy logic was motivated in large measure by the need for a conceptual framework which can address the issue of uncertainty and lexical imprecision.

Some of the essential characteristics of fuzzy logic relate to the following [242].

- In fuzzy logic, exact reasoning is viewed as a limiting case of approximate reasoning.
- In fuzzy logic, everything is a matter of degree.
- In fuzzy logic, knowledge is interpreted a collection of elastic or, equivalently, fuzzy constraint on a collection of variables.
- Inference is viewed as a process of propagation of elastic constraints.
- Any logical system can be fuzzified.

There are two main characteristics of fuzzy systems that give them better performance for specific applications.

- Fuzzy systems are suitable for uncertain or approximate reasoning, especially for the system with a mathematical model that is difficult to derive.
- Fuzzy logic allows decision making with estimated values under incomplete or uncertain information.

Artificial neural systems can be considered as simplified mathematical models of brain-like systems and they function as parallel distributed computing networks. However, in contrast to conventional computers, which are programmed to perform specific task, most neural networks must be taught, or trained. They can learn new associations, new functional dependencies and new patterns.

The study of brain-style computation has its roots over 50 years ago in the work of McCulloch and Pitts (1943) and slightly later in Hebb's famous *Organization of Behavior* (1949). The early work in artificial intelligence was torn between those who believed that intelligent systems could best be built on computers modeled after brains, and those like Minsky and Papert who believed that intelligence was fundamentally symbol processing of the kind readily modeled on the *von Neumann computer*. For a variety of reasons, the symbol-processing approach became the dominant theme in artifcial intelligence. The 1980s showed a rebirth in interest in neural computing: Hopfield (1985) provided the mathematical foundation for understanding the dynamics of an important class of networks; Rumelhart and McClelland (1986) introduced the backpropagation learning algorithm for complex, multi-layer networks and thereby provided an answer to one of the most severe criticisms of the original perceptron work.

Perhaps the most important advantage of neural networks is their adaptivity. Neural networks can automatically adjust their weights to optimize their behavior as pattern recognizers, decision makers, system controllers, predictors, etc. Adaptivity allows the neural network to perform well even when the environment or the system being controlled varies over time. There are many control problems that can benefit from continual nonlinear modeling and adaptation.

While fuzzy logic performs an inference mechanism under cognitive uncertainty, computational neural networks offer exciting advantages, such as learning, adaptation, fault-tolerance, parallelism and generalization. A brief comparative study between fuzzy systems and neural networks in their operations in the context of knowledge acquisition, uncertainty, reasoning and adaptation is presented in the following table [93]:

To enable a system to deal with cognitive uncertainties in a manner more like humans, one may incorporate the concept of fuzzy logic into the neural networks. The resulting *hybrid system* is called fuzzy neural, neural fuzzy, neuro-fuzzy or fuzzy-neuro network.

Neural networks are used to *tune* membership functions of fuzzy systems that are employed as decision-making systems for controlling equipment. Although fuzzy logic can encode expert knowledge directly using rules with linguistic labels, it usually takes a lot of time to design and tune the membership functions which quantitatively define these linquistic labels. Neural network learning techniques can automate this process and substantially reduce development time and cost while improving performance.

Skills		Fuzzy Systems	Neural Nets
Knowledge	Inputs	Human experts	Sample sets
acquisition	Tools	Interaction	Algorithms
Uncertainty	Information	Quantitive and	Quantitive
		Qualitive	
	Cognition	Decision making	Perception
Reasoning	Mechanism	Heuristic search	Parallel computat.
	Speed	Low	High
Adaption	Fault-tolerance	Low	Very high
	Learning	Induction	Adjusting weights
Natural	Implementation	Explicit	Implicit
language	Flexibility	High	Low

Table 0.1. Properties of fuzzy systems and neural networks.

In theory, neural networks, and fuzzy systems are equivalent in that they are convertible, yet in practice each has its own advantages and disadvantages. For neural networks, the knowledge is automatically acquired by the backpropagation algorithm, but the learning process is relatively slow and analysis of the trained network is difficult (black box). Neither is it possible to extract structural knowledge (rules) from the trained neural network, nor can we integrate special information about the problem into the neural network in order to simplify the learning procedure.

Fuzzy systems are more favorable in that their behavior can be explained based on fuzzy rules and thus their performance can be adjusted by tuning the rules. But since, in general, knowledge acquisition is difficult and also the universe of discourse of each input variable needs to be divided into several intervals, applications of fuzzy systems are restricted to the fields where expert knowledge is available and the number of input variables is small.

To overcome the problem of knowledge acquisition, neural networks are extended to automatically *extract fuzzy rules from numerical data.*

Cooperative approaches use neural networks to optimize certain parameters of an ordinary fuzzy system, or to preprocess data and extract fuzzy (control) rules from data.

The basic processing elements of neural networks are called *artificial neurons,* or simply *neurons.* The signal flow from of neuron inputs, x_j, is considered to be unidirectional as indicated by arrows, as is a neuron's output signal flow. Consider a simple neural net in Figure 0.1. All signals and weights are real numbers. The input neurons do not change the input signals so their output is the same as their input. The signal x_i interacts with the weight w_i to produce the product $p_i = w_i x_i$, $i = 1, \ldots, n$. The input information p_i is aggregated, by addition, to produce the input

$$\text{net} = p_1 + \cdots + p_n = w_1 x_1 + \cdots + w_n x_n$$

to the neuron. The neuron uses its transfer function f, which could be a sigmoidal function,

$$f(t) = \frac{1}{1 + e^{-t}}$$

to compute the output

$$y = f(\text{net}) = f(w_1 x_1 + \cdots + w_n x_n).$$

This simple neural net, which employs multiplication, addition, and sigmoidal f, will be called as regular (or standard) neural net.

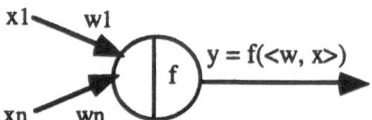

Fig. 0.1. A simple neural net.

If we employ other operations like a t-norm, or a t-conorm, to combine the incoming data to a neuron we obtain what we call a *hybrid neural net*. These modifications lead to a fuzzy neural architecture based on fuzzy arithmetic operations. A hybrid neural net may not use multiplication, addition, or a sigmoidal function (because the results of these operations are not necesserily are in the unit interval).

A *hybrid neural net* is a neural net with crisp signals and weights and crisp transfer function. However, (i) we can combine x_i and w_i using a t-norm, t-conorm, or some other continuous operation; (ii) we can aggregate the p_i's with a t-norm, t-conorm, or any other continuous function; (iii) f can be any continuous function from input to output.

We emphasize here that all inputs, outputs and the weights of a hybrid neural net are real numbers taken from the unit interval $[0, 1]$. A processing element of a hybrid neural net is called *fuzzy neuron*.

It is well-known that regular nets are universal approximators, i.e. they can approximate any continuous function on a compact set to arbitrary accuracy. In a discrete fuzzy expert system one inputs a discrete approximation to the fuzzy sets and obtains a discrete approximation to the output fuzzy set. Usually discrete fuzzy expert systems and fuzzy controllers are continuous mappings. Thus we can conclude that given a continuous fuzzy expert system, or continuous fuzzy controller, there is a regular net that can uniformly approximate it to any degree of accuracy on compact sets. The problem with this result that it is non-constructive and does not tell you how to build the net.

Hybrid neural nets can be used to implement fuzzy IF-THEN rules in a constructive way. Though hybrid neural nets can not use directly the standard error backpropagation algorithm for learning, they can be trained by steepest descent methods to learn the parameters of the membership functions representing the linguistic terms in the rules.

The direct fuzzification of conventional neural networks is to extend connection weigths and/or inputs and/or fuzzy desired outputs (or targets) to fuzzy numbers. This extension is summarized in the next table.

Fuzzy neural net	Weights	Inputs	Targets
Type 1	crisp	fuzzy	crisp
Type 2	crisp	fuzzy	fuzzy
Type 3	fuzzy	fuzzy	fuzzy
Type 4	fuzzy	crisp	fuzzy
Type 5	crisp	crisp	fuzzy
Type 6	fuzzy	crisp	crisp
Type 7	fuzzy	fuzzy	crisp

Table 0.2. Direct fuzzification of neural networks.

Fuzzy neural networks of *Type 1* are used in classification problem of a fuzzy input vector to a crisp class. The networks of *Type 2, 3* and *4* are used to implement fuzzy IF-THEN rules. However, the last three types in Table 0.2 are unrealistic.

- In *Type 5*, outputs are always real numbers because both inputs and weights are real numbers.
- In *Type 6* and *7*, the fuzzification of weights is not necessary because targets are real numbers.

A regular fuzzy neural network is a neural network with fuzzy signals and/or fuzzy weights, sigmoidal transfer function and all the operations are defined by Zadeh's extension principle. Consider a simple regular fuzzy neural net in Figure 0.2.

All signals and weights are fuzzy numbers. The input neurons do not change the input signals so their output is the same as their input. The signal X_i interacts with the weight W_i to produce the product $P_i = W_i X_i$, $i = 1, \ldots, n$, where we use the extension principle to compute P_i. The input information P_i is aggregated, by standard extended addition, to produce the input

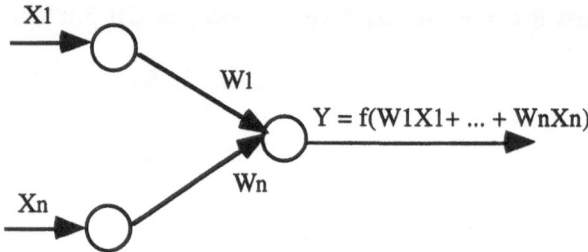

Fig. 0.2. Simple regular fuzzy neural net.

$$net = P_1 + \cdots + P_n = W_1 X_1 + \cdots + W_n X_n$$

to the neuron. The neuron uses its transfer function f, which is a sigmoidal function, to compute the output

$$Y = f(net) = f(W_1 X_1 + \cdots + W_n X_n)$$

where f is a sigmoidal function and the membership function of the output fuzzy set Y is computed by the extension principle.

The main disadvantage of regular fuzzy neural network that they are not universal approximators. Therefore we must abandon the extension principle if we are to obtain a universal approximator.

A hybrid fuzzy neural network is a neural network with fuzzy signals and/or fuzzy weights. However, (i) we can combine X_i and W_i using a t-norm, t-conorm, or some other continuous operation; (ii) we can aggregate the P_i's with a t-norm, t-conorm, or any other continuous function; (iii) f can be *any function* from input to output.

Buckley and Hayashi [31] showed that *hybrid fuzzy neural networks* are universal approximators, i.e. they can approximate any continuous fuzzy functions on a compact domain.

This Lecture Notes is organized in four Chapters. The First Chapter is dealing with inference mechanisms in fuzzy expert systems. The Second Chapter provides a brief description of learning rules of feedforward multi-layer supervised neural networks, and Kohonen's unsupervised learning algorithm for classification of input patterns. In the Third Chapter we explain the basic principles of fuzzy neural hybrid systems. In the Fourth Chapter we present a case study and some excercises for the Reader.

This book is a textbook on neuro-fuzzy systems used for a course taught by the author at Åbo Akademi University between 1995 and 1999.

Contents

1. Fuzzy systems

1.1 An introduction to fuzzy logic

Fuzzy sets were introduced by Zadeh [235] as a means of representing and manipulating data that was not precise, but rather fuzzy.

It was specifically designed to mathematically represent uncertainty and vagueness and to provide formalized tools for dealing with the imprecision intrinsic to many problems. However, the story of fuzzy logic started much more earlier (see James F. Brule's tutorial, [3] for details) ...

To devise a concise theory of logic, and later mathematics, *Aristotle* posited the so-called "Laws of Thought" (see [162]). One of these, the "Law of the Excluded Middle," states that every proposition must either be *True* (**T**) or *False* (**F**). Even when *Parminedes* proposed the first version of this law (around 400 Before Christ) there were strong and immediate objections: for example, *Heraclitus* proposed that things could be simultaneously *True* and *not True*. It was *Plato* who laid the foundation for what would become fuzzy logic, indicating that there was a third region (beyond **T** and **F**) where these opposites "tumbled about."

A systematic alternative to the bi-valued logic of Aristotle was first proposed by *Lukasiewicz* (see [180]) around 1920, when he described a three-valued logic, along with the mathematics to accompany it. The third value he proposed can best be translated as the term "possible," and he assigned it a numeric value between **T** and **F**. Eventually, he proposed an entire notation and axiomatic system from which he hoped to derive modern mathematics.

Later, he explored four-valued logics, five-valued logics, and then declared that in principle there was nothing to prevent the derivation of an infinite-valued logic.Lukasiewicz felt that three- and infinite-valued logics were the most intriguing, but he ultimately settled on a four-valued logic because it seemed to be the most easily adaptable to Aristotelian logic.

It should be noted that *Knuth* also proposed a three-valued logic similar to Lukasiewicz's, from which he speculated that mathematics would become even more elegant than in traditional bi-valued logic.

The notion of an infinite-valued logic was introduced in Zadeh's seminal work "Fuzzy Sets" [235] where he described the mathematics of fuzzy set theory, and by extension fuzzy logic. This theory proposed making the membership function (or the values **F** and **T**) operate over the range of real numbers $[0, 1]$. New operations for the calculus of logic were proposed, and showed to be in principle at least a generalization of classic logic.

There is a strong relationship between *Boolean* logic and the concept of a subset, there is a similar strong relationship between fuzzy logic and fuzzy subset theory.

In classical set theory, a subset A of a set X can be defined by its characteristic function χ_A as a mapping from the elements of X to the elements of the set $\{0, 1\}$,

$$\chi_A : X \rightarrow \{0, 1\}.$$

This mapping may be represented as a set of ordered pairs, with exactly one ordered pair present for each element of X. The first element of the ordered pair is an element of the set X, and the second element is an element of the set $\{0, 1\}$. The value zero is used to represent non-membership, and the value one is used to represent membership. The truth or falsity of the statement

$$"x \text{ is in } A"$$

is determined by the ordered pair $(x, \chi_A(x))$. The statement is true if the second element of the ordered pair is 1, and the statement is false if it is 0.

Similarly, a *fuzzy subset* A of a set X can be defined as a set of ordered pairs, each with the first element from X, and the second element from the interval $[0, 1]$, with exactly one ordered pair present for each element of X.

This defines a mapping, μ_A, between elements of the set X and values in the interval $[0, 1]$. The value zero is used to represent complete non-membership, the value one is used to represent complete membership, and values in between are used to represent intermediate degrees of membership. The set X is referred to as the universe of discourse for the fuzzy subset A. Frequently, the mapping μ_A is described as a function, the membership function of A. The degree to which the statement

$$"x \text{ is in } A"$$

is true is determined by finding the ordered pair $(x, \mu_A(x))$. The degree of truth of the statement is the second element of the ordered pair. It should be noted that the terms *membership function* and *fuzzy subset* get used interchangeably.

Definition 1.1.1 *[235] Let X be a nonempty set. A fuzzy set A in X is characterized by its membership function*

$$\mu_A : X \rightarrow [0, 1]$$

and $\mu_A(x)$ is interpreted as the degree of membership of element x in fuzzy set A for each $x \in X$.

It is clear that A is completely determined by the set of tuples

$$A = \{(x, \mu_A(x)) | x \in X\}$$

Frequently we will write simply $A(x)$ instead of $\mu_A(x)$. The family of all fuzzy (sub)sets in X is denoted by $\mathcal{F}(X)$. Fuzzy subsets of the real line are called *fuzzy quantities*.

If $X = \{x_1, \dots, x_n\}$ is a finite set and A is a fuzzy set in X then we often use the notation

$$A = \mu_1/x_1 + \cdots + \mu_n/x_n$$

where the term μ_i/x_i, $i = 1, \dots, n$ signifies that μ_i is the grade of membership of x_i in A and the plus sign represents the union.

Example 1.1.1 *Suppose we want to define the set of natural numbers "close to 1". This can be expressed by*

$$A = 0.0/-2 + 0.3/-1 + 0.6/0 + 1.0/1 + 0.6/2 + 0.3/3 + 0.0/4.$$

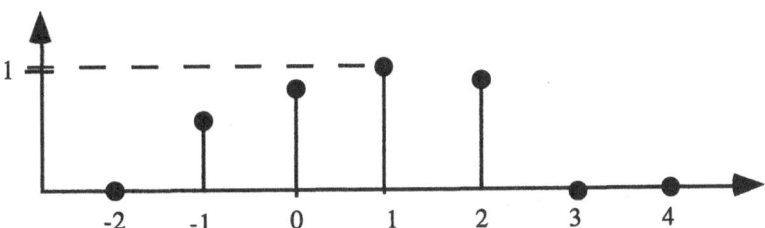

Fig. 1.1. A discrete membership function for "x is close to 1".

Example 1.1.2 *The membership function of the fuzzy set of real numbers "close to 1", is can be defined as (see Fig.1.2)*

$$A(t) = \exp(-\beta(t-1)^2)$$

where β is a positive real number.

Example 1.1.3 *Assume someone wants to buy a cheap car. Cheap can be represented as a fuzzy set on a universe of prices, and depends on his purse. For instance, from Fig. 1.3. cheap is roughly interpreted as follows:*

- *Below 3000\$ cars are considered as cheap, and prices make no real difference to buyer's eyes.*
- *Between 3000\$ and 4500\$, a variation in the price induces a weak preference in favor of the cheapest car.*

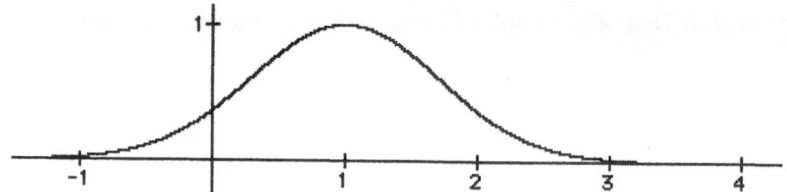

Fig. 1.2. A membership function for "x is close to 1".

- *Between 4500$ and 6000$, a small variation in the price induces a clear preference in favor of the cheapest car.*
- *Beyond 6000$ the costs are too high (out of consideration).*

Definition 1.1.2 *(support) Let A be a fuzzy subset of X; the support of A, denoted* supp(A), *is the crisp subset of X whose elements all have nonzero membership grades in A.*

$$\text{supp}(A) = \{x \in X | A(x) > 0\}.$$

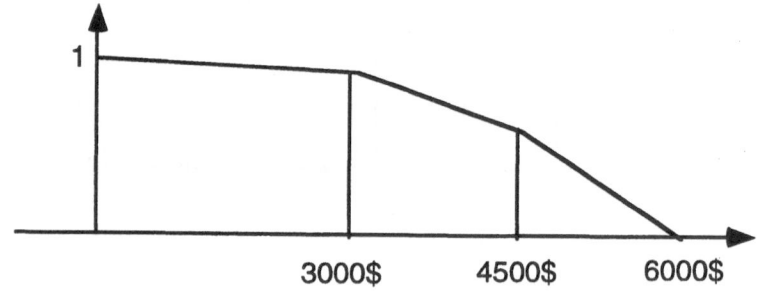

Fig. 1.3. Membership function of "cheap".

Definition 1.1.3 *(normal fuzzy set) A fuzzy subset A of a classical set X is called* normal *if there exists an $x \in X$ such that $A(x) = 1$. Otherwise A is subnormal.*

Definition 1.1.4 *(α-cut) An α-level set of a fuzzy set A of X is a non-fuzzy set denoted by $[A]^\alpha$ and is defined by*

$$[A]^\alpha = \begin{cases} \{t \in X | A(t) \geq \alpha\} & \text{if } \alpha > 0 \\ \text{cl}(\text{supp}A) & \text{if } \alpha = 0 \end{cases}$$

where cl(supp$A$) *denotes the closure of the support of A.*

Example 1.1.4 *Assume* $X = \{-2, -1, 0, 1, 2, 3, 4\}$ *and*

$$A = 0.0/-2 + 0.3/-1 + 0.6/0 + 1.0/1 + 0.6/2 + 0.3/3 + 0.0/4,$$

in this case

$$[A]^\alpha = \begin{cases} \{-1, 0, 1, 2, 3\} & \text{if } 0 \le \alpha \le 0.3 \\ \{0, 1, 2\} & \text{if } 0.3 < \alpha \le 0.6 \\ \{1\} & \text{if } 0.6 < \alpha \le 1 \end{cases}$$

Definition 1.1.5 *(convex fuzzy set) A fuzzy set A of X is called* convex *if $[A]^\alpha$ is a convex subset of X, $\forall \alpha \in [0, 1]$.*

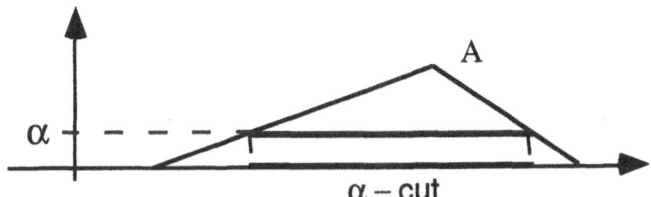

Fig. 1.4. An α-cut of a triangular fuzzy number.

In many situations people are only able to characterize numeric information imprecisely. For example, people use terms such as, about 5000, near zero, or essentially bigger than 5000. These are examples of what are called *fuzzy numbers*. Using the theory of fuzzy subsets we can represent these fuzzy numbers as fuzzy subsets of the set of real numbers. More exactly,

Definition 1.1.6 *(fuzzy number) A fuzzy number A is a fuzzy set of the real line with a normal, (fuzzy) convex and continuous membership function of bounded support. The family of fuzzy numbers will be denoted by $\mathcal{F}$.*

Definition 1.1.7 *(quasi fuzzy number) A quasi fuzzy number A is a fuzzy set of the real line with a normal, fuzzy convex and continuous membership function satisfying the limit conditions*

$$\lim_{t \to \infty} A(t) = 0, \quad \lim_{t \to -\infty} A(t) = 0.$$

Let A be a fuzzy number. Then $[A]^\gamma$ is a closed convex (compact) subset of $I\!R$ for all $\gamma \in [0, 1]$. Let us introduce the notations

$$a_1(\gamma) = \min[A]^\gamma, \quad a_2(\gamma) = \max[A]^\gamma$$

In other words, $a_1(\gamma)$ denotes the left-hand side and $a_2(\gamma)$ denotes the right-hand side of the γ-cut. It is easy to see that

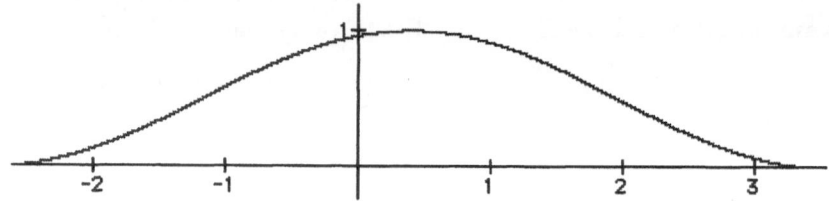

Fig. 1.5. A fuzzy number.

If $\alpha \le \beta$ then $[A]^{\alpha} \supset [A]^{\beta}$

Furthermore, the left-hand side function

$$a_1 : [0,1] \to I\!R$$

is monoton increasing and lower semicontinuous, and the right-hand side function

$$a_2 : [0,1] \to I\!R$$

is monoton decreasing and upper semicontinuous. We shall use the notation

$$[A]^{\gamma} = [a_1(\gamma), a_2(\gamma)].$$

The support of A is the open interval $(a_1(0), a_2(0))$.

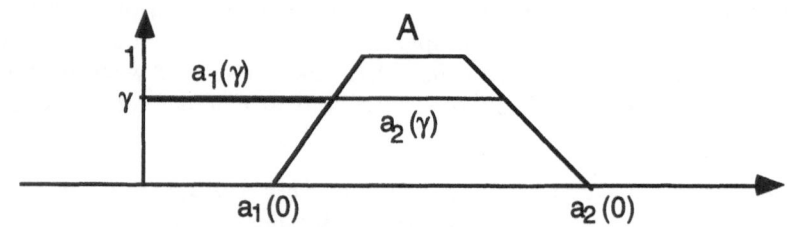

Fig. 1.6. The support of A is $(a_1(0), a_2(0))$.

If A is not a fuzzy number then there exists an $\gamma \in [0,1]$ such that $[A]^{\gamma}$ is not a convex subset of $I\!R$.

Definition 1.1.8 *(triangular fuzzy number) A fuzzy set A is called triangular fuzzy number with peak (or center) a, left width $\alpha > 0$ and right width $\beta > 0$ if its membership function has the following form*

$$A(t) = \begin{cases} 1 - \dfrac{a-t}{\alpha} & \text{if } a - \alpha \le t \le a \\[2mm] 1 - \dfrac{t-a}{\beta} & \text{if } a \le t \le a + \beta \\[2mm] 0 & \text{otherwise} \end{cases}$$

and we use the notation $A = (a, \alpha, \beta)$. It can easily be verified that

$$[A]^\gamma = [a - (1 - \gamma)\alpha, a + (1 - \gamma)\beta], \ \forall \gamma \in [0, 1].$$

The support of A is $(a - \alpha, b + \beta)$.

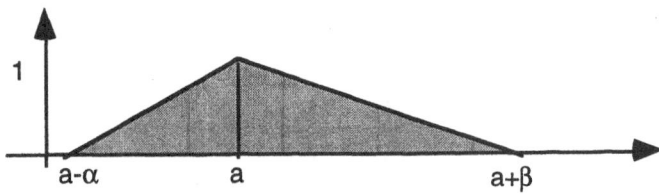

Fig. 1.7. Triangular fuzzy number.

A triangular fuzzy number with center a may be seen as a fuzzy quantity

"x is approximately equal to a".

Definition 1.1.9 *(trapezoidal fuzzy number) A fuzzy set A is called trapezoidal fuzzy number with tolerance interval $[a, b]$, left width α and right width β if its membership function has the following form*

$$A(t) = \begin{cases} 1 - \dfrac{a - t}{\alpha} & \text{if } a - \alpha \leq t \leq a \\ 1 & \text{if } a \leq t \leq b \\ 1 - \dfrac{t - b}{\beta} & \text{if } a \leq t \leq b + \beta \\ 0 & \text{otherwise} \end{cases}$$

and we use the notation $A = (a, b, \alpha, \beta)$. It can easily be shown that

$$[A]^\gamma = [a - (1 - \gamma)\alpha, b + (1 - \gamma)\beta], \ \forall \gamma \in [0, 1].$$

The support of A is $(a - \alpha, b + \beta)$.

A trapezoidal fuzzy number may be seen as a fuzzy quantity

"x is approximately in the interval $[a, b]$".

Definition 1.1.10 *(LR-representation of fuzzy numbers) Any fuzzy number $A \in \mathcal{F}$ can be described as*

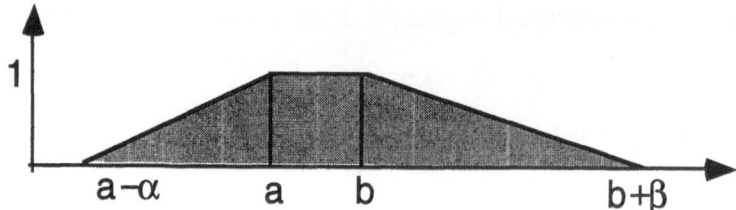

Fig. 1.8. Trapezoidal fuzzy number.

$$
A(t) = \begin{cases} L\left(\dfrac{a-t}{\alpha}\right) & \text{if } t \in [a-\alpha, a] \\ 1 & \text{if } t \in [a, b] \\ R\left(\dfrac{t-b}{\beta}\right) & \text{if } t \in [b, b+\beta] \\ 0 & \text{otherwise} \end{cases}
$$

where $[a, b]$ is the peak *or* core *of A,*

$$
L\colon [0,1] \to [0,1], \quad R\colon [0,1] \to [0,1]
$$

are continuous and non-increasing shape functions with $L(0) = R(0) = 1$ and $R(1) = L(1) = 0$. We call this fuzzy interval of LR-type and refer to it by

$$
A = (a, b, \alpha, \beta)_{LR}
$$

The support of A is $(a - \alpha, b + \beta)$.

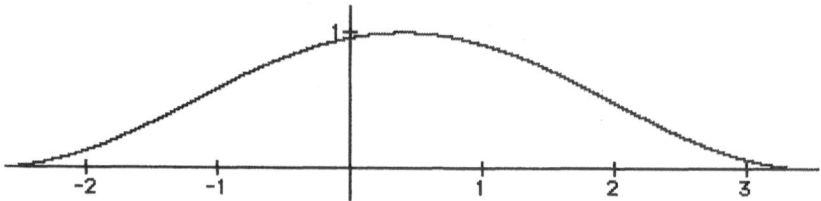

Fig. 1.9. Fuzzy number of type LR with nonlinear reference functions.

Definition 1.1.11 *(quasi fuzzy number of type LR) Any quasi fuzzy number $A \in \mathcal{F}(\mathbb{R})$ can be described as*

$$A(t) = \begin{cases} L\left(\dfrac{a-t}{\alpha}\right) & \text{if } t \leq a, \\ 1 & \text{if } t \in [a,b], \\ R\left(\dfrac{t-b}{\beta}\right) & \text{if } t \geq b, \end{cases}$$

where $[a,b]$ is the peak *or* core *of A,*

$$L\colon [0,\infty) \to [0,1], \quad R\colon [0,\infty) \to [0,1]$$

are continuous and non-increasing shape functions with $L(0) = R(0) = 1$ and

$$\lim_{t\to\infty} L(t) = 0, \quad \lim_{t\to\infty} R(t) = 0.$$

Let $A = (a,b,\alpha,\beta)_{LR}$ be a fuzzy number of type LR. If $a = b$ then we use the notation

$$A = (a,\alpha,\beta)_{LR}$$

and say that A is a quasi-triangular fuzzy number. Furthermore if $L(x) = R(x) = 1 - x$ then instead of $A = (a,b,\alpha,\beta)_{LR}$ we simply write

$$A = (a,b,\alpha,\beta).$$

Definition 1.1.12 *(subsethood) Let A and B are fuzzy subsets of a classical set X. We say that A is a subset of B if $A(t) \leq B(t)$, $\forall t \in X$.*

Definition 1.1.13 *(equality of fuzzy sets) Let A and B are fuzzy subsets of a classical set X. A and B are said to be equal, denoted $A = B$, if $A \subset B$ and $B \subset A$. We note that $A = B$ if and only if $A(x) = B(x)$ for $x \in X$.*

Example 1.1.5 *Let A and B be fuzzy subsets of $X = \{-2,-1,0,1,2,3,4\}$.*

$$A = 0.0/-2 + 0.3/-1 + 0.6/0 + 1.0/1 + 0.6/2 + 0.3/3 + 0.0/4$$

$$B = 0.1/-2 + 0.3/-1 + 0.9/0 + 1.0/1 + 1.0/2 + 0.3/3 + 0.2/4$$

It is easy to check that $A \subset B$ holds.

Definition 1.1.14 *(empty fuzzy set) The empty fuzzy subset of X is defined as the fuzzy subset $\emptyset$ of X such that $\emptyset(x) = 0$ for each $x \in X$.*

It is easy to see that $\emptyset \subset A$ holds for any fuzzy subset A of X.

Definition 1.1.15 *The largest fuzzy set in X, called universal fuzzy set in X, denoted by 1_X, is defined by $1_X(t) = 1$, $\forall t \in X$.*

It is easy to see that $A \subset 1_X$ holds for any fuzzy subset A of X.

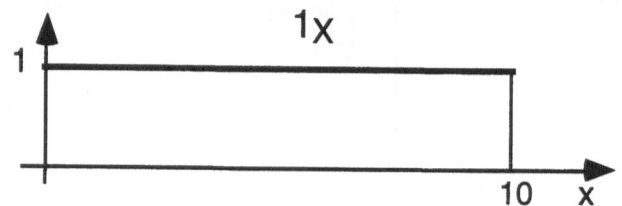

Fig. 1.10. The graph of the universal fuzzy subset in $X = [0, 10]$.

Definition 1.1.16 *(Fuzzy point) Let A be a fuzzy number. If $supp(A) = \{x_0\}$ then A is called a fuzzy point and we use the notation $A = \bar{x}_0$.*

Let $A = \bar{x}_0$ be a fuzzy point. It is easy to see that $[A]^\gamma = [x_0, x_0] = \{x_0\}$, $\forall \gamma \in [0, 1]$.

Exercise 1.1.1 *Let $X = [0, 2]$ be the universe of discourse of fuzzy number A defined by the membership function $A(t) = 1 - t$ if $t \in [0, 1]$ and $A(t) = 0$, otherwise. Interpret A linguistically.*

Exercise 1.1.2 *Let $A = (a, b, \alpha, \beta)_{LR}$ and $A' = (a', b', \alpha', \beta')_{LR}$ be fuzzy numbers of type LR. Give necessary and sufficient conditions for the subsethood of A in A'.*

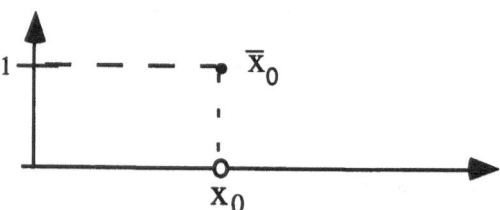

Fig. 1.11. Fuzzy point.

Exercise 1.1.3 *Let $A = (a, \alpha)$ be a symmetrical triangular fuzzy number. Calculate $[A]^\gamma$ as a function of a and α.*

Exercise 1.1.4 *Let $A = (a, \alpha, \beta)$ be a triangular fuzzy number. Calculate $[A]^\gamma$ as a function of a, α and β.*

Exercise 1.1.5 *Let $A = (a, b, \alpha, \beta)_{LR}$ be a fuzzy number of LR-type. Calculate $[A]^\gamma$ as a function of a, b, α, β, L and R.*

1.2 Operations on fuzzy sets

In this section we extend the classical set theoretic operations from ordinary set theory to fuzzy sets. We note that all those operations which are extensions of crisp concepts reduce to their usual meaning when the fuzzy subsets have membership degrees that are drawn from $\{0, 1\}$. For this reason, when extending operations to fuzzy sets we use the same symbol as in set theory.

Let A and B are fuzzy subsets of a nonempty (crisp) set X.

Definition 1.2.1 *(intersection) The intersection of A and B is defined as*

$$(A \cap B)(t) = \min\{A(t), B(t)\} = A(t) \wedge B(t), \ \forall t \in X$$

Example 1.2.1 *Let A and B be fuzzy subsets of $X = \{-2, -1, 0, 1, 2, 3, 4\}$.*

$$A = 0.6/-2 + 0.3/-1 + 0.6/0 + 1.0/1 + 0.6/2 + 0.3/3 + 0.4/4$$

$$B = 0.1/-2 + 0.3/-1 + 0.9/0 + 1.0/1 + 1.0/2 + 0.3/3 + 0.2/4$$

Then $A \cap B$ has the following form

$$A \cup B = 0.1/-2 + 0.3/-1 + 0.6/0 + 1.0/1 + 0.6/2 + 0.3/3 + 0.2/4.$$

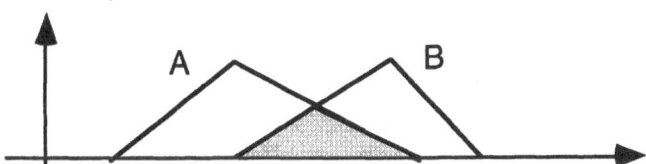

Fig. 1.12. Intersection of two triangular fuzzy numbers.

Definition 1.2.2 *(union) The union of A and B is defined as*

$$(A \cup B)(t) = \max\{A(t), B(t)\} = A(t) \vee B(t), \ \forall t \in X$$

Example 1.2.2 *Let A and B be fuzzy subsets of $X = \{-2, -1, 0, 1, 2, 3, 4\}$.*

$$A = 0.6/-2 + 0.3/-1 + 0.6/0 + 1.0/1 + 0.6/2 + 0.3/3 + 0.4/4$$

$$B = 0.1/-2 + 0.3/-1 + 0.9/0 + 1.0/1 + 1.0/2 + 0.3/3 + 0.2/4$$

Then $A \cup B$ has the following form

$$A \cup B = 0.6/-2 + 0.3/-1 + 0.9/0 + 1.0/1 + 1.0/2 + 0.3/3 + 0.4/4.$$

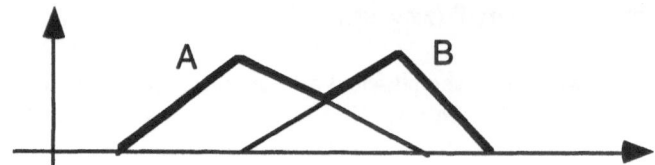

Fig. 1.13. Union of two triangular fuzzy numbers.

Definition 1.2.3 *(complement) The complement of a fuzzy set A is defined* as

$$(\neg A)(t) = 1 - A(t)$$

A closely related pair of properties which hold in ordinary set theory are *the law of excluded middle*

$$A \vee \neg A = X$$

and *the law of noncontradiction principle*

$$A \wedge \neg A = \emptyset$$

It is clear that $\neg 1_X = \emptyset$ and $\neg \emptyset = 1_X$, however, the laws of excluded middle and noncontradiction are not satisfied in fuzzy logic.

Lemma 1.2.1 *The law of excluded middle is not valid. Let* $A(t) = 1/2$, $\forall t \in \mathbb{R}$, *then it is easy to see that*

$$(\neg A \vee A)(t) = \max\{\neg A(t), A(t)\} = \max\{1 - 1/2, 1/2\} = 1/2 \neq 1$$

Lemma 1.2.2 *The law of noncontradiction is not valid. Let* $A(t) = 1/2$, $\forall t \in \mathbb{R}$, *then it is easy to see that*

$$(\neg A \wedge A)(t) = \min\{\neg A(t), A(t)\} = \min\{1 - 1/2, 1/2\} = 1/2 \neq 0$$

However, fuzzy logic does satisfy *De Morgan*'s laws

$$\neg(A \wedge B) = \neg A \vee \neg B, \quad \neg(A \vee B) = \neg A \wedge \neg B$$

Fig. 1.14. A and its complement.

Triangular norms were introduced by Schweizer and Sklar [203] to model distances in probabilistic metric spaces. In fuzzy sets theory triangular norms are extensively used to model logical connective *and.*

Definition 1.2.4 *(Triangular norm.) A mapping*

$$T: [0,1] \times [0,1] \to [0,1]$$

is a triangular norm (t-norm for short) iff it is symmetric, associative, non-decreasing in each argument and $T(a,1) = a$, for all $a \in [0,1]$. In other words, any t-norm T satisfies the properties:

Symmetricity:

$$T(x,y) = T(y,x), \ \forall x, y \in [0,1].$$

Associativity:

$$T(x, T(y,z)) = T(T(x,y), z), \ \forall x, y, z \in [0,1].$$

Monotonicity:

$$T(x,y) \leq T(x',y') \text{ if } x \leq x' \text{ and } y \leq y'.$$

One identy:

$$T(x,1) = x, \ \forall x \in [0,1].$$

These axioms attempt to capture the basic properties of set intersection. The basic t-norms are:

- minimum: $\min(a,b) = \min\{a,b\}$,
- Łukasiewicz: $T_L(a,b) = \max\{a + b - 1, 0\}$
- product: $T_P(a,b) = ab$
- weak:

$$T_W(a,b) = \begin{cases} \min\{a,b\} & \text{if } \max\{a,b\} = 1 \\ 0 & \text{otherwise} \end{cases}$$

- Hamacher [95]:

$$H_\gamma(a,b) = \frac{ab}{\gamma + (1-\gamma)(a+b-ab)}, \ \gamma \geq 0 \qquad (1.1)$$

- Dubois and Prade:

$$D_\alpha(a,b) = \frac{ab}{\max\{a,b,\alpha\}}, \ \alpha \in (0,1)$$

- Yager:

$$Y_p(a,b) = 1 - \min\{1, \sqrt[p]{[(1-a)^p + (1-b)^p]}\}, \ p > 0$$

- Frank [70]:

$$F_\lambda(a,b) = \begin{cases} \min\{a,b\} & \text{if } \lambda = 0 \\ T_P(a,b) & \text{if } \lambda = 1 \\ T_L(a,b) & \text{if } \lambda = \infty \\ 1 - \log_\lambda\left[1 + \dfrac{(\lambda^a - 1)(\lambda^b - 1)}{\lambda - 1}\right] & \text{otherwise} \end{cases}$$

All t-norms may be extended, through associativity, to $n > 2$ arguments. The minimum t-norm is automatically extended and

$$T_P(a_1, \dots, a_n) = a_1 \times a_2 \times \cdots \times a_n,$$

$$T_L(a_1, \dots a_n) = \max\left\{ \sum_{i=1}^{n} a_i - n + 1, 0 \right\}$$

A t-norm T is called strict if T is strictly increasing in each argument. A t-norm T is said to be Archimedean iff T is continuous and $T(x, x) < x$ for all $x \in (0, 1)$. Every Archimedean t-norm T is representable by a continuous and decreasing function $f \colon [0, 1] \rightarrow [0, \infty]$ with $f(1) = 0$ and

$$T(x, y) = f^{-1}(\min\{f(x) + f(y), f(0)\}).$$

The function f is the additive generator of T. A t-norm T is said to be nilpotent if $T(x, y) = 0$ holds for some $x, y \in (0, 1)$. Let T_1, T_2 be t-norms. We say that T_1 is weaker than T_2 (and write $T_1 \leq T_2$) if $T_1(x, y) \leq T_2(x, y)$ for each $x, y \in [0, 1]$.

Triangular conorms are extensively used to model logical connective *or*.

Definition 1.2.5 *(Triangular conorm.) A mapping*

$$S \colon [0, 1] \times [0, 1] \rightarrow [0, 1],$$

is a triangular co-norm (t-conorm) if it is symmetric, associative, non-decreasing in each argument and $S(a, 0) = a$, for all $a \in [0, 1]$. In other words, any t-conorm S satisfies the properties:

$S(x, y) = S(y, x)$ *(symmetricity)*
$S(x, S(y, z)) = S(S(x, y), z)$ *(associativity)*
$S(x, y) \leq S(x', y')$ *if $x \leq x'$ and $y \leq y'$ (monotonicity)*
$S(x, 0) = x$, $\forall x \in [0, 1]$ *(zero identy)*

If T is a t-norm then the equality

$$S(a, b) := 1 - T(1 - a, 1 - b),$$

defines a t-conorm and we say that S is derived from T. The basic t-conorms are:

- maximum: $\max(a, b) = \max\{a, b\}$
- Lukasiewicz: $S_L(a, b) = \min\{a + b, 1\}$
- probabilistic: $S_P(a, b) = a + b - ab$
- strong:

$$STRONG(a, b) = \begin{cases} \max\{a, b\} & \text{if } \min\{a, b\} = 0 \\ 1 & \text{otherwise} \end{cases}$$

- Hamacher:

$$HOR_\gamma(a, b) = \frac{a + b - (2 - \gamma)ab}{1 - (1 - \gamma)ab}, \ \gamma \geq 0$$

- Yager:

$$YOR_p(a, b) = \min\{1, \sqrt[p]{a^p + b^p}\}, \ p > 0.$$

Lemma 1.2.3 *Let T be a t-norm. Then the following statement holds*

$$T_W(x, y) \leq T(x, y) \leq \min\{x, y\}, \ \forall x, y \in [0, 1].$$

Proof. From monotonicity, symmetricity and the extremal condition we get

$$T(x, y) \leq T(x, 1) \leq x, \ T(x, y) = T(y, x) \leq T(y, 1) \leq y.$$

This means that $T(x, y) \leq \min\{x, y\}$.

Lemma 1.2.4 *Let S be a t-conorm. Then the following statement holds*

$$\max\{a, b\} \leq S(a, b) \leq STRONG(a, b), \ \forall a, b \in [0, 1]$$

Proof. From monotonicity, symmetricity and the extremal condition we get

$$S(x, y) \geq S(x, 0) \geq x, \ S(x, y) = S(y, x) \geq S(y, 0) \geq y$$

This means that $S(x, y) \geq \max\{x, y\}$.

Lemma 1.2.5 *$T(a, a) = a$ holds for any $a \in [0, 1]$ if and only if T is the minimum norm.*

Proof. If $T(a, b) = \min(a, b)$ then $T(a, a) = a$ holds obviously. Suppose $T(a, a) = a$ for any $a \in [0, 1]$, and $a \leq b \leq 1$. We can obtain the following expression using monotonicity of T

$$a = T(a, a) \leq T(a, b) \leq \min\{a, b\}.$$

From commutativity of T it follows that

$$a = T(a, a) \leq T(b, a) \leq \min\{b, a\}.$$

These equations show that $T(a, b) = \min\{a, b\}$ for any $a, b \in [0, 1]$.

Lemma 1.2.6 *The distributive law of t-norm T on the max operator holds for any $a, b, c \in [0, 1]$.*

$$T(\max\{a, b\}, c) = \max\{T(a, c), T(b, c)\}.$$

The operation *intersection* can be defined by the help of triangular norms.

Definition 1.2.6 *(t-norm-based intersection) Let T be a t-norm. The T-intersection of A and B is defined as*

$$(A \cap B)(t) = T(A(t), B(t)), \ \forall t \in X.$$

Example 1.2.3 *Let $T(x, y) = T_L(x, y) = \max\{x + y - 1, 0\}$ be the Lukasiewicz t-norm. Then we have*

$$(A \cap B)(t) = \max\{A(t) + B(t) - 1, 0\} \ \forall t \in X.$$

Let A and B be fuzzy subsets of $X = \{-2, -1, 0, 1, 2, 3, 4\}$.

$$A = 0.0/-2 + 0.3/-1 + 0.6/0 + 1.0/1 + 0.6/2 + 0.3/3 + 0.0/4,$$

$$B = 0.1/-2 + 0.3/-1 + 0.9/0 + 1.0/1 + 1.0/2 + 0.3/3 + 0.2/4.$$

Then $A \cap B$ has the following form

$$A \cap B = 0.0/-2 + 0.0/-1 + 0.5/0 + 1.0/1 + 0.6/2 + 0.0/3 + 0.2/4.$$

The operation *union* can be defined by the help of triangular conorms.

Definition 1.2.7 *(t-conorm-based union) Let S be a t-conorm. The S-union of A and B is defined as*

$$(A \cup B)(t) = S(A(t), B(t)), \ \forall t \in X.$$

Example 1.2.4 *Let $S(x, y) = \min\{x + y, 1\}$ be the Lukasiewicz t-conorm. Then we have*

$$(A \cup B)(t) = \min\{A(t) + B(t), 1\}, \ \forall t \in X.$$

Let A and B be fuzzy subsets of $X = \{-2, -1, 0, 1, 2, 3, 4\}$.

$$A = 0.0/-2 + 0.3/-1 + 0.6/0 + 1.0/1 + 0.6/2 + 0.3/3 + 0.0/4,$$

$$B = 0.1/-2 + 0.3/-1 + 0.9/0 + 1.0/1 + 1.0/2 + 0.3/3 + 0.0/4.$$

Then $A \cup B$ has the following form

$$A \cup B = 0.1/-2 + 0.6/-1 + 1.0/0 + 1.0/1 + 1.0/2 + 0.6/3 + 0.2/4.$$

In general, *the law of the excluded middle* and the *noncontradiction principle* properties are not satisfied by t-norms and t-conorms defining the intersection and union operations. However, the Łukasiewicz t-norm and t-conorm do satisfy these properties.

Lemma 1.2.7 *If $T(x, y) = T_L(x, y) = \max\{x + y - 1, 0\}$ then the law of noncontradiction is valid.*

Proof. Let A be a fuzzy set in X. Then from the definition of t-norm-based intersection we get

$$(A \cap \neg A)(t) = T_L(A(t), 1 - A(t)) = (A(t) + 1 - A(t) - 1) \vee 0 = 0, \ \forall t \in X.$$

Lemma 1.2.8 *If* $S(x,y) = S_L(x,y) = \min\{1, x + y\}$, *then the law of excluded middle is valid.*

Proof. Let A be a fuzzy set in X. Then from the definition of t-conorm-based union we get

$$(A \cup \neg A)(t) = S_L(A(t), 1 - A(t)) = (A(t) + 1 - A(t)) \wedge 1 = 1,$$

for all $t \in X$.

Exercise 1.2.1 *Let A and B be fuzzy subsets of $X = \{-2, -1, 0, 1, 2, 3, 4\}$.*

$$A = 0.5/-2 + 0.4/-1 + 0.6/0 + 1.0/1 + 0.6/2 + 0.3/3 + 0.4/4$$

$$B = 0.1/-2 + 0.7/-1 + 0.9/0 + 1.0/1 + 1.0/2 + 0.3/3 + 0.2/4$$

Suppose that their intersection is defined by the probabilistic t-norm $T_P(a, b) = ab$. What is then the membership function of $A \cap B$?

Exercise 1.2.2 *Let A and B be fuzzy subsets of $X = \{-2, -1, 0, 1, 2, 3, 4\}$.*

$$A = 0.5/-2 + 0.4/-1 + 0.6/0 + 1.0/1 + 0.6/2 + 0.3/3 + 0.4/4$$

$$B = 0.1/-2 + 0.7/-1 + 0.9/0 + 1.0/1 + 1.0/2 + 0.3/3 + 0.2/4$$

Suppose that their union is defined by the probabilistic t-conorm $T_P(a, b) = a + b - ab$. What is then the membership function of $A \cup B$?

Exercise 1.2.3 *Let A and B be fuzzy subsets of $X = \{-2, -1, 0, 1, 2, 3, 4\}$.*

$$A = 0.7/-2 + 0.4/-1 + 0.6/0 + 1.0/1 + 0.6/2 + 0.3/3 + 0.4/4$$

$$B = 0.1/-2 + 0.2/-1 + 0.9/0 + 1.0/1 + 1.0/2 + 0.3/3 + 0.2/4$$

Suppose that their intersection is defined by the Hamacher's t-norm with $\gamma = 0$. What is then the membership function of $A \cap B$?

Exercise 1.2.4 *Let A and B be fuzzy subsets of $X = \{-2, -1, 0, 1, 2, 3, 4\}$.*

$$A = 0.7/-2 + 0.4/-1 + 0.6/0 + 1.0/1 + 0.6/2 + 0.3/3 + 0.4/4$$

$$B = 0.1/-2 + 0.2/-1 + 0.9/0 + 1.0/1 + 1.0/2 + 0.3/3 + 0.2/4$$

Suppose that their intersection is defined by the Hamacher's t-conorm with $\gamma = 0$. What is then the membership function of $A \cup B$?

Exercise 1.2.5 *Show that if $\gamma \leq \gamma'$ then $H_\gamma(x, y) \geq H_{\gamma'}(x, y)$ holds for all $x, y \in [0, 1]$, i.e. the family H_γ is monoton decreasing.*

1.3 Fuzzy relations

A classical relation can be considered as a set of tuples, where a tuple is an ordered pair. A binary tuple is denoted by (u, v), an example of a ternary tuple is (u, v, w) and an example of n-ary tuple is $(x_1, \ldots, x_n)$.

Definition 1.3.1 *(classical n-ary relation) Let $X_1, \ldots, X_n$ be classical sets. The subsets of the Cartesian product $X_1 \times \cdots \times X_n$ are called n-ary relations. If $X_1 = \cdots = X_n$ and $R \subset X^n$ then R is called an n-ary relation in X.*

Let R be a binary relation in R. Then the characteristic function of R is defined as

$$\chi_R(u, v) = \begin{cases} 1 \text{ if } (u, v) \in R \\ 0 \text{ otherwise} \end{cases}$$

Example 1.3.1 *Let X be the domain of men {John, Charles, James} and Y the domain of women {Diana, Rita, Eva}, then the relation "married to" on $X \times Y$ is, for example*

$$\{(Charles, Diana), (John, Eva), (James, Rita)\}$$

Example 1.3.2 *Consider the following relation $(u, v) \in R$ iff $u \in [a, b]$ and $v \in [0, c]$:*

$$\chi_R(u, v) = \begin{cases} 1 \quad \text{if } (u, v) \in [a, b] \times [0, c] \\ 0 \text{ otherwise} \end{cases}$$

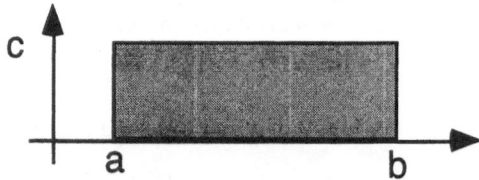

Fig. 1.15. Graph of a crisp relation.

Let R be a binary relation in a classical set X. Then

Definition 1.3.2 *(reflexivity) R is reflexive if $\forall u \in U : (u, u) \in R$*

Definition 1.3.3 *(anti-reflexivity) R is anti-reflexive if $\forall u \in U : (u, u) \notin R$*

Definition 1.3.4 *(symmetricity) R is symmetric if from $(u, v) \in R$ then $(v, u) \in R$*

Definition 1.3.5 *(anti-symmetricity) R is anti-symmetric if $(u, v) \in R$ and $(v, u) \in R$ then $u = v$*

Definition 1.3.6 *(transitivity) R is transitive if $(u,v) \in R$ and $(v,w) \in R$ then $(u,w) \in R$, $\forall u, v, w \in U$*

Example 1.3.3 *Consider the classical inequality relations on the real line R. It is clear that $\leq$ is reflexive, anti-symmetric and transitive, while $<$ is anti-reflexive, anti-symmetric and transitive.*

Other important classes of binary relations are the following:

Definition 1.3.7 *(equivalence) R is an equivalence relation if, R is reflexive, symmetric and transitive*

Definition 1.3.8 *(partial order) R is a partial order relation if it is reflexive, anti-symmetric and transitive*

Definition 1.3.9 *(total order) R is a total order relation if it is partial order and $(u,v) \in R$ or $(v,u) \in R$ hold for any u and v.*

Example 1.3.4 *Let us consider the binary relation "subset of". It is clear that it is a partial order relation. The relation $\leq$ on natural numbers is a total order relation.*

Example 1.3.5 *Consider the relation "mod 3" on natural numbers*

$$\{(m,n) \mid (n-m) \bmod 3 \equiv 0\}$$

This is an equivalence relation.

Definition 1.3.10 *(fuzzy relation) Let X and Y be nonempty sets. A fuzzy relation R is a fuzzy subset of $X \times Y$. In other words, $R \in \mathcal{F}(X \times Y)$. If $X = Y$ then we say that R is a binary fuzzy relation in X.*

Let R be a binary fuzzy relation on $\mathbb{R}$. Then $R(u,v)$ is interpreted as the degree of membership of (u,v) in R.

Example 1.3.6 *A simple example of a binary fuzzy relation on $U = \{1,2,3\}$, called "approximately equal" can be defined as*

$$R(1,1) = R(2,2) = R(3,3) = 1,$$
$$R(1,2) = R(2,1) = R(2,3) = R(3,2) = 0.8,$$
$$R(1,3) = R(3,1) = 0.3.$$

The membership function of R is given by

$$R(u,v) = \begin{cases} 1 & \text{if } u = v \\ 0.8 & \text{if } |u-v| = 1 \\ 0.3 & \text{if } |u-v| = 2 \end{cases}$$

In matrix notation it can be represented as

$$R = \begin{pmatrix} & 1 & 2 & 3 \\ 1 & 1 & 0.8 & 0.3 \\ 2 & 0.8 & 1 & 0.8 \\ 3 & 0.3 & 0.8 & 1 \end{pmatrix}$$

Fuzzy relations are very important because they can describe interactions between variables. Let R and S be two binary fuzzy relations on $X \times Y$.

Definition 1.3.11 *(intersection) The intersection of R and G is defined by*

$$(R \cap G)(u, v) = \min\{R(u, v), G(u, v)\} = R(u, v) \wedge G(u, v), \ (u, v) \in X \times Y.$$

Note that $R \colon X \times Y \to [0, 1]$, i.e. the domain of R is the whole Cartesian product $X \times Y$.

Definition 1.3.12 *(union) The union of R and S is defined by*

$$(R \cup G)(u, v) = \max\{R(u, v), G(u, v)\} = R(u, v) \vee G(u, v), \ (u, v) \in X \times Y.$$

Example 1.3.7 *Let us define two binary relations R = "x is considerable smaller than y" and G = "x is very close to y"*

$$R = \begin{pmatrix} & y_1 & y_2 & y_3 & y_4 \\ x_1 & 0.5 & 0.1 & 0.1 & 0.7 \\ x_2 & 0 & 0.8 & 0 & 0 \\ x_3 & 0.9 & 1 & 0.7 & 0.8 \end{pmatrix} \qquad G = \begin{pmatrix} & y_1 & y_2 & y_3 & y_4 \\ x_1 & 0.4 & 0 & 0.9 & 0.6 \\ x_2 & 0.9 & 0.4 & 0.5 & 0.7 \\ x_3 & 0.3 & 0 & 0.8 & 0.5 \end{pmatrix}$$

The intersection of R and G means that "x is considerable smaller than y" **and** "x is very close to y".

$$(R \cap G)(x, y) = \begin{pmatrix} & y_1 & y_2 & y_3 & y_4 \\ x_1 & 0.4 & 0 & 0.1 & 0.6 \\ x_2 & 0 & 0.4 & 0 & 0 \\ x_3 & 0.3 & 0 & 0.7 & 0.5 \end{pmatrix}$$

The union of R and G means that "x is considerable smaller than y" **or** "x is very close to y".

$$(R \cup G)(x, y) = \begin{pmatrix} & y_1 & y_2 & y_3 & y_4 \\ x_1 & 0.5 & 0.1 & 0.9 & 0.7 \\ x_2 & 0.9 & 0.8 & 0.5 & 0.7 \\ x_3 & 0.9 & 1 & 0.8 & 0.8 \end{pmatrix}$$

Consider a classical relation R on $\mathbb{R}$.

$$R(u, v) = \begin{cases} 1 & \text{if } (u, v) \in [a, b] \times [0, c] \\ 0 & \text{otherwise} \end{cases}$$

It is clear that the *projection* (or **shadow**) of R on the X-axis is the closed interval $[a, b]$ and its projection on the Y-axis is $[0, c]$.

Definition 1.3.13 *(projection of classical relations)*
Let R be a classical relation on $X \times Y$. The projection of R on X, denoted by $\Pi_X(R)$, is defined as

$$\Pi_X(R) = \{x \in X \mid \exists y \in Y \text{ such that } (x, y) \in R\}$$

similarly, the projection of R on Y, denoted by $\Pi_Y(R)$, is defined as

$$\Pi_Y(R) = \{y \in Y \mid \exists x \in X \text{ such that } (x, y) \in R\}$$

Definition 1.3.14 *(projection of binary fuzzy relations)*
Let R be a binary fuzzy relation on $X \times Y$. The projection of R on X is a fuzzy subset of X, denoted by $\Pi_X(R)$, defined as

$$\Pi_X(R)(x) = \sup\{R(x, y) \mid y \in Y\}$$

and the projection of R on Y is a fuzzy subset of Y, denoted by $\Pi_Y(R)$, defined as

$$\Pi_Y(R)(y) = \sup\{R(x, y) \mid x \in X\}$$

If R is fixed then instead of $\Pi_X(R)(x)$ we write simply $\Pi_X(x)$.

Example 1.3.8 *Consider the fuzzy relation $R = $ "x is considerable smaller than y"*

$$R = \begin{pmatrix} & y_1 & y_2 & y_3 & y_4 \\ x_1 & 0.5 & 0.1 & 0.1 & 0.7 \\ x_2 & 0 & 0.8 & 0 & 0 \\ x_3 & 0.9 & 1 & 0.7 & 0.8 \end{pmatrix}$$

then the projection on X means that

- x_1 *is assigned the highest membership degree from the tuples (x_1, y_1), (x_1, y_2), (x_1, y_3), (x_1, y_4), i.e. $\Pi_X(x_1) = 0.7$, which is the maximum of the first row.*
- x_2 *is assigned the highest membership degree from the tuples (x_2, y_1), (x_2, y_2), (x_2, y_3), (x_2, y_4), i.e. $\Pi_X(x_2) = 0.8$, which is the maximum of the second row.*
- x_3 *is assigned the highest membership degree from the tuples (x_3, y_1), (x_3, y_2), (x_3, y_3), (x_3, y_4), i.e. $\Pi_X(x_3) = 1$, which is the maximum of the third row.*

Definition 1.3.15 *(Cartesian product of fuzzy sets)*
   The Cartesian product of two fuzzy sets $A \in \mathcal{F}(X)$ and $B \in \mathcal{F}(Y)$ is defined by

$$(A \times B)(u, v) = \min\{A(u), B(v)\}, \ (u, v) \in X \times Y.$$

It is clear that the Cartesian product of two fuzzy sets $A \in \mathcal{F}(X)$ and $B \in \mathcal{F}(Y)$ is a binary fuzzy relation in $X \times Y$, i.e.

$$A \times B \in \mathcal{F}(X \times Y).$$

Assume A and B are normal fuzzy sets. An interesting property of $A \times B$ is that
$$\Pi_Y(A \times B) = B$$
and
$$\Pi_X(A \times B) = A.$$
Really,
$$\Pi_X(x) = \sup\{(A \times B)(x, y) \,|\, y \in Y\}$$
$$= \sup\{\min\{A(x), B(y)\} \,|\, y \in Y\} =$$
$$\min\{A(x), \sup\{B(y)\} \,|\, y \in Y\}\} = A(x).$$

Similarly to the one-dimensional case, intersection and union operations on fuzzy relations can be defined via t-norms and t-conorms, respectively.

Definition 1.3.16 *(t-norm-based intersection) Let T be a t-norm and let R and G be binary fuzzy relations in $X \times Y$. Their T-intersection is defined by*

$$(R \cap S)(u, v) = T(R(u, v), G(u, v)), \ (u, v) \in X \times Y.$$

Definition 1.3.17 *(t-conorm-based union) Let S be a t-conorm and let R and G be binary fuzzy relations in $X \times Y$. Their S-union is defined by*

$$(R \cup S)(u, v) = S(R(u, v), G(u, v)), \ (u, v) \in X \times Y.$$

Definition 1.3.18 *(sup-min composition) Let $R \in \mathcal{F}(X \times Y)$ and $G \in \mathcal{F}(Y \times Z)$. The sup-min composition of R and G, denoted by $R \circ G$ is defined as*
$$(R \circ S)(u, w) = \sup_{v \in Y} \min\{R(u, v), S(v, w)\}$$

It is clear that $R \circ G$ is a binary fuzzy relation in $X \times Z$.

Example 1.3.9 *Consider two fuzzy relations $R = $ "x is considerable smaller than y" and $G = $ "y is very close to z"*

$$R = \begin{pmatrix} & y_1 & y_2 & y_3 & y_4 \\ x_1 & 0.5 & 0.1 & 0.1 & 0.7 \\ x_2 & 0 & 0.8 & 0 & 0 \\ x_3 & 0.9 & 1 & 0.7 & 0.8 \end{pmatrix} \qquad G = \begin{pmatrix} & z_1 & z_2 & z_3 \\ y_1 & 0.4 & 0.9 & 0.3 \\ y_2 & 0 & 0.4 & 0 \\ y_3 & 0.9 & 0.5 & 0.8 \\ y_4 & 0.6 & 0.7 & 0.5 \end{pmatrix}$$

Then their sup − min composition is

$$R \circ G = \begin{pmatrix} & z_1 & z_2 & z_3 \\ x_1 & 0.6 & 0.8 & 0.5 \\ x_2 & 0 & 0.4 & 0 \\ x_3 & 0.7 & 0.9 & 0.7 \end{pmatrix}$$

Formally,

$$\begin{pmatrix} & y_1 & y_2 & y_3 & y_4 \\ x_1 & 0.5 & 0.1 & 0.1 & 0.7 \\ x_2 & 0 & 0.8 & 0 & 0 \\ x_3 & 0.9 & 1 & 0.7 & 0.8 \end{pmatrix} \circ \begin{pmatrix} & z_1 & z_2 & z_3 \\ y_1 & 0.4 & 0.9 & 0.3 \\ y_2 & 0 & 0.4 & 0 \\ y_3 & 0.9 & 0.5 & 0.8 \\ y_4 & 0.6 & 0.7 & 0.5 \end{pmatrix} = \begin{pmatrix} & z_1 & z_2 & z_3 \\ x_1 & 0.6 & 0.7 & 0.5 \\ x_2 & 0 & 0.4 & 0 \\ x_3 & 0.7 & 0.9 & 0.7 \end{pmatrix}$$

i.e., the composition of R and G is nothing else, but the classical product of the matrices R and G with the difference that instead of addition we use maximum and instead of multiplication we use minimum operator. For example,

$$(R \circ G)(x_1, z_1) = \max\{0.5 \wedge 0.4, 0.1 \wedge 0, 0.1 \wedge 0.9, 0.7 \wedge 0.6\} = 0.6$$

$$(R \circ G)(x_1, z_2) = \max\{0.5 \wedge 0.9, 0.1 \wedge 0.4, 0.1 \wedge 0.5, 0.7 \wedge 0.7\} = 0.7$$

$$(R \circ G)(x_1, z_3) = \max\{0.5 \wedge 0.3, 0.1 \wedge 0, 0.1 \wedge 0.8, 0.7 \wedge 0.5\} = 0.5$$

Definition 1.3.19 *(sup-T composition) Let T be a t-norm and let $R \in \mathcal{F}(X \times Y)$ and $G \in \mathcal{F}(Y \times Z)$. The sup-T composition of R and G, denoted by $R \circ G$ is defined as*

$$(R \circ S)(u, w) = \sup_{v \in Y} T(R(u, v), S(v, w))$$

Following Zadeh [237] we can define the sup-min composition of a fuzzy set and fuzzy relation as follows

Definition 1.3.20 *Let $C \in \mathcal{F}(X)$ and $R \in \mathcal{F}(X \times Y)$. The membership function of the composition of a fuzzy set C and a fuzzy relation R is defined by*

$$(C \circ R)(y) = \sup_{x \in X} \min\{C(x), R(x, y)\}, \quad \forall y \in Y.$$

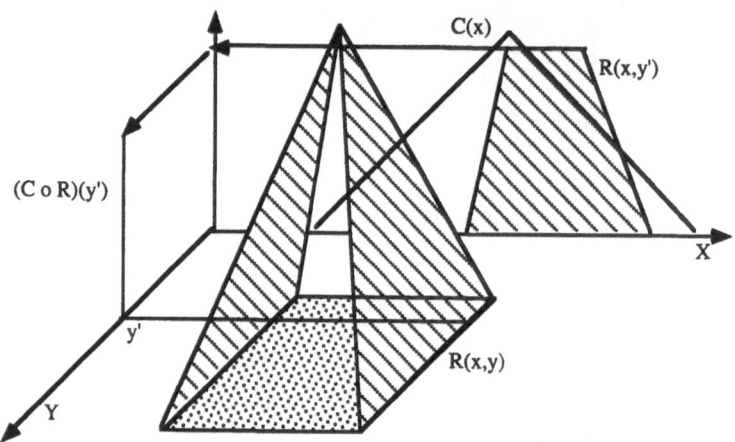

Fig. 1.16. Composition of a fuzzy number and a fuzzy relation.

The composition of a fuzzy set C and a fuzzy relation R can be considered as the shadow of the relation R on the fuzzy set C.

In the above definition we can use any t-norm for modeling the compositional operator.

Definition 1.3.21 *Let T be a t-norm $C \in \mathcal{F}(X)$ and $R \in \mathcal{F}(X \times Y)$. The membership function of the composition of a fuzzy set C and a fuzzy relation R is defined by*

$$(C \circ R)(y) = \sup_{x \in X} T(C(x), R(x, y)),$$

for all $y \in Y$.

For example, if $T_P(x, y) = xy$ is the product t-norm then the sup-T compositiopn of a fuzzy set C and a fuzzy relation R is defined by

$$(C \circ R)(y) = \sup_{x \in X} T_P(C(x), R(x, y)) = \sup_{x \in X} C(x)R(x, y)$$

and if $T_L(x, y) = \max\{0, x + y - 1\}$ is the Łukasiewicz t-norm then we get

$$(C \circ R)(y) = \sup_{x \in X} T_L(C(x), R(x, y)) = \sup_{x \in X} \max\{0, C(x) + R(x, y) - 1\}$$

for all $y \in Y$.

Example 1.3.10 *Let A and B be fuzzy numbers and let $R = A \times B$ a fuzzy relation. Observe the following property of composition*

$$A \circ R = A \circ (A \times B) = B, \quad B \circ R = B \circ (A \times B) = A.$$

This fact can be interpreted as: if A and B have relation $A \times B$ and then the composition of A and $A \times B$ is exactly B, and then the composition of B and $A \times B$ is exactly A.

Example 1.3.11 *Let* C *be a fuzzy set in the universe of discourse* $\{1, 2, 3\}$ *and let* R *be a binary fuzzy relation in* $\{1, 2, 3\}$. *Assume that*

$$C = 0.2/1 + 1/2 + 0.2/3$$

and

$$R = \begin{pmatrix} & 1 & 2 & 3 \\ 1 & 1 & 0.8 & 0.3 \\ 2 & 0.8 & 1 & 0.8 \\ 3 & 0.3 & 0.8 & 1 \end{pmatrix}$$

Using Definition 1.3.20 we get

$$C \circ R = (0.2/1 + 1/2 + 0.2/3) \circ \begin{pmatrix} & 1 & 2 & 3 \\ 1 & 1 & 0.8 & 0.3 \\ 2 & 0.8 & 1 & 0.8 \\ 3 & 0.3 & 0.8 & 1 \end{pmatrix} =$$

$$0.8/1 + 1/2 + 0.8/3.$$

Example 1.3.12 *Let* C *be a fuzzy set in the universe of discourse* $[0, 1]$ *and let* R *be a binary fuzzy relation in* $[0, 1]$. *Assume that* $C(x) = x$ *and* $R(x, y) = 1 - |x - y|$. *Using the definition of sup-min composition (1.3.20) we get*

$$(C \circ R)(y) = \sup_{x \in [0,1]} \min\{x, 1 - |x - y|\} = \frac{1 + y}{2}$$

for all $y \in [0, 1]$.

Example 1.3.13 *Let* C *be a fuzzy set in the universe of discourse* $\{1, 2, 3\}$ *and let* R *be a binary fuzzy relation in* $\{1, 2, 3\}$. *Assume that*

$$C = 1/1 + 0.2/2 + 1/3$$

and

$$R = \begin{pmatrix} & 1 & 2 & 3 \\ 1 & 0.4 & 0.8 & 0.3 \\ 2 & 0.8 & 0.4 & 0.8 \\ 3 & 0.3 & 0.8 & 0 \end{pmatrix}$$

Then the sup-T_P composition of C *and* R *is calculated by*

$$C \circ R = (1/1 + 0.2/2 + 1/3) \circ \begin{pmatrix} & 1 & 2 & 3 \\ 1 & 0.4 & 0.8 & 0.3 \\ 2 & 0.8 & 0.4 & 0.8 \\ 3 & 0.3 & 0.8 & 0 \end{pmatrix}$$

$$= 0.4/1 + 0.8/2 + 0.3/3.$$

1.4 The extension principle

In order to use fuzzy numbers and relations in any intellgent system we must be able to perform arithmetic operations with these fuzzy quantities. In particular, we must be able to to *add, subtract, multiply* and *divide* with fuzzy quantities. The process of doing these operations is called *fuzzy arithmetic.*

We shall first introduce an important concept from fuzzy set theory called the *extension principle.* We then use it to provide for these arithmetic operations on fuzzy numbers.

In general the extension principle pays a fundamental role in enabling us to extend any point operations to operations involving fuzzy sets. In the following we define this principle.

Definition 1.4.1 *(extension principle) Assume X and Y are crisp sets and let f be a mapping from X to Y,*

$$f : X \to Y$$

such that for each $x \in X, f(x) = y \in Y$. Assume A is a fuzzy subset of X, using the extension principle, we can define $f(A)$ as a fuzzy subset of Y such that

$$f(A)(y) = \begin{cases} \sup_{x \in f^{-1}(y)} A(x) & \text{if } f^{-1}(y) \neq \emptyset \\ 0 & \text{otherwise} \end{cases} \qquad (1.2)$$

where $f^{-1}(y) = \{x \in X \mid f(x) = y\}$.

It should be noted that if f is strictly increasing (or strictly decreasing) then (1.2) turns into

$$f(A)(y) = \begin{cases} A(f^{-1}(y)) & \text{if } y \in Range(f) \\ 0 & \text{otherwise} \end{cases}$$

where $Range(f) = \{y \in Y \mid \exists x \in X \text{ such that } f(x) = y\}$.

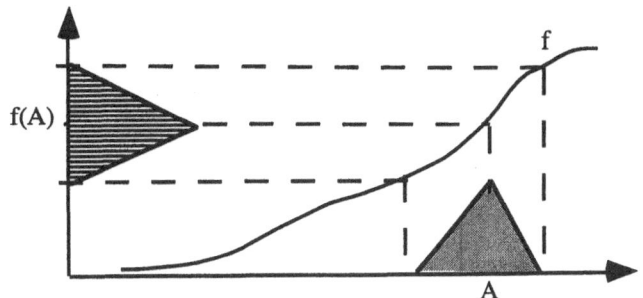

Fig. 1.17. Extension of a monoton increasing function.

Example 1.4.1 *Let* $f(x) = x^2$ *and let* $A \in \mathcal{F}$ *be a symmetric triangular fuzzy number with membership function*

$$A(x) = \begin{cases} 1 - \dfrac{|a - x|}{\alpha} & \text{if } |a - x| \leq \alpha \\ 0 & \text{otherwise} \end{cases}$$

Then using the extension principle we get

$$f(A)(y) = \begin{cases} A(\sqrt{y}) & \text{if } y \geq 0 \\ 0 & \text{otherwise} \end{cases}$$

that is

$$f(A)(y) = \begin{cases} 1 - \dfrac{|a - \sqrt{y}|}{\alpha} & \text{if } |a - \sqrt{y}| \leq \alpha \text{ and } y \geq 0 \\ 0 & \text{otherwise} \end{cases}$$

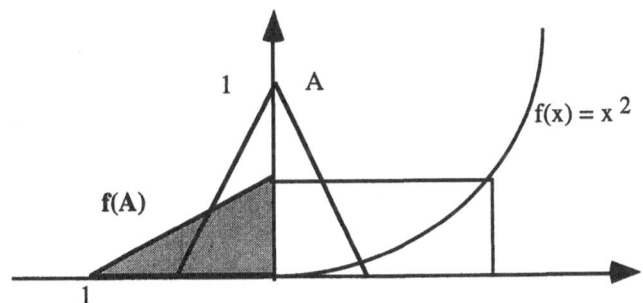

Fig. 1.18. The quadratic image of a symmetric triangular fuzzy number.

Example 1.4.2 *Let*

$$f(x) = \frac{1}{1 + e^{-x}}$$

be a sigmoidal function and let A be a fuzzy number. Then from

$$f^{-1}(y) = \begin{cases} \ln(y/(1-y)) & \text{if } 0 \le y \le 1 \\ 0 & \text{otherwise} \end{cases}$$

it follows that

$$f(A)(y) = \begin{cases} A(\ln(y/(1-y))) & \text{if } 0 \le y \le 1 \\ 0 & \text{otherwise} \end{cases}$$

Example 1.4.3 *Let $\lambda \ne 0$ be a real number and let $f(x) = \lambda x$ be a linear function. Suppose $A \in \mathcal{F}$ is a fuzzy number. Then using the extension principle we obtain*

$$f(A)(y) = \sup\{A(x) \mid \lambda x = y\} = A\left(\frac{y}{\lambda}\right).$$

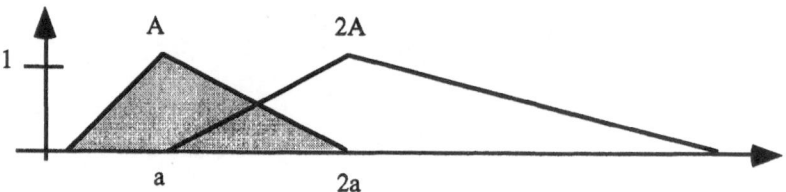

Fig. 1.19. The fuzzy number λA for $\lambda = 2$.

For $\lambda = 0$ then we get

$$f(A)(y) = (0 \times A)(y) = \sup\{A(x) \mid 0x = y\} = \begin{cases} 0 \text{ if } y \ne 0 \\ 1 \text{ if } y = 0 \end{cases}$$

That is $0 \times A = \bar{0}$ for all $A \in \mathcal{F}$.

If $f(x) = \lambda x$ and $A \in \mathcal{F}$ then we will write $f(A) = \lambda A$. Especially, if $\lambda = -1$ then we have

$$(-1A)(x) = (-A)(x) = A(-x), \ x \in \mathbb{R}$$

It should be noted that Zadeh's extension principle is nothing else but a straightforward generalization of set-valued functions (see [181] for details). Namely, let $f \colon X \to Y$ be a function. Then the image of a (crisp) subset $A \subset X$ by f is defined by

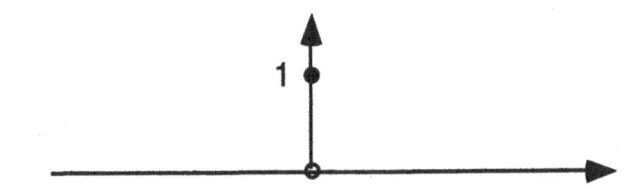

Fig. 1.20. $0 \times A$ is equal to $\bar{0}$.

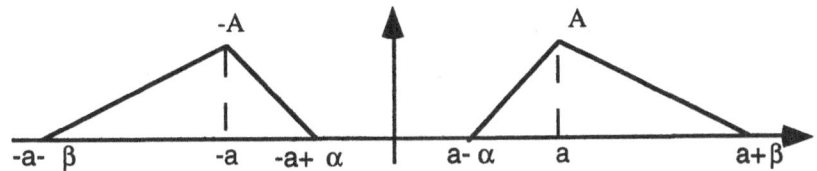

Fig. 1.21. Fuzzy number A and $-A$.

$$f(A) = \{f(x) \mid x \in A\}$$

and the characteristic function of $f(A)$ is

$$\chi_{f(A)}(y) = \sup\{\chi_A(x) \mid x \in f^{-1}(y)\}$$

Then replacing χ_A by a fuzzy set μ_A we get Zadeh's extension principle (1.2).

1.5 The extension principle for n-place functions

Definition 1.5.1 *(sup-min extension n-place functions) Let $X_1, X_2, \ldots, X_n$ and Y be a family of (nonempty) sets. Assume f is a mapping from the Cartesian product*

$$X_1 \times X_2 \times \cdots \times X_n$$

into Y, that is, for each n-tuple $(x_1, \ldots, x_n)$ such that $x_i \in X_i$, we have

$$f(x_1, x_2, \ldots, x_n) = y \in Y.$$

Let $A_1, \ldots, A_n$ be fuzzy subsets of $X_1, \ldots, X_n$, respectively; then the extension principle allows for the evaluation of $f(A_1, \ldots, A_n)$. In particular, $f(A_1, \ldots, A_n) = B$, where B is a fuzzy subset of Y such that

$$B(y) = \begin{cases} \sup\{\min\{A_1(x_1), \ldots, A_n(x_n)\} \mid x \in f^{-1}(y)\} & \text{if } f^{-1}(y) \neq \emptyset \quad (1.3) \\ 0 & \text{otherwise.} \end{cases}$$

For $n = 2$ then the extension principle reads

$$f(A_1, A_2)(y) = \sup\{A_1(x_1) \wedge A_2(x_2) \mid f(x_1, x_2) = y\}$$

The extension principle for n-place functions is also a straightforward generalization of set-valued functions. Namely, let $f\colon X_1 \times X_2 \to Y$ be a function. Then the image of a (crisp) subset $(A_1, A_2) \subset X_1 \times X_2$ by f is defined by

$$f(A_1, A_2) = \{f(x_1, x_2) \mid x_1 \in A \text{ and } x_2 \in A_2\}$$

and the characteristic function of $f(A_1, A_2)$ is

$$\chi_{f(A_1, A_2)}(y) = \sup\{\min\{\chi_{A_1}(x), \chi_{A_2}(x)\} \mid x \in f^{-1}(y)\}.$$

Then replacing the characteristic functions by fuzzy sets we get Zadeh's extension principle for n-place functions (1.3).

Example 1.5.1 (*extended addition*) *Let $f\colon X \times X \to X$ be defined as*

$$f(x_1, x_2) = x_1 + x_2,$$

i.e. f is the addition operator. Suppose A_1 and A_2 are fuzzy subsets of X. Then using the extension principle we get

$$f(A_1, A_2)(y) = \sup_{x_1 + x_2 = y} \min\{A_1(x_1), A_2(x_2)\}$$

and we use the notation $f(A_1, A_2) = A_1 + A_2$.

Example 1.5.2 (*extended subtraction*) *Let $f\colon X \times X \to X$ be defined as*

$$f(x_1, x_2) = x_1 - x_2,$$

i.e. f is the subtraction operator. Suppose A_1 and A_2 are fuzzy subsets of X. Then using the extension principle we get

$$f(A_1, A_2)(y) = \sup_{x_1 - x_2 = y} \min\{A_1(x_1), A_2(x_2)\}$$

and we use the notation $f(A_1, A_2) = A_1 - A_2$.

We note that from the equality

$$\sup_{x_1 - x_2 = y} \min\{A_1(x_1), A_2(x_2)\} = \sup_{x_1 + x_2 = y} \min\{A_1(x_1), A_2(-x_2)\}$$

it follows that $A_1 - A_2 = A_1 + (-A_2)$ holds. However, if $A \in \mathcal{F}$ is a fuzzy number then

$$(A - A)(y) = \sup_{x_1 - x_2 = y} \min\{A(x_1), A(x_2)\}, \ y \in \mathbb{R}$$

is not equal to the fuzzy number $\bar{0}$, where $\bar{0}(t) = 1$ if $t = 0$ and $\bar{0}(t) = 0$ otherwise.

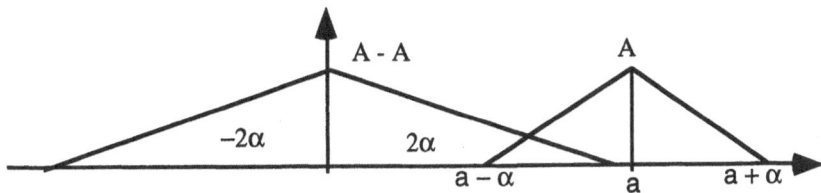

Fig. 1.22. The membership function of $A - A$.

Example 1.5.3 *Let* $f\colon X \times X \to X$ *be defined as*

$$f(x_1, x_2) = \lambda_1 x_1 + \lambda_2 x_2,$$

Suppose A_1 *and* A_2 *are fuzzy subsets of* X. *Then using the extension principle we get*

$$f(A_1, A_2)(y) = \sup_{\lambda_1 x_1 + \lambda_2 x_2 = y} \min\{A_1(x_1), A_2(x_2)\}$$

and we use the notation $f(A_1, A_2) = \lambda_1 A_1 + \lambda_2 A_2$.

Example 1.5.4 *(extended multiplication) Let* $f\colon X \times X \to X$ *be defined as*

$$f(x_1, x_2) = x_1 x_2,$$

i.e. f *is the multiplication operator. Suppose* A_1 *and* A_2 *are fuzzy subsets of* X. *Then using the extension principle we get*

$$f(A_1, A_2)(y) = \sup_{x_1 x_2 = y} \min\{A_1(x_1), A_2(x_2)\}$$

and we use the notation $f(A_1, A_2) = A_1 A_2$.

Example 1.5.5 *(extended division) Let* $f\colon X \times X \to X$ *be defined as*

$$f(x_1, x_2) = x_1 / x_2,$$

i.e. f *is the division operator. Suppose* A_1 *and* A_2 *are fuzzy subsets of* X. *Then using the extension principle we get*

$$f(A_1, A_2)(y) = \sup_{x_1 / x_2 = y,\ x_2 \neq 0} \min\{A_1(x_1), A_2(x_2)\}$$

and we use the notation $f(A_1, A_2) = A_1 / A_2$.

Definition 1.5.2 *Let* $X \neq \emptyset$ *and* $Y \neq \emptyset$ *be crisp sets and let* f *be a function from* $\mathcal{F}(X)$ *to* $\mathcal{F}(Y)$. *Then* f *is called a fuzzy function (or mapping) and we use the notation*

$$f\colon \mathcal{F}(X) \to \mathcal{F}(Y).$$

It should be noted, however, that a fuzzy function is not necessarily defined by Zadeh's extension principle. It can be any function which maps a fuzzy set $A \in \mathcal{F}(X)$ into a fuzzy set $B := f(A) \in \mathcal{F}(Y)$.

Definition 1.5.3 *Let $X \neq \emptyset$ and $Y \neq \emptyset$ be crisp sets. A fuzzy mapping $f \colon \mathcal{F}(X) \to \mathcal{F}(Y)$ is said to be monoton increasing if from $A, A' \in \mathcal{F}(X)$ and $A \subset A'$ it follows that $f(A) \subset f(A')$.*

Theorem 1.5.1 *Let $X \neq \emptyset$ and $Y \neq \emptyset$ be crisp sets. Then every fuzzy mapping $f \colon \mathcal{F}(X) \to \mathcal{F}(Y)$ defined by the extension principle is monoton increasing.*

Proof Let $A, A' \in \mathcal{F}(X)$ such that $A \subset A'$. Then using the definition of sup-min extension principle we get

$$f(A)(y) = \sup_{x \in f^{-1}(y)} A(x) \leq \sup_{x \in f^{-1}(y)} A'(x) = f(A')(y)$$

for all $y \in Y$.

Lemma 1.5.1 *Let $A, B \in \mathcal{F}$ be fuzzy numbers and let $f(A, B) = A + B$ be defined by sup-min extension principle. Then f is monoton increasing.*

Proof. Let $A, A', B, B' \in \mathcal{F}$ such that $A \subset A'$ and $B \subset B'$. Then using the definition of sup-min extension principle we get

$$(A + B)(z) = \sup_{x+y=z} \min\{A(x), B(y)\}$$
$$\leq \sup_{x+y=z} \min\{A'(x), B'(y)\} = (A' + B')(z)$$

Which ends the proof.

The following lemma can be proved in a similar way.

Lemma 1.5.2 *Let $A, B \in \mathcal{F}$ be fuzzy numbers, let λ_1, λ_2 be real numbers and let*

$$f(A, B) = \lambda_1 A + \lambda_2 B$$

be defined by sup-min extension principle. Then f is a monoton increasing fuzzy function.

Let $A = (a_1, a_2, \alpha_1, \alpha_2)_{LR}$ and $B = (b_1, b_2, \beta_1, \beta_2)_{LR}$ be fuzzy numbers of LR-type.

Using the (sup-min) extension principle we can verify the following rule for addition of fuzzy numbers of LR-type.

$$A + B = (a_1 + b_1, a_2 + b_2, \alpha_1 + \beta_1, \alpha_2 + \beta_2)_{LR}$$

furthermore, if $\lambda \in I\!R$ is a real number then λA can be represented as

$$\lambda A = \begin{cases} (\lambda a_1, \lambda a_2, \alpha_1, \alpha_2)_{LR} & \text{if } \lambda \geq 0 \\ (\lambda a_2, \lambda a_1, |\lambda|\alpha_2, |\lambda|\alpha_1)_{RL} & \text{if } \lambda < 0 \end{cases}$$

In particular if $A = (a_1, a_2, \alpha_1, \alpha_2)$ and $B = (b_1, b_2, \beta_1, \beta_2)$ are fuzzy numbers of trapezoidal form then

$$A + B = (a_1 + b_1, a_2 + b_2, \alpha_1 + \beta_1, \alpha_2 + \beta_2),$$

$$A - B = (a_1 - b_2, a_2 - b_1, \alpha_1 + \beta_2, \alpha_2 + \beta_1).$$

If $A = (a, \alpha_1, \alpha_2)$ and $B = (b, \beta_1, \beta_2)$ are fuzzy numbers of triangular form then

$$A + B = (a + b, \alpha_1 + \beta_1, \alpha_2 + \beta_2),$$

$$A - B = (a - b, \alpha_1 + \beta_2, \alpha_2 + \beta_1)$$

and if $A = (a, \alpha)$ and $B = (b, \beta)$ are fuzzy numbers of symmetrical triangular form then

$$A + B = (a + b, \alpha + \beta),$$

$$A - B = (a - b, \alpha + \beta),$$

$$\lambda A = (\lambda a, |\lambda|\alpha).$$

The above results can be generalized to linear combinations of fuzzy numbers.

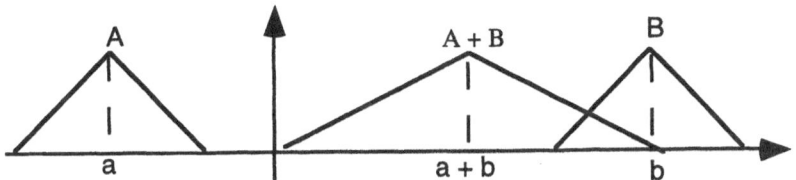

Fig. 1.23. Addition of triangular fuzzy numbers.

Lemma 1.5.3 *Let $A_i = (a_i, \alpha_i)$ be a fuzzy number of symmetrical triangular form and let λ_i be a real number, $i = 1, \ldots, n$. Then their linear combination*

$$\sum_{i=1}^{n} \lambda_i A_i := \lambda_1 A_1 + \cdots + \lambda_n A_n$$

can be represented as

$$\sum_{i=1}^{n} \lambda_i A_i = (\lambda_1 a_1 + \cdots + \lambda_n a_n, |\lambda_1|\alpha_1 + \cdots + |\lambda_n|\alpha_n)$$

Assume $A_i = (a_i, \alpha)$, $i = 1, \ldots, n$ are fuzzy numbers of symmetrical triangular form and $\lambda_i \in [0, 1]$, such that $\lambda_1 + \ldots + \lambda_n = 1$.

Then their *convex linear combination* can be represented as

$$\sum_{i=1}^{n} \lambda_i A_i =$$

$$(\lambda_1 a_1 + \cdots + \lambda_n a_n, \ \lambda_1 \alpha + \cdots + \lambda_n \alpha) = (\lambda_1 a_1 + \cdots + \lambda_n a_n, \ \alpha).$$

Let A and B be fuzzy numbers with $[A]^\alpha = [a_1(\alpha), a_2(\alpha)]$ and $[B]^\alpha = [b_1(\alpha), b_2(\alpha)]$. Then it can easily be shown that

$$[A + B]^\alpha = [a_1(\alpha) + b_1(\alpha), a_2(\alpha) + b_2(\alpha)]$$

$$[-A]^\alpha = [-a_2(\alpha), -a_1(\alpha)]$$

$$[A - B]^\alpha = [a_1(\alpha) - b_2(\alpha), a_2(\alpha) - b_1(\alpha)]$$

$$[\lambda A]^\alpha = [\lambda a_1(\alpha), \lambda a_2(\alpha)], \ \lambda \geq 0$$

$$[\lambda A]^\alpha = [\lambda a_2(\alpha), \lambda a_1(\alpha)], \ \lambda < 0$$

for all $\alpha \in [0, 1]$, i.e. any α-level set of the extended sum of two fuzzy numbers is equal to the sum of their α-level sets. The following two theorems show that this property is valid for any continuous function.

Theorem 1.5.2 *[193] Let $f : X \to X$ be a continuous function and let A be fuzzy numbers. Then*

$$[f(A)]^\alpha = f([A]^\alpha)$$

where $f(A)$ is defined by the extension principle (1.2) and

$$f([A]^\alpha) = \{f(x) \mid x \in [A]^\alpha\}.$$

If $[A]^\alpha = [a_1(\alpha), a_2(\alpha)]$ and f is monoton increasing then from the above theorem we get

$$[f(A)]^\alpha = f([A]^\alpha) = f([a_1(\alpha), a_2(\alpha)]) = [f(a_1(\alpha)), f(a_2(\alpha))].$$

Theorem 1.5.3 *[193] Let $f : X \times X \to X$ be a continuous function and let A and B be fuzzy numbers. Then*

$$[f(A, B)]^\alpha = f([A]^\alpha, [B]^\alpha)$$

where

$$f([A]^\alpha, [B]^\alpha) = \{f(x_1, x_2) \mid x_1 \in [A]^\alpha, x_2 \in [B]^\alpha\}.$$

Let $f(x, y) = xy$ and let $[A]^\alpha = [a_1(\alpha), a_2(\alpha)]$ and $[B]^\alpha = [b_1(\alpha), b_2(\alpha)]$ be two fuzzy numbers. Applying Theorem 1.5.3 we get

$$[f(A, B)]^\alpha = f([A]^\alpha, [B]^\alpha) = [A]^\alpha[B]^\alpha.$$

However the equation

$$[AB]^\alpha = [A]^\alpha[B]^\alpha = [a_1(\alpha)b_1(\alpha), a_2(\alpha)b_2(\alpha)]$$

holds if and only if A and B are both nonnegative, i.e. $A(x) = B(x) = 0$ for $x \leq 0$.

If B is nonnegative then we have

$$[A]^\alpha[B]^\alpha = [\min\{a_1(\alpha)b_1(\alpha), a_1(\alpha)b_2(\alpha)\}, \max\{a_2(\alpha)b_1(\alpha), a_2(\alpha)b_2(\alpha)\}]$$

In general case we obtain a very complicated expression for the α level sets of the product AB

$$[A]^\alpha[B]^\alpha = [\min\{a_1(\alpha)b_1(\alpha), a_1(\alpha)b_2(\alpha), a_2(\alpha)b_1(\alpha), a_2(\alpha)b_2(\alpha)\},$$

$$\max\{a_1(\alpha)b_1(\alpha), a_1(\alpha)b_2(\alpha), a_2(\alpha)b_1(\alpha), a_2(\alpha)b_2(\alpha)\}]$$

The above properties of extended operations *addition*, *subtraction* and *multiplication by scalar* of fuzzy fuzzy numbers of type LR are often used in *fuzzy neural networks*.

Definition 1.5.4 *(fuzzy max)*
Let $f(x, y) = \max\{x, y\}$ and let $[A]^\alpha = [a_1(\alpha), a_2(\alpha)]$ and $[B]^\alpha = [b_1(\alpha), b_2(\alpha)]$ be two fuzzy numbers. Applying Theorem 1.5.3 we get

$$[f(A, B)]^\alpha = f([A]^\alpha, [B]^\alpha) = \max\{[A]^\alpha, [B]^\alpha\}$$

$$= [a_1(\alpha) \vee b_1(\alpha), a_2(\alpha) \vee b_2(\alpha)]$$

and we use the notation $\max\{A, B\}$.

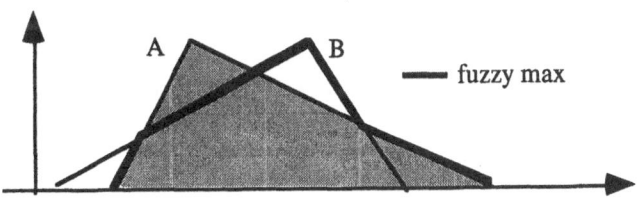

Fig. 1.24. Fuzzy max of triangular fuzzy numbers.

Definition 1.5.5 *(fuzzy min)*

Let $f(x, y) = \min\{x, y\}$ and let $[A]^\alpha = [a_1(\alpha), a_2(\alpha)]$ and $[B]^\alpha = [b_1(\alpha), b_2(\alpha)]$ be two fuzzy numbers. Applying Theorem 1.5.3 we get

$$
\begin{aligned}
[f(A, B)]^\alpha &= f([A]^\alpha, [B]^\alpha) \\
&= \min\{[A]^\alpha, [B]^\alpha\} \\
&= [a_1(\alpha) \wedge b_1(\alpha), a_2(\alpha) \wedge b_2(\alpha)],
\end{aligned}
$$

and we will use the notation $\min\{A, B\}$.

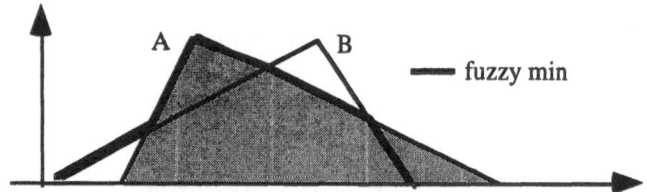

Fig. 1.25. Fuzzy min of triangular fuzzy numbers.

The fuzzy max and min are commutative and associative operations. Furthermore, if A, B and C are fuzzy numbers then

$$\max\{A, \min\{B, C\}\} = \min\{\max\{A, B\}, \max\{A, C\}\}$$

$$\min\{A, \max\{B, C\}\} = \max\{\min\{A, B\}, \min\{A, C\}\}$$

i.e. min and max are distributive.

In the definition of the extension principle one can use any t-norm for modeling the compositional operator.

Definition 1.5.6 *(sup-T extension principle) Let T be a t-norm and let f be a mapping from $X_1 \times X_2 \times \cdots \times X_n$ to Y, Assume $f(A_1, \ldots, A_n)$ is a fuzzy subset of $X_1 \times X_2 \times \cdots \times X_n$, using the extension principle, we can define $f(A)$ as a fuzzy subset of Y such that*

$$
f(A_1, \ldots, A_n)(y) =
\begin{cases}
\sup\{T(A_1(x), \ldots, A_n(x)) \mid x \in f^{-1}(y)\} & \text{if } f^{-1}(y) \neq \emptyset \\
0 & \text{otherwise.}
\end{cases}
$$

Example 1.5.6 *Let $T_P(u, v) = uv$ be the product t-norm and let $f(x_1, x_2) = x_1 + x_2$ be the addition operation on the real line. If A and B are fuzzy numbers then their sup-T extended sum, denoted by $A \oplus B$, is defined by*

$$f(A, B)(y) = \sup_{x_1 + x_2 = y} T_P(A_1(x_1), A_2(x_2)) = \sup_{x_1 + x_2 = y} A_1(x_1) A_2(x_2)$$

Example 1.5.7 *Let* $T(u,v) = \max\{0, u + v - 1\}$ *be the Lukasiewicz t-norm and let* $f(x_1, x_2) = x_1 + x_2$ *be the addition operation on the real line. If* A *and* B *are fuzzy numbers then their sup-T extended sum, denoted by* $A \oplus B$, *is defined by*

$$f(A,B)(y) = \sup_{x_1 + x_2 = y} T_L(A_1(x_1), A_2(x_2))$$
$$= \sup_{x_1 + x_2 = y} \max\{0, A_1(x_1) + A_2(x_2) - 1\}.$$

The reader can find some results on t-norm-based operations on fuzzy numbers in [71, 72, 78].

The following theorem illustrates that, if instead of min-norm in Zadeh's extension principle, we use an arbitrary t-norm, then we obtain results similar to those of Nguyen (see Theorem 1.5.3).

Theorem 1.5.4 *[71] Let* $X \neq \emptyset$, $Y \neq \emptyset$, $Z \neq \emptyset$ *be sets and let* T *be a t-norm. If* $f : X \times Y \rightarrow Z$ *is a two-place function and* $A \in \mathcal{F}(X)$, $B \in \mathcal{F}(Y)$ *then a necessary and sufficient condition for the equality*

$$[f(A,B)]^\alpha = \bigcup_{T(\xi,\eta) \geq \alpha} f([A]^\xi, [B]^\eta), \quad \alpha \in (0,1], \tag{1.4}$$

is, that for each $z \in Z$,

$$\sup_{f(x,y)=z} T(A(x), B(y))$$

is attained.

The next theorem shows that the equality (1.4) holds for all upper semicontinuous T and continuous f in the class of upper semicontinuous and compactly-supported fuzzy subsets. In the following, X, Y, Z are locally compact topological speces.

Theorem 1.5.5 *[71] If* $f : X \times Y \rightarrow Z$ *is continuous and the t-norm* T *is upper semicontinuous, then*

$$[f(A,B)]^\alpha = \bigcup_{T(\xi,\eta) \geq \alpha} f([A]^\xi, [B]^\eta), \quad \alpha \in (0,1],$$

holds for each $A \in \mathcal{F}(X, \mathcal{K})$ *and* $B \in \mathcal{F}(Y, \mathcal{K})$.

The following examples illustrate that the α-cuts of the fuzzy set $f(A,B)$ can be generated in a simple way when the t-norm in question has a simple form.

Example 1.5.8 *If* $T(x,y) = \min(x,y)$, *then using the fact that* $\xi \geq \alpha$ *and* $\eta \geq \alpha$ *implies*

$$f([A]^\xi, [B]^\eta) \subset f([A]^\alpha, [B]^\alpha),$$

equation (1.4) is reduced to the well-known form of Nguyen:

$$[f(A, B)]^\alpha = f([A]^\alpha, [B]^\alpha) \;\; \alpha \in (0, 1],$$

Example 1.5.9 If $T(x, y) = T_W(x, y)$, where

$$T_W(x, y) = \begin{cases} \min\{x, y\} & \text{if } \max\{x, y\} = 1 \\ 0 & \text{otherwise} \end{cases}$$

is the weak t-norm, then (1.4) turns into

$$[f(A, B)]^\alpha = f([A]^1, [B]^\alpha) \cup f([A]^\alpha, [B]^1) \;\; \alpha \in (0, 1],$$

since $T_W(\xi, \eta) \geq \alpha > 0$ holds only if $\xi = 1$ or $\eta = 1$.

Thus if $[A]^1 = \emptyset$ or $[B]^1 = \emptyset$, then $[f(A, B)]^\alpha = \emptyset$, $\forall \alpha \in (0, 1]$. If there exist unique x_0 and y_0 such that $A(x_0) = B(y_0) = 1$, then we obtain

$$[f(A, B)]^\alpha = f(x_0, [B]^\alpha) \cup f([A]^\alpha, y_0) \;\; \alpha \in (0, 1],$$

Example 1.5.10 If $T(x, y) = xy$, then the equation (1.4) yields

$$[f(A, B)]^\alpha = \bigcup_{\xi \in [\alpha, 1]} f([A]^\xi, [B]^{\alpha/\xi}), \;\; \alpha \in (0, 1].$$

Example 1.5.11 If $T(x, y) = \max\{0, x + y - 1\}$, then

$$[f(A, B)]^\alpha = \bigcup_{\xi \in [\alpha, 1]} f([A]^\xi, [B]^{\alpha+1-\xi}), \;\; \alpha \in (0, 1].$$

Exercise 1.5.1 Let $A_1 = (a_1, \alpha)$ and $A_2 = (a_2, \alpha)$ be fuzzy numbers of symmetric triangular form. Compute analytically the membership function of their product-sum, $A_1 \oplus A_2$, defined by

$$(A_1 \oplus A_2)(y) = \sup_{x_1 + x_2 = y} T_P(A_1(x_1), A_2(x_2)) = \sup_{x_1 + x_2 = y} A_1(x_1) A_2(x_2).$$

Exercise 1.5.2 Let $A_1 = (a_1, \alpha)$ and $A_2 = (a_2, \alpha)$ be fuzzy numbers of symmetric triangular form. Compute analytically the membership function of their H_0-sum , $A_1 \oplus A_2$, defined by

$$(A_1 \oplus A_2)(y) = \sup_{x_1 + x_2 = y} H_0(A_1(x_1), A_2(x_2))$$

$$= \sup_{x_1 + x_2 = y} \frac{A_1(x_1) A_2(x_2)}{A_1(x_1) + A_2(x_2) - A_1(x_1) A_2(x_2)}.$$

1.6 Metrics for fuzzy numbers

Let A and B be fuzzy numbers with $[A]^\alpha = [a_1(\alpha), a_2(\alpha)]$ and $[B]^\alpha = [b_1(\alpha), b_2(\alpha)]$. We metricize the set of fuzzy numbers by the metrics

- **Hausdorff distance**

$$D(A, B) = \sup_{\alpha \in [0,1]} \max\{|a_1(\alpha) - b_1(\alpha)|, |a_2(\alpha) - b_2(\alpha)|\}.$$

i.e. $D(A, B)$ is the maximal distance between the α-level sets of A and B. For example, if $A = (a, \alpha)$ and $B = (b, \alpha)$ are fuzzy numbers of symmetric triangular form with the same width $\alpha > 0$ then

$$D(A, B) = |a - b|,$$

and if $A = (a, \alpha)$ and $B = (b, \beta)$ then

$$D(A, B) = |a - b| + |\alpha - \beta|.$$

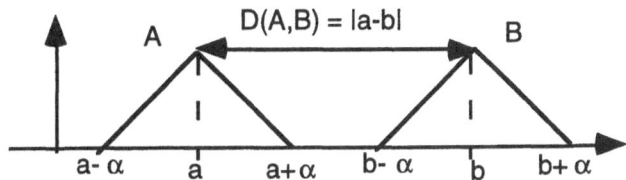

Fig. 1.26. Hausdorff distance between triangular fuzzy numbers A and B.

- C_∞ **distance**

$$C_\infty(A, B) = \|A - B\|_\infty = \sup\{|A(u) - B(u)| : u \in \mathbb{R}\}.$$

i.e. $C_\infty(A, B)$ is the maximal distance between the membership grades of A and B. The following statement holds $0 \le C_\infty(A, B) \le 1$.

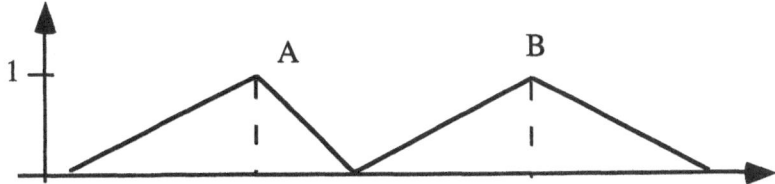

Fig. 1.27. $C(A, B) = 1$ whenever the supports of A and B are disjunctive.

- **Hamming distance** Suppose A and B are fuzzy sets in X. Then their Hamming distance, denoted by $H(A, B)$, is defined by

$$H(A, B) = \int_X |A(x) - B(x)| \, dx.$$

- **Discrete Hamming distance** Suppose A and B are discrete fuzzy sets

$$A = \mu_1/x_1 + \ldots + \mu_n/x_n, \quad B = \nu_1/x_1 + \ldots + \nu_n/x_n$$

Then their Hamming distance is defined by

$$H(A, B) = \sum_{j=1}^{n} |\mu_j - \nu_j|.$$

It should be noted that $D(A, B)$ is a better measure of *similarity* than $C_\infty(A, B)$, because $C_\infty(A, B) \leq 1$ holds even though the supports of A and B are very far from each other.

Definition 1.6.1 *Let f be a fuzzy function from $\mathcal{F}$ to $\mathcal{F}$. Then f is said to be continuous in metric D if $\forall \epsilon > 0$ there exists $\delta > 0$ such that if*

$$D(A, B) \leq \delta$$

then

$$D(f(A), f(B)) \leq \epsilon$$

In a similar way we can define the continuity of fuzzy functions in metric C_∞.

Definition 1.6.2 *Let f be a fuzzy function from $\mathcal{F}(\mathbb{R})$ to $\mathcal{F}(\mathbb{R})$. Then f is said to be continuous in metric C_∞ if $\forall \epsilon > 0$ there exists $\delta > 0$ such that if*

$$C_\infty(A, B) \leq \delta$$

then

$$C_\infty(f(A), f(B)) \leq \epsilon.$$

We note that in the definition of continuity in metric C_∞ the domain and the range of f can be the family of all fuzzy subsets of the real line, while in the case of continuity in metric D the the domain and the range of f is the set of fuzzy numbers.

Exercise 1.6.1 *Let $f(x) = \sin x$ and let $A = (a, \alpha)$ be a fuzzy number of symmetric triangular form. Calculate the membership function of the fuzzy set $f(A)$.*

Exercise 1.6.2 *Let $B_1 = (b_1, \beta_1)$ and $B_2 = (b_2, \beta_2)$ be fuzzy number of symmetric triangular form. Calculate the α-level set of their product $B_1 B_2$.*

Exercise 1.6.3 *Let $B_1 = (b_1, \beta_1)$ and $B_2 = (b_2, \beta_2)$ be fuzzy number of symmetric triangular form. Calculate the α-level set of their fuzzy max $\max\{B_1, B_2\}$.*

Exercise 1.6.4 *Let $B_1 = (b_1, \beta_1)$ and $B_2 = (b_2, \beta_2)$ be fuzzy number of symmetric triangular form. Calculate the α-level set of their fuzzy min $\min\{B_1, B_2\}$.*

Exercise 1.6.5 *Let $A = (a, \alpha)$ and $B = (b, \beta)$ be fuzzy numbers of symmetrical triangular form. Calculate the distances $D(A, B)$, $H(A, B)$ and $C_\infty(A, B)$ as a function of a, b, α and β.*

Exercise 1.6.6 *Let $A = (a, \alpha_1, \alpha_2)$ and $B = (b, \beta_1, \beta_2)$ be fuzzy numbers of triangular form. Calculate the distances $D(A, B)$, $H(A, B)$ and $C_\infty(A, B)$ as a function of a, b, α_1, α_2, β_1 and β_2.*

Exercise 1.6.7 *Let $A = (a_1, a_2, \alpha_1, \alpha_2)$ and $B = (b_1, b_2, \beta_1, \beta_2)$ be fuzzy numbers of trapezoidal form. Calculate the distances $D(A, B)$, $H(A, B)$ and $C_\infty(A, B)$.*

Exercise 1.6.8 *Let $A = (a_1, a_2, \alpha_1, \alpha_2)_{LR}$ and $B = (b_1, b_2, \beta_1, \beta_2)_{LR}$ be fuzzy numbers of type LR. Calculate the distances $D(A, B)$, $H(A, B)$ and $C_\infty(A, B)$.*

Exercise 1.6.9 *Let A and B be discrete fuzzy subsets of*

$$X = \{-2, -1, 0, 1, 2, 3, 4\}$$

defined by

$$A = 0.7/-2 + 0.4/-1 + 0.6/0 + 1.0/1 + 0.6/2 + 0.3/3 + 0.4/4$$

$$B = 0.1/-2 + 0.2/-1 + 0.9/0 + 1.0/1 + 1.0/2 + 0.3/3 + 0.2/4$$

Calculate the Hamming distance between A and B.

1.7 Measures of possibility and necessity

Fuzzy numbers can also be considered as possibility distributions [61]. If $A \in \mathcal{F}$ is a fuzzy number and $x \in \mathbb{R}$ a real number then $A(x)$ can be interpreted as the degree of possiblity of the statement "x is A".

Let $A, B \in \mathcal{F}$ be fuzzy numbers. The degree of possibility that the proposition "A is less than or equal to B" is true denoted by $\text{Pos}[A \leq B]$ and defined by the extension principle as

$$\text{Pos}[A \leq B] = \sup_{x \leq y} \min\{A(x), B(y)\} = \sup_{z \leq 0}(A - B)(z), \qquad (1.5)$$

In a similar way, the degree of possibility that the proposition "A is greater than or equal to B" is true, denoted by $\text{Pos}[A \geq B]$, is defined by

$$\text{Pos}[A \geq B] = \sup_{x \geq y} \min\{A(x), B(y)\} = \sup_{z \geq 0}(A - B)(z). \qquad (1.6)$$

Finally, the degree of possibility that the proposition is true "A is equal to

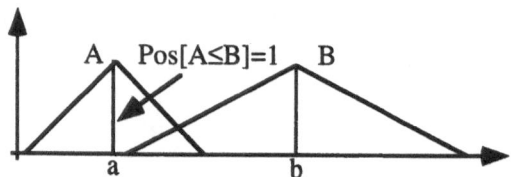

Fig. 1.28. $\text{Pos}[A \leq B] = 1$, because $a \leq b$.

B" and denoted by $\text{Pos}[A = B]$, is defined by

$$\text{Pos}[A = B] = \sup_{x} \min\{A(x), B(x)\} = (A - B)(0), \qquad (1.7)$$

Let $A = (a, \alpha)$ and $B = (b, \beta)$ fuzzy numbers of symmetric triangular form. It is easy to compute that,

$$\text{Pos}[A \leq B] = \begin{cases} 1 & \text{if } a \leq b \\ 1 - \dfrac{a - b}{\alpha + \beta} & \text{otherwise} \\ 0 & \text{if } a \geq b + \alpha + \beta \end{cases} \qquad (1.8)$$

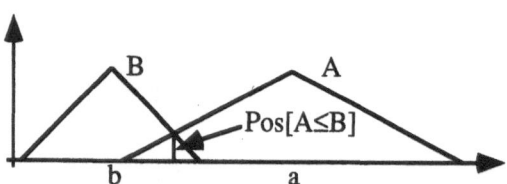

Fig. 1.29. $\text{Pos}[A \leq B] < 1$, because $a > b$.

The degree of necessity that the proposition "A is less than or equal to B" is true, denoted by $\text{Pos}[A \leq B]$, is defined as

$$\text{Nes}[A \leq B] = 1 - \text{Pos}[A \geq B].$$

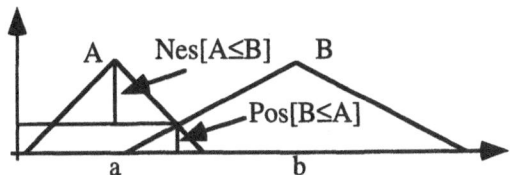

Fig. 1.30. Nes$[A \leq B] < 1$, $(a < b, A \cap B \neq \emptyset)$.

If $A = (a, \alpha)$ and $B = (b, \beta)$ are fuzzy numbers of symmetric triangular form then

$$\text{Nes}[A \leq B] = \begin{cases} 1 & \text{if } a \leq b - \alpha - \beta \\ \dfrac{b-a}{\alpha + \beta} & \text{otherwise} \\ 0 & \text{if } a \geq b \end{cases} \qquad (1.9)$$

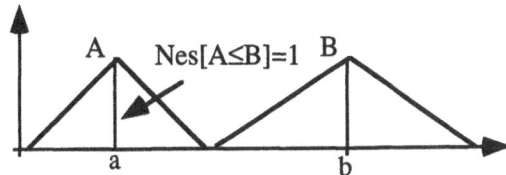

Fig. 1.31. Nes$[A \leq B] = 1$, $(a < b$ and $A \cap B = \emptyset)$.

Let $\xi \in \mathcal{F}$ be a fuzzy number. Given a subset $D \subset \mathbb{R}$, the grade of possibility of the statement "D contains the value of ξ" is defined by

$$\text{Pos}(\xi|D) = \sup_{x \in D} \xi(x) \qquad (1.10)$$

The quantity $1 - \text{Pos}(\xi|\bar{D})$, where $\bar{D}$ is the complement of D, is denoted by

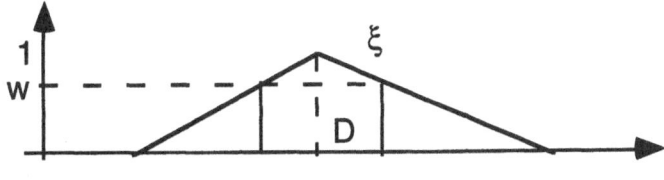

Fig. 1.32. Pos$(\xi|D) = 1$ és Nes$(\xi|D) = 1 - w$.

$Nes(\xi \mid D)$ and is interpreted as the grade of necessity of the statement "D contains the value of ξ". It satisfies dual property with respect to (1.10):

$$\text{Nes}(\xi|D) = 1 - \text{Pos}(\xi|\bar{D}).$$

If $D = [a, b] \subset I\!R$ then instead of $\text{Nes}(\xi \mid [a, b])$ we shall write $\text{Nes}(a \leq \xi \leq b)$ and if $D = \{x\}, x \in I\!R$ we write $\text{Nes}(\xi = x)$.

Let $\xi_1, \xi_2, \ldots$ be a sequence of fuzzy numbers. We say that $\{\xi_n\}$ converges pointwise to a fuzzy set ξ (and write $\lim_{n\to\infty} \xi_n = \xi$) if

$$\lim_{n\to\infty} \xi_n(x) = \xi(x),$$

for all $x \in I\!R$.

Definition 1.7.1 *[47, 48] Let $A \in \mathcal{F}$ be a fuzzy number with*

$$[A]^\gamma = [a_1(\gamma), a_2(\gamma)], \ \gamma \in [0, 1].$$

The mean (or expected) value of A can be defined via the Goetschel-Voxman defuzzification method [92] as

$$E(A) = \frac{\int_0^1 \gamma \cdot \dfrac{a_1(\gamma) + a_2(\gamma)}{2} \, d\gamma}{\int_0^1 \gamma \, d\gamma} = \int_0^1 \gamma(a_1(\gamma) + a_2(\gamma))d\gamma, \qquad (1.11)$$

i.e. the weight of the arithmetic mean of $a_1(\gamma)$ and $a_2(\gamma)$ is just γ.

Definition 1.7.2 *[47, 48] The variance of $A \in \mathcal{F}$ is defined by*

$$\text{Var}(A) = \frac{1}{2} \int_0^1 \gamma(a_2(\gamma) - a_1(\gamma))^2 d\gamma. \qquad (1.12)$$

Exercise 1.7.1 *Let $A = (a, \alpha, \beta)$ be a triangular fuzzy number. Show that,*

$$E(A) = a + \frac{\beta - \alpha}{6}, \quad \text{Var}(A) = \frac{(\alpha + \beta)^2}{24}.$$

Exercise 1.7.2 *Let A be the characteristic function of the crisp interval $[a, b]$. Show that,*

$$\text{Var}(A) = \left(\frac{b - a}{2}\right)^2.$$

Exercise 1.7.3 *Let A and B be fuzzy numbers and let $\lambda \in I\!R$ be a real number. Prove that the expected value of fuzzy numbers defined by (1.11) is a linear function on $\mathcal{F}$, that is,*

$$E(A + B) = E(A) + E(B), \quad E(\lambda A) = \lambda E(A),$$

where the addition and multiplication by a scalar of fuzzy numbers is defined by the sup-min extension principle.

1.8 Fuzzy implications

Let $p =$ "x is in A" and $q =$ "y is in B" be crisp propositions, where A and B are crisp sets for the moment. The implication $p \to q$ is interpreted as

$$\neg(p \wedge \neg q).$$

The full interpretation of the material implication $p \to q$ is that the degree of truth of $p \to q$ quantifies to what extend q is at least as true as p, i.e.

$$\tau(p \to q) = \begin{cases} 1 \text{ if } \tau(p) \le \tau(q) \\ 0 \text{ otherwise} \end{cases}$$

where $\tau(.)$ denotes the truth value of a proposition.

$\tau(p)$	$\tau(q)$	$\tau(p \to q)$
1	1	1
0	1	1
0	0	1
1	0	0

Table 1.1. Truth table for the material implication.

Example 1.8.1 *Let $p =$ "x is bigger than 10" and $q =$ "x is bigger than 9". It is easy to see that $p \to q$ is true, because it can never happen that x is bigger than 10 and at the same time x is not bigger than 9.*

Consider the implication statement: if "pressure is high" then "volume is small". The membership function of the fuzzy set $A =$ "big pressure",

$$A(u) = \begin{cases} 1 & \text{if } u \ge 5 \\ 1 - \dfrac{5 - u}{4} & \text{if } 1 \le u \le 5 \\ 0 & \text{otherwise} \end{cases}$$

can be interpreted as

- x is in the fuzzy set *big pressure* with grade of membership zero, for all $0 \le x \le 1$
- 2 is in the fuzzy set *big pressure* with grade of membership 0.25
- 4 is in the fuzzy set *big pressure* with grade of membership 0.75
- x is in the fuzzy set *big pressure* with grade of membership one, for all $x \ge 5$

The membership function of the fuzzy set B, *small volume*,

$$B(v) = \begin{cases} 1 & \text{if } v \leq 1 \\ 1 - \dfrac{v-1}{4} & \text{if } 1 \leq v \leq 5 \\ 0 & \text{otherwise} \end{cases}$$

can be interpreted as

- y is in the fuzzy set *small volume* with grade of membership zero, for all $y \geq 5$
- 4 is in the fuzzy set *small volume* with grade of membership 0.25
- 2 is in the fuzzy set *small volume* with grade of membership 0.75
- y is in the fuzzy set *small volume* with grade of membership one, for all $y \leq 1$

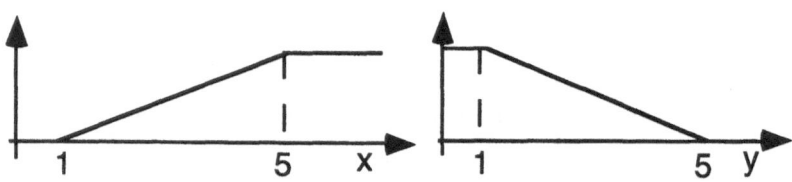

Fig. 1.33. "x is big pressure" and "y is small volume".

If p is a proposition of the form "x is A" where A is a fuzzy set, for example, "big pressure" and q is a proposition of the form "y is B" for example, "small volume" then one encounters the following problem: *How to define the membership function of the fuzzy implication $A \to B$?* It is clear that $(A \to B)(x, y)$ should be defined *pointwise* i.e. $(A \to B)(x, y)$ should be a function of $A(x)$ and $B(y)$. That is

$$(A \to B)(u, v) = I(A(u), B(v)).$$

We shall use the notation

$$(A \to B)(u, v) = A(u) \to B(v).$$

In our interpretation $A(u)$ is considered as the truth value of the proposition "u is big pressure", and $B(v)$ is considered as the truth value of the proposition "v is small volume".

$$u \text{ is big pressure} \to v \text{ is small volume} \equiv A(u) \to B(v)$$

One possible extension of material implication to implications with intermediate truth values is

$$A(u) \to B(v) = \begin{cases} 1 \text{ if } A(u) \le B(v) \\ 0 \text{ otherwise} \end{cases}$$

This implication operator is called *Standard Strict*.

"4 is big pressure" → "1 is small volume" $= A(4) \to B(1) = 0.75 \to 1 = 1$

However, it is easy to see that this fuzzy implication operator is not appropriate for real-life applications. Namely, let $A(u) = 0.8$ and $B(v) = 0.8$. Then we have

$$A(u) \to B(v) = 0.8 \to 0.8 = 1$$

Let us suppose that there is a small error of measurement or small rounding error of digital computation in the value of $B(v)$, and instead 0.8 we have to proceed with 0.7999. Then from the definition of Standard Strict implication operator it follows that

$$A(u) \to B(v) = 0.8 \to 0.7999 = 0$$

This example shows that small changes in the input can cause a big deviation in the output, i.e. our system is very sensitive to rounding errors of digital computation and small errors of measurement.

A smoother extension of material implication operator can be derived from the equation

$$X \to Y = \sup\{Z | X \cap Z \subset Y\}$$

where X, Y and Z are classical sets.

Using the above principle we can define the following fuzzy implication operator

$$A(u) \to B(v) = \sup\{z | \min\{A(u), z\} \le B(v)\}$$

that is,

$$A(u) \to B(v) = \begin{cases} 1 & \text{if } A(u) \le B(v) \\ B(v) & \text{otherwise} \end{cases}$$

This operator is called *Gödel* implication. Using the definitions of negation and union of fuzzy subsets the material implication $p \to q = \neg p \vee q$ can be extended by

$$A(u) \to B(v) = \max\{1 - A(u), B(v)\}$$

This operator is called *Kleene-Dienes* implication.

In many practical applications one uses Mamdani's implication operator to model causal relationship between fuzzy variables. This operator simply takes the minimum of truth values of fuzzy predicates

$$A(u) \to B(v) = \min\{A(u), B(v)\}$$

It is easy to see this is not a correct extension of material implications, because $0 \to 0$ yields zero. However, in knowledge-based systems, we are usually not interested in rules, in which the antecedent part is false.

There are three important classes of fuzzy implication operators:

- S-**implications**: defined by

$$x \to y = S(n(x), y)$$

where S is a t-conorm and n is a negation on $[0,1]$. These implications arise from the Boolean formalism $p \to q = \neg p \vee q$. Typical examples of S-implications are the Łukasiewicz and Kleene-Dienes implications.

- R-**implications**: obtained by residuation of continuous t-norm T, i.e.

$$x \to y = \sup\{z \in [0,1] \,|\, T(x, z) \leq y\}$$

These implications arise from the *Intutionistic Logic* formalism. Typical examples of R-implications are the Gödel and Gaines implications.

- **t-norm implications**: if T is a t-norm then

$$x \to y = T(x, y)$$

Although these implications do not verify the properties of material implication they are used as model of implication in many applications of fuzzy logic. Typical examples of t-norm implications are the Mamdani $(x \to y = \min\{x, y\})$ and Larsen $(x \to y = xy)$ implications.

The most often used fuzzy implication operators are listed in the following table.

Name	Definition
Early Zadeh	$x \to y = \max\{1 - x, \min(x, y)\}$
Łukasiewicz	$x \to y = \min\{1, 1 - x + y\}$
Mamdani	$x \to y = \min\{x, y\}$
Larsen	$x \to y = xy$
Standard Strict	$x \to y = \begin{cases} 1 \text{ if } x \leq y \\ 0 \text{ otherwise} \end{cases}$
Gödel	$x \to y = \begin{cases} 1 \text{ if } x \leq y \\ y \text{ otherwise} \end{cases}$
Gaines	$x \to y = \begin{cases} 1 \quad \text{ if } x \leq y \\ y/x \text{ otherwise} \end{cases}$
Kleene-Dienes	$x \to y = \max\{1 - x, y\}$
Kleene-Dienes-Łukasiewicz	$x \to y = 1 - x + xy$
Yager	$x \to y = y^x$

Table 1.2. Fuzzy implication operators.

1.9 Linguistic variables

The use of fuzzy sets provides a basis for a systematic way for the manipulation of vague and imprecise concepts. In particular, we can employ fuzzy sets to represent linguistic variables. A linguistic variable can be regarded either as a variable whose value is a fuzzy number or as a variable whose values are defined in linguistic terms.

Definition 1.9.1 *(linguistic variable) A linguistic variable is characterized by a quintuple*

$$(x, T(x), U, G, M)$$

in which

- *x is the name of variable;*
- *T(x) is the term set of x, that is, the set of names of linguistic values of x with each value being a fuzzy number defined on U;*
- *G is a syntactic rule for generating the names of values of x;*
- *and M is a semantic rule for associating with each value its meaning.*

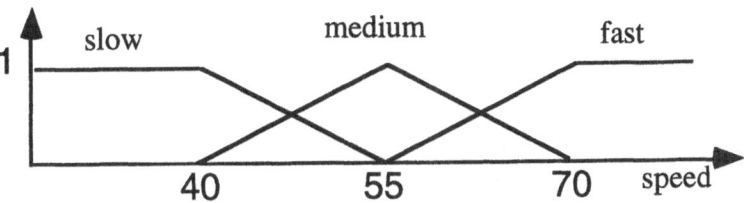

Fig. 1.34. Values of linguistic variable *speed*.

For example, if *speed* is interpreted as a linguistic variable, then its term set T (speed) could be

$$T = \{\text{slow, moderate, fast, very slow, more or less fast, sligthly slow, } \dots \}$$

where each term in T (speed) is characterized by a fuzzy set in a universe of discourse $U = [0, 100]$. We might interpret

- *slow* as "a speed below about 40 mph"
- *moderate* as "a speed close to 55 mph"
- *fast* as "a speed above about 70 mph"

These terms can be characterized as fuzzy sets whose membership functions are shown in the figure below.

In many practical applications we normalize the domain of inputs and use the following type of fuzzy partition

- NB (Negative Big),
- NM (Negative Medium)
- NS (Negative Small),
- ZE (Zero)
- PS (Positive Small),
- PM (Positive Medium)
- PB (Positive Big)

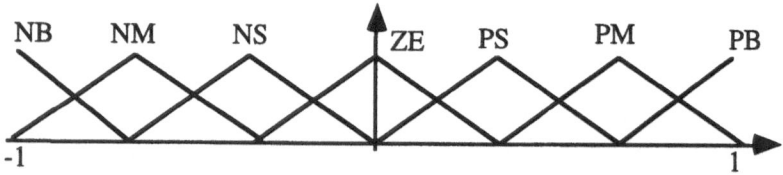

Fig. 1.35. A possible fuzzy partition of $[-1, 1]$.

If A a fuzzy set in X then we can modify the meaning of A with the help of words such as *very, more or less, slightly*, etc. For example, the membership function of fuzzy sets "very A" and "more or less A" can be defined by

$$(\text{very } A)(x) = (A(x))^2,$$

$$(\text{more or less } A)(x) = \sqrt{A(x)}, \ \forall x \in X$$

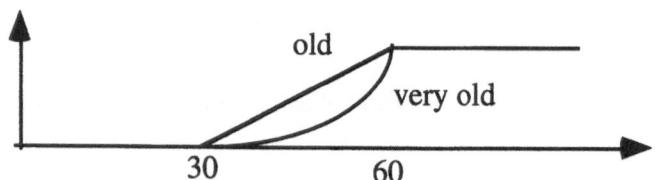

Fig. 1.36. Membership functions of fuzzy sets *old* and *very old*.

1.9.1 The linguistic variable *Truth*

Truth also can be interpreted as linguistic variable with a possible term set

$$T = \{\text{Absolutely false, Very false, False, Fairly true, True, Very true,} \\ \text{Absolutely true}\}$$

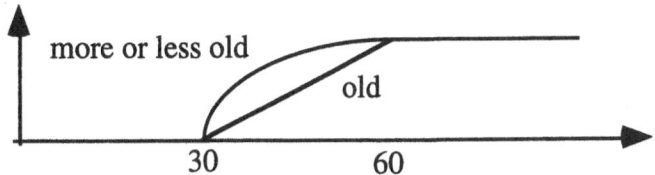

Fig. 1.37. Membership function of fuzzy sets *old* and *more or less old*.

One may define the membership function of linguistic terms of truth as

$$\text{True}(u) = u, \quad \text{False}(u) = 1 - u$$

for each $u \in [0, 1]$, and

$$\text{Absolutely false}(u) = \begin{cases} 1 \text{ if } u = 0 \\ 0 \text{ otherwise} \end{cases}$$

$$\text{Absolutely true}(u) = \begin{cases} 1 \text{ if } u = 1 \\ 0 \text{ otherwise} \end{cases}$$

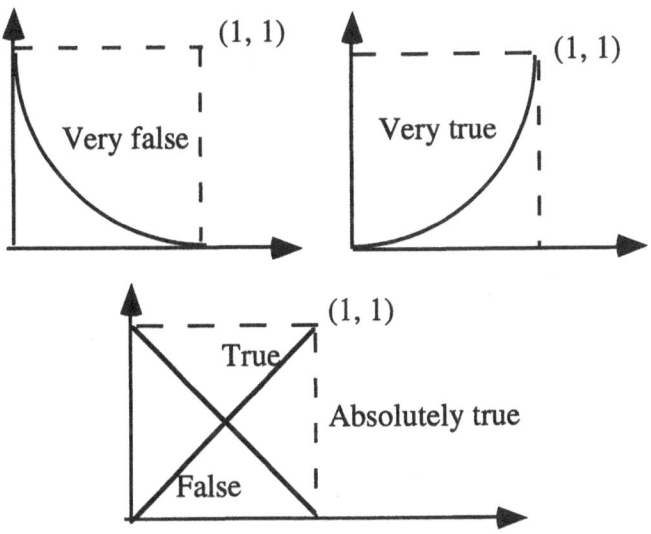

Fig. 1.38. Some values of lingusitic variable *Truth*.

The words "Fairly" and "Very" are interpreted as

$$\text{Fairly true}(u) = \sqrt{u}, \quad \text{Very true}(u) = u^2,$$

Fairly false$(u) = \sqrt{1-u}$,   Very false$(u) = (1-u)^2$

for each $u \in [0,1]$. Suppose we have the fuzzy statement "x is A". Let τ be a term of linguistic variable *Truth*. Then the statement "x is A is τ" is interpreted as "x is $\tau \circ A$". Where

$$(\tau \circ A)(u) = \tau(A(u))$$

for each $u \in [0,1]$. For example, let $\tau = $ "true". Then "x is A is true" is defined by "x is $\tau \circ A$" $= $ "x is A" because

$$(\tau \circ A)(u) = \tau(A(u)) = A(u)$$

for each $u \in [0,1]$. It is why "everything we write is considered to be true".

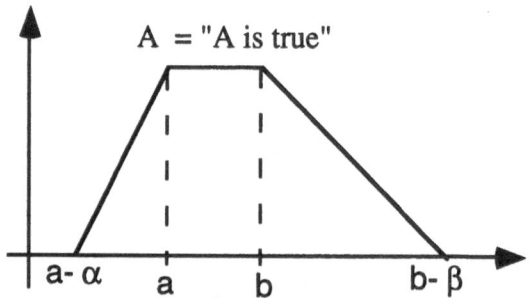

Fig. 1.39. "A is true".

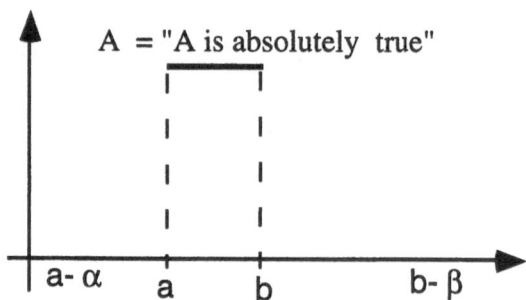

Fig. 1.40. "A is absolutely true".

Let $\tau = $ "absolutely true". Then the statement "x is A is Absolutely true" is defined by "x is $\tau \circ A$", where

$$(\tau \circ A)(x) = \begin{cases} 1 \text{ if } A(x) = 1 \\ 0 \text{ otherwise} \end{cases}$$

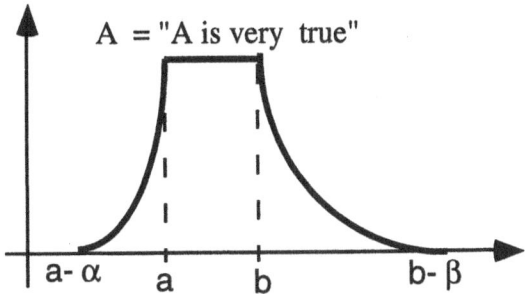

Fig. 1.41. ”A is very true”.

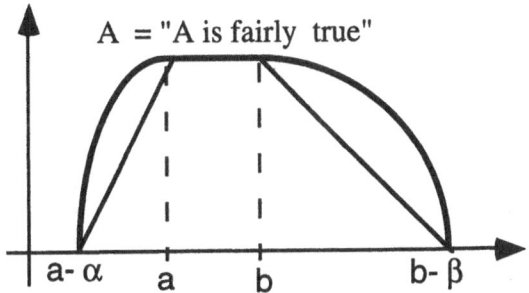

Fig. 1.42. ”A is fairly true”.

Let $\tau = $ ”fairly true”. Then the statement ”x is A is Fairly true” is defined by ”x is $\tau \circ A$”, where

$$(\tau \circ A)(x) = \sqrt{A(x)}$$

Let $\tau = $ ”very true”. Then the statement ”x is A is Fairly true” is defined by ”x is $\tau \circ A$”, where

$$(\tau \circ A)(x) = (A(x))^2$$

Let $\tau = $ ”false”. Then the statement ”x is A is false” is defined by ”x is $\tau \circ A$”, where

$$(\tau \circ A)(x) = 1 - A(x)$$

Let $\tau = $ ”absolutely false”. Then the statement ”x is A is Absolutely false” is defined by ”x is $\tau \circ A$”, where

$$(\tau \circ A)(x) = \begin{cases} 1 \text{ if } A(x) = 0 \\ 0 \text{ otherwise} \end{cases}$$

1.10 The theory of approximate reasoning

In 1979 *Zadeh* introduced the theory of approximate reasoning [240]. This theory provides a powerful framework for reasoning in the face of imprecise

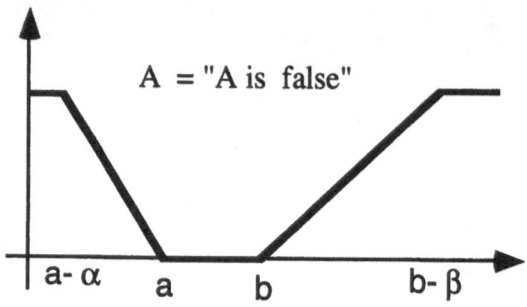

Fig. 1.43. "*A* is false".

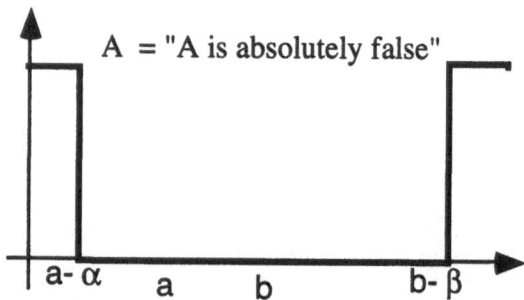

Fig. 1.44. "*A* is absolutely false".

and uncertain information. Central to this theory is the representation of propositions as statements assigning fuzzy sets as values to variables.

Suppose we have two interactive variables $x \in X$ and $y \in Y$ and the causal relationship between x and y is completely known. Namely, we know that y is a function of x

$$y = f(x)$$

Then we can make inferences easily

premise	$y = f(x)$
fact	$x = x'$

$$\text{consequence} \quad y = f(x')$$

This inference rule says that if we have $y = f(x), \forall x \in X$ and we observe that $x = x'$ then y takes the value $f(x')$.

More often than not we do not know the complete causal link f between x and y, only we now the values of $f(x)$ for some particular values of x,

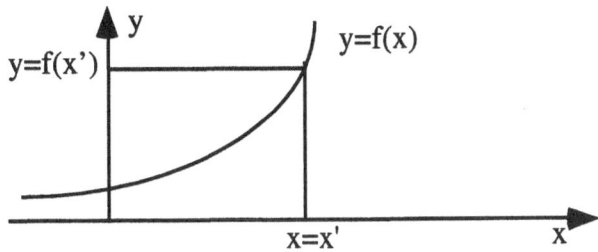

Fig. 1.45. Simple crisp inference.

$$\Re_1 : \quad \text{if } x = x_1 \text{ then } y = y_1$$
$$\Re_2 : \quad \text{if } x = x_2 \text{ then } y = y_2$$
$$\ldots$$
$$\Re_n : \quad \text{If } x = x_n \text{ then } y = y_n$$

Suppose that we are given an $x' \in X$ and want to find an $y' \in Y$ which correponds to x' under the rule-base $\{\Re_1, \ldots, \Re_n\}$,

$\Re_1 :$	if	$x = x_1$ then	$y = y_1$
$\Re_2 :$	If	$x = x_2$ then	$y = y_2$
$\ldots$			
$\Re_n :$	if	$x = x_n$ then	$y = y_n$
fact:		$x = x'$	
consequence:			$y = y'$

This problem is frequently quoted as interpolation.

Let x and y be linguistic variables, e.g. "x is high" and "y is small". The basic problem of approximate reasoning is to find the membership function of the consequence C from the rule-base $\{\Re_1, \ldots, \Re_n\}$ and the fact A,

$\Re_1 :$	if x is A_1 then y is C_1,
$\Re_2 :$	if x is A_2 then y is C_2,
$\ldots\ldots\ldots\ldots$	
$\Re_n :$	if x is A_n then y is C_n
fact:	x is A
consequence:	y is C

In [240] Zadeh introduces a number of translation rules which allow us to represent some common linguistic statements in terms of propositions in our language. In the following we describe some of these translation rules.

Definition 1.10.1 *Entailment rule:*

$$x \text{ is } A \qquad\qquad Mary \text{ is very young}$$
$$A \subset B \qquad\qquad very\ young \subset young$$
$$\overline{x \text{ is } B} \qquad\qquad \overline{Mary \text{ is young}}$$

Definition 1.10.2 *Conjuction rule:*

$$x \text{ is } A \qquad\qquad pressure\ is\ not\ very\ high$$
$$x \text{ is } B \qquad\qquad pressure\ is\ not\ very\ low$$
$$\overline{x \text{ is } A \cap B} \qquad \overline{pressure\ is\ not\ very\ high\ and\ not\ very\ low}$$

Definition 1.10.3 *Disjunction rule:*

$$\qquad x \text{ is } A \qquad\qquad\qquad pressure\ is\ not\ very\ high\ vspace4pt$$
$$or \quad x \text{ is } B \qquad\qquad or \quad pressure\ is\ not\ very\ low$$
$$\overline{x \text{ is } A \cup B} \qquad\qquad \overline{pressure\ is\ not\ very\ high\ or\ not\ very\ low}$$

Definition 1.10.4 *Projection rule:*

$$(x, y) \text{ have relation } R \qquad\qquad (x, y) \text{ have relation } R$$
$$\overline{x \text{ is } \Pi_X(R)} \qquad\qquad\qquad \overline{y \text{ is } \Pi_Y(R)}$$

$$(x, y) \text{ is close to } (3, 2) \qquad\qquad (x, y) \text{ is close to } (3, 2)$$
$$\overline{x \text{ is close to } 3} \qquad\qquad\qquad \overline{y \text{ is close to } 2}$$

Definition 1.10.5 *Negation rule:*

$$not\ (x \text{ is } A) \qquad\qquad not\ (x \text{ is high})$$
$$\overline{x \text{ is } \neg A} \qquad\qquad \overline{x \text{ is not high}}$$

In fuzzy logic and approximate reasoning, the most important fuzzy implication inference rule is the *Generalized Modus Ponens* (GMP). The classical *Modus Ponens* inference rule says:

premise	if p then q
fact	p
consequence	q

This inference rule can be interpreted as: If p is true and $p \to q$ is true then q is true.

The fuzzy implication inference is based on the compositional rule of inference for approximate reasoning suggested by Zadeh [237].

Definition 1.10.6 *(compositional rule of inference)*

premise	if x is A then	y is B
fact	x is A'	
consequence:		y is B'

where the consequence B' is determined as a composition of the fact and the fuzzy implication operator

$$B' = A' \circ (A \to B)$$

that is,

$$B'(v) = \sup_{u \in U} \min\{A'(u), (A \to B)(u,v)\}, \ v \in V.$$

The consequence B' is nothing else but the shadow of $A \to B$ on A'.

The *Generalized Modus Ponens*, which reduces to calssical modus ponens when $A' = A$ and $B' = B$, is closely related to the forward data-driven inference which is particularly useful in the *Fuzzy Logic Control*.

In many practical cases instead of sup-min composition we use sup-T composition, where T is a t-norm.

Definition 1.10.7 *(sup-T compositional rule of inference)*

premise	if x is A then	y is B
fact	x is A'	
consequence:		y is B'

where the consequence B' is determined as a composition of the fact and the fuzzy implication operator

$$B' = A' \circ (A \to B)$$

that is,

$$B'(v) = \sup\{T(A'(u), (A \to B)(u,v)) \mid u \in U\}, \ v \in V.$$

It is clear that T can not be chosen independently of the implication operator.

The classical *Modus Tollens inference rule* says: If $p \to q$ is true and q is false then p is false. The *Generalized Modus Tollens*,

premise	if x is A then	y is B
fact		y is B'
consequence:	x is A'	

which reduces to "Modus Tollens" when $B = \neg B$ and $A' = \neg A$, is closely related to the backward goal-driven inference which is commonly used in expert systems, especially in the realm of *medical diagnosis*.

Suppose that A, B and A' are fuzzy numbers. The Generalized Modus Ponens should satisfy some rational properties

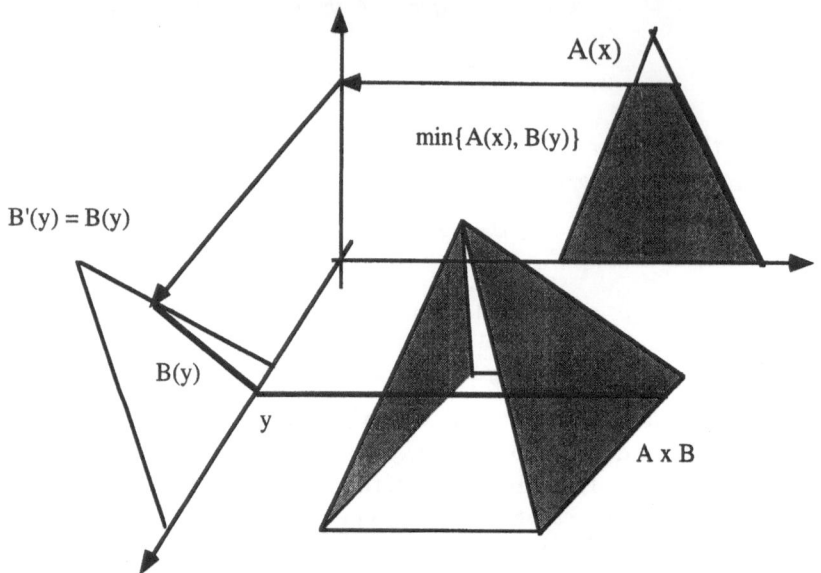

Fig. 1.46. $A \circ A \times B = B$.

Property 1.10.1 *Basic property:*

$$\frac{\begin{array}{ll} if & x \ is \ A \ then \quad y \ is \ B \\ & x \ is \ A \end{array}}{\qquad\qquad\qquad y \ is \ B}$$

$$\frac{\begin{array}{ll} if & \text{pressure is big } then \quad \text{volume is small} \\ & \text{pressure is big} \end{array}}{\qquad\qquad\qquad \text{volume is small}}$$

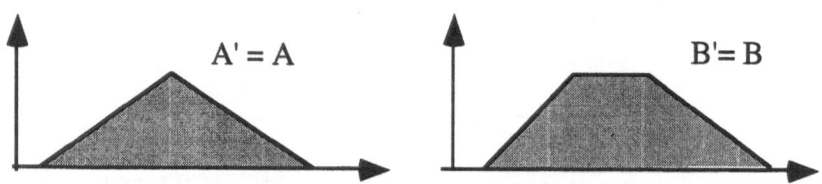

Fig. 1.47. Basic property.

Property 1.10.2 *Total indeterminance:*

$$\frac{\begin{array}{ll} \textit{if } x \textit{ is } A \textit{ then} & y \textit{ is } B \\ x \textit{ is } \neg A & \end{array}}{y \textit{ is unknown}}$$

$$\frac{\begin{array}{ll} \textit{if } \text{ pressure is big} & \textit{then } \text{volume is small} \\ \text{pressure is not big} & \end{array}}{\text{volume is unknown}}$$

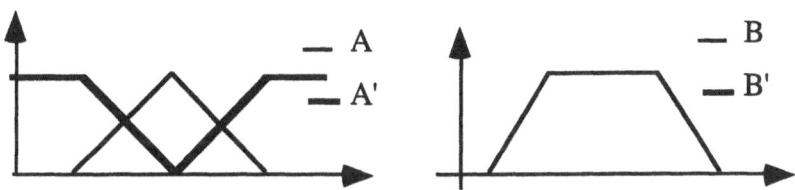

Fig. 1.48. Total indeterminance.

Property 1.10.3 *Subset:*

$$\frac{\begin{array}{ll} \textit{if } x \textit{ is } A \textit{ then} & y \textit{ is } B \\ x \textit{ is } A' \subset A & \end{array}}{y \textit{ is } B}$$

$$\frac{\begin{array}{ll} \textit{if } \text{ pressure is big } \textit{then} & \text{volume is small} \\ \text{pressure is very big} & \end{array}}{\text{volume is small}}$$

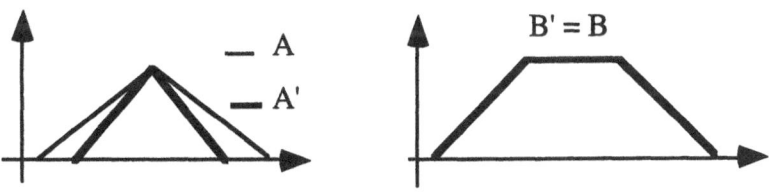

Fig. 1.49. Subset property.

Property 1.10.4 *Superset:*

$$\text{if} \quad x \text{ is } A \text{ then} \quad y \text{ is } B$$
$$x \text{ is } A'$$

$$y \text{ is } B' \supset B$$

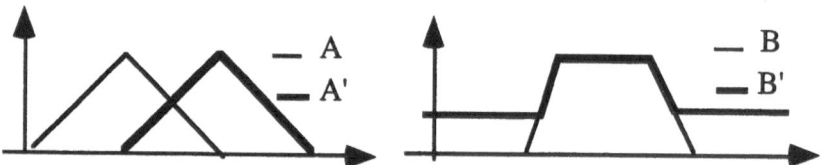

Fig. 1.50. Superset property.

Suppose that A, B and A' are fuzzy numbers. We show that the Generalized Modus Ponens with Mamdani's implication operator does not satisfy all the four properties listed above.

Example 1.10.1 *(The GMP with Mamdani implication)*

$$\text{if} \quad x \text{ is } A \text{ then} \quad y \text{ is } B$$
$$x \text{ is } A'$$

$$y \text{ is } B'$$

where the membership function of the consequence B' is defined by

$$B'(y) = \sup\{A'(x) \wedge A(x) \wedge B(y) | x \in I\!R\}, \ y \in I\!R.$$

Basic property. Let $A' = A$ and let $y \in I\!R$ be arbitrarily fixed. Then we have

$$B'(y) = \sup_x \min\{A(x), \min\{A(x), B(y)\}$$

$$= \sup_x \min\{A(x), B(y)\}$$

$$= \min\{B(y), \sup_x A(x)\} = \min\{B(y), 1\} = B(y).$$

So the basic property is satisfied.

Total indeterminance. Let $A' = \neg A = 1 - A$ and let $y \in I\!R$ be arbitrarily fixed. Then we have

$$B'(y) = \sup_x \min\{1 - A(x), \min\{A(x), B(y)\}$$

$$= \sup_x \min\{A(x), 1 - A(x), B(y)\}$$

$$= \min\{B(y), \sup_x \min\{A(x), 1 - A(x)\}\}$$

$$= \min\{B(y), 1/2\} = 1/2 B(y) < 1$$

this means that the total indeterminance property is not satisfied.

Subset. Let $A' \subset A$ and let $y \in \mathbb{R}$ be arbitrarily fixed. Then we have

$$B'(y) = \sup_x \min\{A'(x), \min\{A(x), B(y)\}$$

$$= \sup_x \min\{A(x), A'(x), B(y)\}$$

$$= \min\{B(y), \sup_x A'(x)\} = \min\{B(y), 1\} = B(y)$$

So the subset property is satisfied.

Superset. Let $y \in \mathbb{R}$ be arbitrarily fixed. Then we have

$$B'(y) = \sup_x \min\{A'(x), \min\{A(x), B(y)\}\} =$$

$$\sup_x \min\{A(x), A'(x), B(y)\} \le B(y).$$

So the superset property of GMP is not satisfied by Mamdani's implication operator.

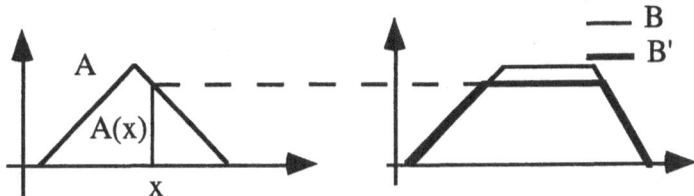

Fig. 1.51. The GMP with Mamdani's implication operator.

Example 1.10.2 *(The GMP with Larsen's product implication)*

$$\begin{array}{ll} if & x\ is\ A\ then\quad y\ is\ B \\ & x\ is\ A' \end{array}$$

$$\rule{6cm}{0.4pt}$$

$$y\ is\ B'$$

where the membership function of the consequence B' is defined by

$$B'(y) = \sup \min\{A'(x), A(x)B(y)|x \in \mathbb{R}\},\ y \in \mathbb{R}.$$

Basic property. Let $A' = A$ and let $y \in \mathbb{R}$ be arbitrarily fixed. Then we have

$$B'(y) = \sup_x \min\{A(x), A(x)B(y)\} = B(y).$$

So the basic property is satisfied.

Total indeterminance. Let $A' = \neg A = 1 - A$ and let $y \in \mathbb{R}$ be arbitrarily fixed. Then we have

$$B'(y) = \sup_x \min\{1 - A(x), A(x)B(y)\} = \frac{B(y)}{1 + B(y)} < 1$$

this means that the total indeterminance property is not satisfied.

Subset. Let $A' \subset A$ and let $y \in I\!\!R$ be arbitrarily fixed. Then we have

$$B'(y) = \sup_x \min\{A'(x), A(x)B(y)\} = \sup_x \min\{A(x), A'(x)B(y)\} = B(y)$$

So the subset property is satisfied.

Superset. Let $y \in I\!\!R$ be arbitrarily fixed. Then we have

$$B'(y) = \sup_x \min\{A'(x), A(x)B(y)\} \le B(y).$$

So, the superset property is not satisfied.

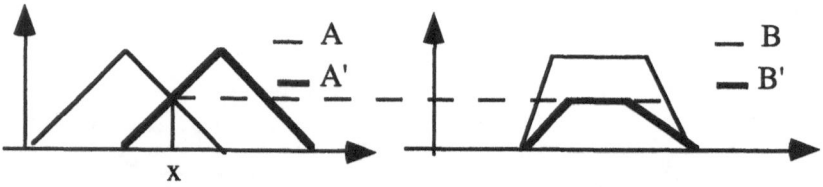

Fig. 1.52. The GMP with Larsen's implication operator.

Suppose we are given one block of fuzzy rules of the form

$$\Re_1 : \qquad \text{if } x \text{ is } A_1 \text{ then } z \text{ is } C_1,$$
$$\Re_2 : \qquad \text{if } x \text{ is } A_2 \text{ then } z \text{ is } C_2,$$
$$\dots\dots\dots\dots$$
$$\Re_n : \qquad \text{if } x \text{ is } A_n \text{ then } z \text{ is } C_n$$
$$\text{fact:} \qquad x \text{ is } A$$

$$\text{consequence:} \qquad\qquad z \text{ is } C$$

The i-th fuzzy rule from this rule-base

$$\Re_i : \text{if } x \text{ is } A_i \text{ then } z \text{ is } C_i$$

is implemented by a *fuzzy implication* R_i and is defined as

$$R_i(u, w) = A_i(u) \rightarrow C_i(w)$$

There are two main approaches to determine the membership function of consequence C.

- **Combine the rules first.** In this approach, we first combine all the rules by an aggregation operator **Agg** into one rule which used to obtain C from A.

$$R = \mathbf{Agg}\left(\Re_1, \Re_2, \cdots, \Re_n\right)$$

If the sentence connective *also* is interpreted as *and* then we get

$$R = \bigcap_{i=1}^{n} R_i$$

that is

$$R(u, w) = \bigcap_{i=1}^{n} R_i(u, w) = \min(A_i(u) \to C_i(w))$$

or by using a t-norm T for modeling the connective *and*

$$R(u, w) = T(R_1(u, w), \dots, R_n(u, w))$$

If the sentence connective *also* is interpreted as *or* then we get

$$R = \bigcup_{i=1}^{n} R_i$$

that is

$$R(u, w) = \bigcup_{i=1}^{n} R_i(u, v, w) = \max(A_i(u) \to C_i(w))$$

or by using a t-conorm S for modeling the connective *or*

$$R(u, w) = S(R_1(u, w), \dots, R_n(u, w))$$

Then we compute C from A by the compositional rule of inference as

$$C = A \circ R = A \circ \mathbf{Agg}\left(R_1, R_2, \cdots, R_n\right)$$

- **Fire the rules first.** Fire the rules individually, given A, and then combine their results into C.
We first compose A with each R_i producing intermediate result

$$C_i' = A \circ R_i$$

for $i = 1, \dots, n$ and then combine the C_i' component wise into C' by some aggregation operator **Agg**

$$C' = \mathbf{Agg}\left(C_1', \dots, C_n'\right) = \mathbf{Agg}\left(A \circ R_1, \dots, A \circ R_n\right).$$

We show that the sup-min compositional operator and the connective *also* interpreted as the union operator are commutative. Thus the consequence, C, inferred from the complete set of rules is equivalent to the aggregated result, C', derived from individual rules.

Lemma 1.10.1 *Let*

$$C = A \circ \bigcup_{i=1}^{n} R_i$$

be defined by standard sup-min composition as

$$C(w) = \sup_u \min\{A(u), \max\{R_1(u, w), \dots, R_n(u, w)\}\}$$

and let

$$C' = \bigcup_{i=1}^{n} A \circ R_i$$

defined by the sup-min composition as

$$C'(w) = \max\{\sup_u A(u) \wedge R_i(u, w), \dots, \sup_u A(u) \wedge R_n(u, w)\}.$$

Then $C(w) = C'(w)$ *for all* w *from the universe of discourse* W.

Proof. Using the distributivity of $\wedge$ over $\vee$ we get

$$C(w) = \sup_u\{A(u) \wedge (R_1(u, w) \vee \dots \vee R_n(u, w))\}$$

$$= \sup_u\{(A(u) \wedge R_1(u, w)) \vee \dots \vee (A(u) \wedge R_n(u, w))$$

$$= \max\{\sup_u A(u) \wedge R_1(u, w), \dots, \sup_u A(u) \wedge R_n(u, w)\}$$

$$= C'(w).$$

Which ends the proof.

Similar statement holds for the sup-product compositional rule of inference, i.e the sup-product compositional operator and the connective *also* as the union operator are commutative.

Lemma 1.10.2 *Let*

$$C = A \circ \bigcup_{i=1}^{n} R_i$$

be defined by sup-product composition as

$$C(w) = \sup_u A(u) \max\{R_1(u, w), \dots, R_n(u, w)\}$$

and let

$$C' = \bigcup_{i=1}^{n} A \circ R_i$$

defined by the sup-product composition as

$$C'(w) = \max\{\sup_u A(u)R_i(u, w), \dots, \sup_u A(u)R_n(u, w)\}$$

Then $C(w) = C'(w)$ *holds for each* w *from the universe of discourse* W.

Proof. Using the distributivity of multiplication over $\vee$ we have

$$C(w) = \sup_u \{A(u)(R_1(u,w) \vee \ldots \vee R_n(u,w))\} =$$

$$\sup_u \{A(u)R_1(u,w) \vee \ldots \vee A(u)R_n(u,w)\}$$

$$= \max\{\sup_u A(u)R_1(u,w), \ldots, \sup_u A(u)R_n(u,w)\} = C'(w).$$

Which ends the proof.

However, the sup-min compositional operator and the connective *also* interpreted as the intersection operator are not usually commutative. In this case, the consequence, C, inferred from the complete set of rules is included in the aggregated result, C', derived from individual rules.

Lemma 1.10.3 *Let*

$$C = A \circ \bigcap_{i=1}^{n} R_i$$

be defined by standard sup-min composition as

$$C(w) = \sup_u \min\{A(u), \min\{R_1(u,w), \ldots, R_n(u,w)\}\}$$

and let

$$C' = \bigcap_{i=1}^{n} A \circ R_i$$

defined by the sup-min composition as

$$C'(w) = \min\{\sup_u \{A(u) \wedge R_i(u,w)\}, \ldots, \sup_u \{A(u) \wedge R_n(u,w)\}\}.$$

Then $C \subset C'$, i.e $C(w) \leq C'(w)$ holds for all w from the universe of discourse W.

Proof. From the relationship

$$A \circ \bigcap_{i=1}^{n} R_i \subset A \circ R_i$$

for each $i = 1, \ldots, n$, we get

$$A \circ \bigcap_{i=1}^{n} R_i \subset \bigcap_{i=1}^{n} A \circ R_i.$$

Which ends the proof.

Similar statement holds for the sup-t-norm compositional rule of inference, i.e the sup-product compositional operator and the connective *also* interpreted as the intersection operator are not commutative. In this case, the consequence, C, inferred from the complete set of rules is included in the aggregated result, C', derived from individual rules.

Lemma 1.10.4 *Let*

$$C = A \circ \bigcap_{i=1}^{n} R_i$$

be defined by sup-T composition as

$$C(w) = \sup_u T(A(u), \min\{R_1(u, w), \ldots, R_n(u, w)\})$$

and let

$$C' = \bigcap_{i=1}^{n} A \circ R_i$$

defined by the sup-T composition as

$$C'(w) = \min\{\sup_u T(A(u), R_i(u, w)), \ldots, \sup_u T(A(u), R_n(u, w))\}.$$

Then $C \subset C'$, i.e $C(w) \leq C'(w)$ holds for all w from the universe of discourse W.

Example 1.10.3 *We illustrate Lemma 1.10.3 by a simple example. Assume we have two fuzzy rules of the form*

$$\Re_1 : \text{if } x \text{ is } A_1 \text{ then } z \text{ is } C_1$$
$$\Re_2 : \text{if } x \text{ is } A_2 \text{ then } z \text{ is } C_2$$

where A_1, A_2 and C_1, C_2 are discrete fuzzy numbers of the universe of discourses $\{x_1, x_2\}$ and $\{z_1, z_2\}$, respectively. Suppose that we input a fuzzy set $A = a_1/x_1 + a_2/x_2$ to the system and let

$$R_1 = \begin{pmatrix} & z_1 & z_2 \\ x_1 & 0 & 1 \\ x_2 & 1 & 0 \end{pmatrix}, \quad R_2 = \begin{pmatrix} & z_1 & z_2 \\ x_1 & 1 & 0 \\ x_2 & 0 & 1 \end{pmatrix}$$

represent the fuzzy rules. We first compute the consequence C by

$$C = A \circ (R_1 \cap R_2).$$

Using the definition of intersection of fuzzy relations we get

$$C = (a_1/x_1 + a_2/x_2) \circ \left[\begin{pmatrix} & z_1 & z_2 \\ x_1 & 0 & 1 \\ x_2 & 1 & 0 \end{pmatrix} \cap \begin{pmatrix} & z_1 & z_2 \\ x_1 & 1 & 0 \\ x_2 & 0 & 1 \end{pmatrix} \right] =$$

$$(a_1/x_1 + a_2/x_2) \circ \begin{pmatrix} & z_1 & z_2 \\ x_1 & 0 & 0 \\ x_2 & 0 & 0 \end{pmatrix} = \emptyset$$

Let us compute now the membership function of the consequence C' by

$$C' = (A \circ R_1) \cap (A \circ R_2)$$

Using the definition of sup-min composition we get

$$A \circ R_1 = (a_1/x_1 + a_2/x_2) \circ \begin{pmatrix} & z_1 & z_2 \\ x_1 & 0 & 1 \\ x_2 & 1 & 0 \end{pmatrix}.$$

Plugging into numerical values

$$(A \circ R_1)(z_1) = \max\{a_1 \wedge 0, a_2 \wedge 1\} = a_2, \; (A \circ R_1)(z_2) = \max\{a_1 \wedge 1, a_2 \wedge 0\} = a_1,$$

So,

$$A \circ R_1 = a_2/z_1 + a_1/z_2$$

and from

$$A \circ R_2 = (a_1/x_1 + a_2/x_2) \circ \begin{pmatrix} & z_1 & z_2 \\ x_1 & 1 & 0 \\ x_2 & 0 & 1 \end{pmatrix} =$$

we get

$$A \circ R_2 = a_1/z_1 + a_2/z_2.$$

Finally,

$$C' = a_2/z_1 + a_1/z_2 \cap a_1/z_1 + a_2/z_2 = a_1 \wedge a_2/z_1 + a_1 \wedge a_2/z_2.$$

Which means that C is a proper subset of C' whenever $\min\{a_1, a_2\} \neq 0$.

Suppose now that the fact of the GMP is given by a fuzzy singleton. Then the process of computation of the membership function of the consequence becomes very simple.

For example, if we use Mamdani's implication operator in the GMP then

rule 1:	if x is A_1 then	z is C_1
fact:	x is $\bar{x}_0$	
consequence:		z is C

where the membership function of the consequence C is computed as

$$C(w) = \sup_u \min\{\bar{x}_0(u), (A_1 \rightarrow C_1)(u, w)\} =$$

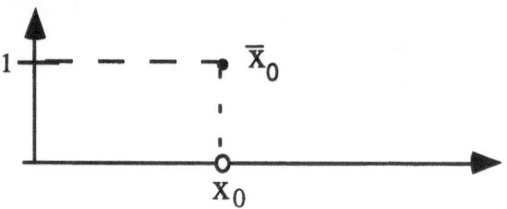

Fig. 1.53. Fuzzy singleton.

$$\sup_{u} \min\{\bar{x}_0(u), \min\{A_1(u), C_1(w)\}\}, \ w \in W.$$

Observing that $\bar{x}_0(u) = 0$, $\forall u \neq x_0$ the supremum turns into a simple minimum

$$C(w) = \min\{\bar{x}_0(x_0) \wedge A_1(x_0) \wedge C_1(w)s\} = \min\{A_1(x_0), C_1(w)\}, \ w \in W.$$

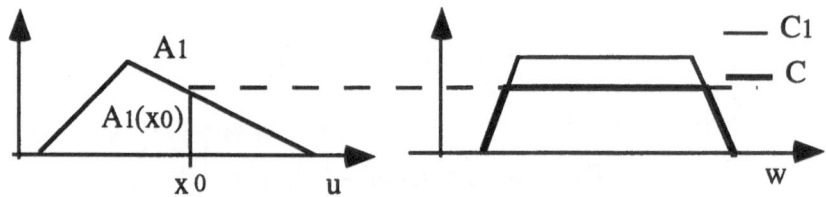

Fig. 1.54. Inference with Mamdani's implication operator.

and if we use Gödel implication operator in the GMP then

$$C(w) = \sup_{u} \min\{\bar{x}_0(u), (A_1 \rightarrow C_1)(u, w)\} = A_1(x_0) \rightarrow C_1(w)$$

So,

$$C(w) = \begin{cases} 1 & \text{if } A_1(x_0) \leq C_1(w) \\ C_1(w) & \text{otherwise} \end{cases}$$

Lemma 1.10.5 *Consider one block of fuzzy rules of the form*

$$\Re_i: \text{if } \dot{x} \text{ is } A_i \text{ then } z \text{ is } C_i, \ 1 \leq i \leq n$$

and suppose that the input to the system is a fuzzy singleton. Then the consequence, C, inferred from the complete set of rules is equal to the aggregated result, C', derived from individual rules. This statements holds for any kind of aggregation operators used to combine the rules.

Proof. Suppose that the input of the system $A = \bar{x}_0$ is a fuzzy singleton. On the one hand we have

$$C(w) = (A \circ \text{Agg}\,[R_1, \dots, R_n])(w)$$

$$= \sup_u \{\bar{x}_0(u) \wedge \text{Agg}\,[R_1, \dots, R_n](u, w)\}$$

$$= \text{Agg}\,[R_1, \dots, R_n](x_0, w) = \text{Agg}\,[R_1(x_0, w), \dots, R_n(x_0, w)].$$

On the other hand

$$C'(w) = \text{Agg}\,[A \circ R_1, \dots, A \circ R_n](w)$$

$$= \text{Agg}\,[\sup_u \min\{\bar{x}_0(u), R_1(u, w)\}, \dots, \sup_u \min\{\bar{x}_0(u), R_1(u, w)\}]$$

$$= \text{Agg}\,[R_1(x_0, w), \dots, R_n(x_0, w)] = C(w).$$

Which ends the proof.

Consider one block of fuzzy rules of the form

$$\Re = \{A_i \to C_i,\ 1 \le i \le n\}$$

where A_i and C_i are fuzzy numbers.

Lemma 1.10.6 *Suppose that in $\Re$ the supports of A_i are pairwise disjunctive:*

$$\text{supp}(A_i) \cap \text{supp}(A_j) = \emptyset, \ \text{for } i \ne j.$$

If the implication operator is defined by

$$x \to z = \begin{cases} 1 \ \text{if } x \le z \\ z \ \text{otherwise} \end{cases}$$

(Gödel implication) then

$$\bigcap_{i=1}^{n} A_i \circ (A_i \to C_i) = C_i$$

holds for $i = 1, \dots, n$

Proof. Since the GMP with Gödel implication satisfies the basic property we get

$$A_i \circ (A_i \to C_i) = A_i.$$

From $supp(A_i) \cap supp(A_j) = \emptyset$, for $i \ne j$ it follows that

$$A_i \circ (A_j \to C_j) = 1, \ i \ne j$$

where **1** is the universal fuzzy set. So,

$$\bigcap_{i=1}^{n} A_i \circ (A_i \to C_i) = C_i \cap \mathbf{1} = C_i$$

Which ends the proof.

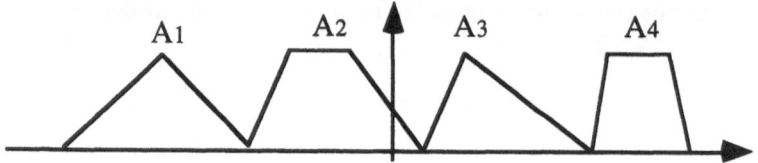

Fig. 1.55. Pairwise disjunctive supports.

Definition 1.10.8 *The rule-base $\Re$ is said to be separated if the core of A_i, defined by*

$$core(A_i) = \{x \mid A_i(x) = 1\},$$

is not contained in $\cap_{j \neq i} supp(A_j)$ for $i = 1, \ldots, n$.

This property means that deleting any of the rules from $\Re$ leaves a point $\hat{x}$ to which no rule applies. It means that every rule is useful.

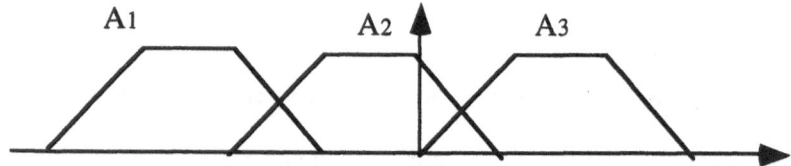

Fig. 1.56. Separated rule-base.

The following theorem shows that Lemma 1.10.6 remains valid for separated rule-bases.

Theorem 1.10.1 *[58] Let $\Re$ be separated. If the implication is modelled by the Gödel implication operator then*

$$\bigcap_{i=1}^{n} A_i \circ (A_i \to C_i) = C_i$$

holds for $i = 1, \ldots, n$

Proof. Since the Gödel implication satisfies the basic property of the GMP we get

$$A_i \circ (A_i \to C_i) = A_i.$$

Since $core(A_i) \cap supp(A_j) \neq \emptyset$, for $i \neq j$ there exists an element $\hat{x}$ such that $\hat{x} \in core(A_i)$ and $\hat{x} \notin supp(A_j)$, $i \neq j$. That is $A_i(\hat{x}) = 1$ and $A_j(\hat{x}) = 0$, $i \neq j$. Applying the compositional rule of inference with Gödel implication operator we get

$$(A_i \circ A_j \to C_j)(z) = \sup_{x} \min\{A_i(x), A_j(x) \to C_j(x))\} \leq$$

$$\min\{A_i(\hat{x}), A_j(\hat{x}) \rightarrow C_j(\hat{x}))\} = \min\{1, 1\} = 1, \ i \neq j$$

for any z. So,

$$\bigcap_{i=1}^{n} A_i \circ (A_i \rightarrow C_i) = C_i \cap 1 = C_i$$

Which ends the proof.

Exercise 1.10.1 *Show that the GMP with Gödel implication operator satisfies properties (1)-(4).*

Exercise 1.10.2 *Show that the GMP with Lukasiewicz implication operator satisfies properties (1)-(4).*

Exercise 1.10.3 *Show that the statement of Lemma 1.10.6 also holds for Lukasiewicz implication operator.*

Exercise 1.10.4 *Show that the statement of Theorem 1.10.1 also holds for Lukasiewicz implication operator.*

1.11 An introduction to fuzzy logic controllers

Conventional controllers are derived from control theory techniques based on mathematical models of the open-loop process, called *system*, to be controlled.

The purpose of the feedback controller is to guarantee a desired response of the output y. The process of keeping the output y close to the setpoint (reference input) y^*, despite the presence disturbances of the system parameters, and noise measurements, is called regulation. The output of the controller (which is the input of the system) is the control action u. The general form of the discrete-time control law is

$$u(k) = f(e(k), e(k-1), \dots, e(k-\tau), u(k-1), \dots, u(k-\tau)) \qquad (1.13)$$

providing a control action that describes the relationship between the input and the output of the controller. In (1.13), e represents the error between the desired setpoint y^* and the output of the system y; parameter τ defines the order of the controller, and f is in general a nonlinear function.

Different control algorithms

- proportional (P)
- integral (I)
- derivative (D)

and their combinations can be derived from control law (1.13) for different values of parameter τ and for different functions f.

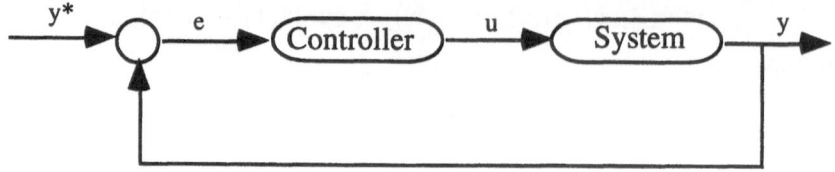

Fig. 1.57. A basic feedback control system..

Example 1.11.1 *A conventional proportional-integral (PI) controller can be described by the function*

$$u = K_p e + K_i \int e \, dt = \int (K_p \dot{e} + K_i e) \, dt$$

or by its differential form

$$du = (K_p \dot{e} + K_i e) \, dt$$

The proportional term provides control action equal to some multiple of the error, while the integral term forces the steady state error to zero.

The discrete-time equivalent expression for the above PI controller is

$$u(k) = K_p e(k) + K_i \sum_{i=1}^{\tau} e(i)$$

where τ defines the order of the controller.

The seminal work by L.A. Zadeh on fuzzy algorithms [237] introduced the idea of formulating the control algorithm by logical rules.

In a fuzzy logic controller (FLC), the dynamic behavior of a fuzzy system is characterized by a set of linguistic description rules based on expert knowledge. The expert knowledge is usually of the form

IF (a set of conditions are satisfied) **THEN**

(a set of consequences can be inferred).

Since the antecedents and the consequents of these IF-THEN rules are associated with fuzzy concepts (linguistic terms), they are often called *fuzzy conditional statements*. In our terminology, a *fuzzy control rule* is a fuzzy conditional statement in which the antecedent is a condition in its application domain and the consequent is a control action for the system under control.

Basically, fuzzy control rules provide a convenient way for expressing control policy and domain knowledge. Furthermore, several linguistic variables might be involved in the antecedents and the conclusions of these rules. When this is the case, the system will be referred to as a multi-input-multi-output

(MIMO) fuzzy system. For example, in the case of two-input-single-output (MISO) fuzzy systems, fuzzy control rules have the form

$$\Re_1 : \text{if } x \text{ is } A_1 \text{ and } y \text{ is } B_1 \text{ then } z \text{ is } C_1$$
also
$$\Re_2 : \text{if } x \text{ is } A_2 \text{ and } y \text{ is } B_2 \text{ then } z \text{ is } C_2$$
also

$\cdots$

also
$$\Re_n : \text{if } x \text{ is } A_n \text{ and } y \text{ is } B_n \text{ then } z \text{ is } C_n$$

where x and y are the process state variables, z is the control variable, A_i, B_i, and C_i are linguistic values of the linguistic variables x, y and z in the universes of discourse U, V, and W, respectively, and an implicit sentence connective *also* links the rules into a rule set or, equivalently, a rule-base. We can represent the FLC in a form similar to the conventional control law (1.13)

$$u(k) = F(e(k), e(k-1), \ldots , e(k-\tau), u(k-1), \ldots , u(k-\tau)) \qquad (1.14)$$

where the function F is described by a fuzzy rule-base. However it does not mean that the FLC is a kind of transfer function or difference equation. The knowledge-based nature of FLC dictates a limited usage of the past values of the error e and control u because it is rather unreasonable to expect meaningful linguistic statements for $e(k-3)$, $e(k-4), \ldots , e(k-\tau)$. A typical FLC describes the relationship between the change of the control

$$\Delta u(k) = u(k) - u(k-1)$$

on the one hand, and the error $e(k)$ and its change

$$\Delta e(k) = e(k) - e(k-1).$$

on the other hand. Such control law can be formalized as

$$\Delta u(k) = F(e(k), \Delta(e(k))) \qquad (1.15)$$

and is a manifestation of the general FLC expression (1.14) with $\tau = 1$. The actual output of the controller $u(k)$ is obtained from the previous value of control $u(k-1)$ that is updated by $\Delta u(k)$

$$u(k) = u(k-1) + \Delta u(k).$$

This type of controller was suggested originally by *Mamdani and Assilian* in 1975 [182] and is called the *Mamdani-type* FLC. A prototypical rule-base of a simple FLC realising the control law (1.15) is listed in the following

$\mathfrak{R}_1$:   **If** e is "positive" and Δe is "near zero" **then** Δu is "positive"

$\mathfrak{R}_2$:   **If** e is "negative" and Δe is "near zero" **then** Δu is "negative"

$\mathfrak{R}_3$:   **If** e is "near zero" and Δe is "near zero" **then** Δu is "near zero"

$\mathfrak{R}_4$:   **If** e is "near zero" and Δe is "positive" **then** Δu is "positive"

$\mathfrak{R}_5$:   **If** e is "near zero" and Δe is "negative" **then** Δu is "negative"

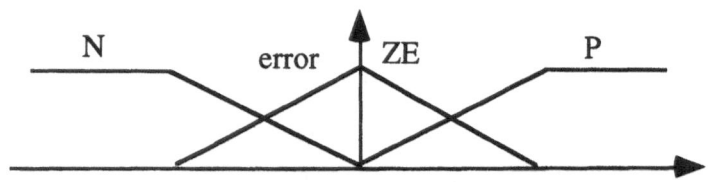

Fig. 1.58. Membership functions for the *error*.

So, our task is the find a crisp control action z_0 from the fuzzy rule-base and from the actual crisp inputs x_0 and y_0:

$\mathfrak{R}_1$:	if x is A_1 and y is B_1 then	z is C_1
also		
$\mathfrak{R}_2$:	if x is A_2 and y is B_2 then	z is C_2
also		
. . .	. . .	
also		
$\mathfrak{R}_n$:	if x is A_n and y is B_n then	z is C_n
input	x is x_0 and y is y_0	
output		z_0

Of course, the inputs of fuzzy rule-based systems should be given by fuzzy sets, and therefore, we have to fuzzify the crisp inputs. Furthermore, the output of a fuzzy system is always a fuzzy set, and therefore to get crisp value we have to defuzzify it.

Fuzzy logic control systems usually consist from four major parts: *Fuzzification interface, Fuzzy rule-base, Fuzzy inference machine* and *Defuzzification interface*.

A fuzzification operator has the effect of transforming crisp data into fuzzy sets. In most of the cases we use fuzzy singletons as fuzzifiers

$$fuzzifier(x_0) := \bar{x}_0$$

where x_0 is a crisp input value from a process.

Suppose now that we have two input variables x and y. A fuzzy control rule

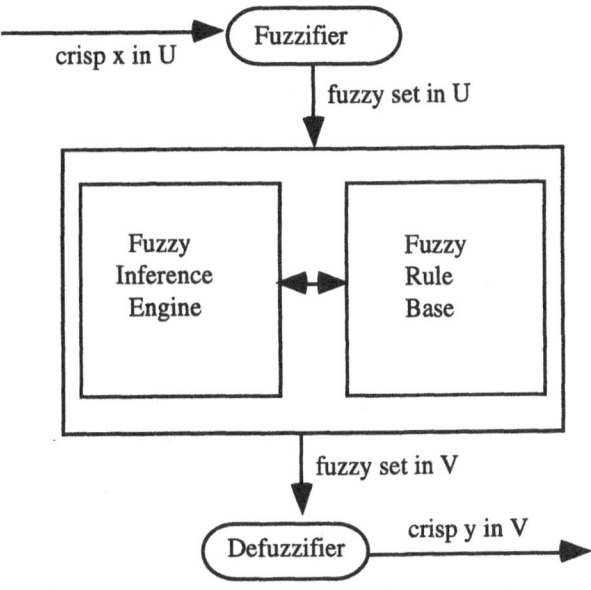

Fig. 1.59. Fuzzy logic controller.

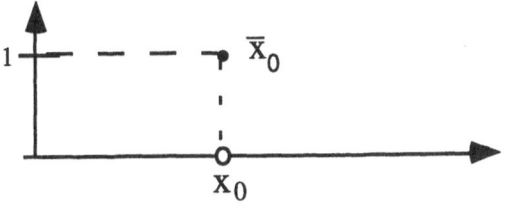

Fig. 1.60. Fuzzy singleton as fuzzifier.

$$\Re_i : \text{if } (x \text{ is } A_i \text{ and } y \text{ is } B_i) \text{ then } (z \text{ is } C_i)$$

is implemented by a *fuzzy implication* R_i and is defined as

$$R_i(u, v, w) = \big[A_i(u) \text{ and } B_i(v)\big] \rightarrow C_i(w)$$

where the logical connective *and* is implemented by Cartesian product, i.e.

$$\big[A_i(u) \text{ and } B_i(v)\big] \rightarrow C_i(w) =$$

$$\big[A_i(u) \times B_i(v)\big] \rightarrow C_i(w) = \min\{A_i(u), B_i(v)\} \rightarrow C_i(w)$$

Of course, we can use any t-norm to model the logical connective *and*.

An FLC consists of a set of fuzzy control rules which are related by the dual concepts of fuzzy implication and the sup–t-norm compositional rule of inference. These fuzzy control rules are combined by using the sentence connective *also*. Since each fuzzy control rule is represented by a fuzzy relation,

the overall behavior of a fuzzy system is characterized by these fuzzy relations. In other words, a fuzzy system can be characterized by a single fuzzy
relation which is the combination in question involves the sentence connective
also. Symbolically, if we have the collection of rules

$$\Re_1 : \text{if } x \text{ is } A_1 \text{ and } y \text{ is } B_1 \text{ then } z \text{ is } C_1$$

also

$$\Re_2 : \text{if } x \text{ is } A_2 \text{ and } y \text{ is } B_2 \text{ then } z \text{ is } C_2$$

also

. . .   . . .

also

$$\Re_n : \text{if } x \text{ is } A_n \text{ and } y \text{ is } B_n \text{ then } z \text{ is } C_n$$

The procedure for obtaining the fuzzy output of such a knowledge base consists from the following three steps:

- Find the firing level of each of the rules.
- Find the output of each of the rules.
- Aggregate the individual rule outputs to obtain the overall system output.

To infer the output z from the given process states x, y and fuzzy relations
R_i, we apply the compositional rule of inference:

$\Re_1$:	if x is A_1 and y is B_1 then z is C_1
$\Re_2$:	if x is A_2 and y is B_2 then z is C_2
. . . .	
$\Re_n$:	if x is A_n and y is B_n then z is C_n
fact :	x is $\bar{x}_0$ and y is $\bar{y}_0$
consequence :	z is C

where the consequence is computed by

$$\text{consequence} = \mathbf{Agg}\,(\text{fact} \circ \Re_1, \ldots, \text{fact} \circ \Re_n).$$

That is,

$$C = \mathbf{Agg}(\bar{x}_0 \times \bar{y}_0 \circ R_1, \ldots, \bar{x}_0 \times \bar{y}_0 \circ R_n)$$

taking into consideration that $\bar{x}_0(u) = 0$, $u \neq x_0$ and $\bar{y}_0(v) = 0$, $v \neq y_0$, the
computation of the membership function of C is very simple:

$$C(w) = \mathbf{Agg}\{A_1(x_0) \times B_1(y_0) \to C_1(w), \ldots, A_n(x_0) \times B_n(y_0) \to C_n(w)\}$$

for all $w \in W$.

The procedure for obtaining the fuzzy output of such a knowledge base
can be formulated as

- The firing level of the i-th rule is determined by

$$A_i(x_0) \times B_i(y_0).$$

- The output of of the i-th rule is calculated by

$$C_i'(w) := A_i(x_0) \times B_i(y_0) \to C_i(w)$$

for all $w \in W$.

- The overall system output, C, is obtained from the individual rule outputs C_i' by

$$C(w) = \mathbf{Agg}\{C_1', \dots, C_n'\}$$

for all $w \in W$.

Example 1.11.2 *If the sentence connective* also *is interpreted as* anding *the rules by using minimum-norm then the membership function of the consequence is computed as*

$$C = (\bar{x}_0 \times \bar{y}_0 \circ R_1) \cap \dots \cap (\bar{x}_0 \times \bar{y}_0 \circ R_n).$$

That is,

$$C(w) = \min\{A_1(x_0) \times B_1(y_0) \to C_1(w), \dots, A_n(x_0) \times B_n(y_0) \to C_n(w)\}$$

for all $w \in W$.

Example 1.11.3 *If the sentence connective* also *is interpreted as* oring *the rules by using minimum-norm then the membership function of the consequence is computed as*

$$C = (\bar{x}_0 \times \bar{y}_0 \circ R_1) \cup \dots \cup (\bar{x}_0 \times \bar{y}_0 \circ R_n).$$

That is,

$$C(w) = \max\{A_1(x_0) \times B_1(y_0) \to C_1(w), \dots, A_n(x_0) \times B_n(y_0) \to C_n(w)\}$$

for all $w \in W$.

Example 1.11.4 *Suppose that the Cartesian product and the implication operator are implemented by the t-norm $T(u,v) = uv$. If the sentence connective* also *is interpreted as* oring *the rules by using minimum-norm then the membership function of the consequence is computed as*

$$C = (\bar{x}_0 \times \bar{y}_0 \circ R_1) \cup \dots \cup (\bar{x}_0 \times \bar{y}_0 \circ R_n).$$

That is,

$$C(w) = \max\{A_1(x_0)B_1(y_0)C_1(w), \dots, A_n(x_0)B_n(y_0)C_n(w)\}$$

for all $w \in W$.

1.12 Defuzzification methods

The output of the inference process so far is a fuzzy set, specifying a possibility distribution of control action. In the on-line control, a nonfuzzy (crisp) control action is usually required. Consequently, one must defuzzify the fuzzy control action (output) inferred from the fuzzy control algorithm, namely:

$$z_0 = defuzzifier(C),$$

where z_0 is the nonfuzzy control output and *defuzzifier* is the defuzzification operator.

Definition 1.12.1 *(defuzzification) Defuzzification is a process to select a representative element from the fuzzy output C inferred from the fuzzy control algorithm.*

The most often used defuzzification operators are

- **Center-of-Area/Gravity.** The defuzzified value of a fuzzy set C is defined as its fuzzy centroid:

$$z_0 = \frac{\int_W zC(z)\,dz}{\int_W C(z)\,dz}.$$

The calculation of the Center-of-Area defuzzified value is simplified if we consider finite universe of discourse W and thus discrete membership function $C(w)$

$$z_0 = \frac{\sum z_j C(z_j)}{\sum C(z_j)}.$$

- **Center-of-Sums, Center-of-Largest-Area**
- **First-of-Maxima.** The defuzzified value of a fuzzy set C is its smallest maximizing element, i.e.

$$z_0 = \min\{z \,|\, C(z) = \max_w C(w)\}.$$

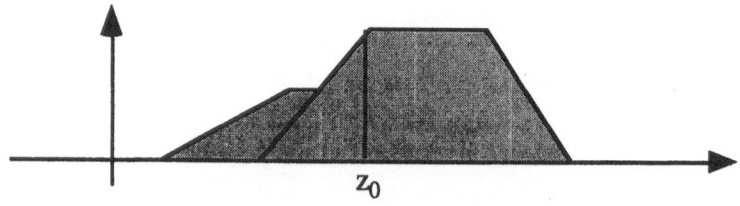

Fig. 1.61. First-of-Maxima defuzzification method.

- **Middle-of-Maxima.** The defuzzified value of a discrete fuzzy set C is defined as a mean of all values of the universe of discourse, having maximal membership grades

$$z_0 = \frac{1}{N} \sum_{j=1}^{N} z_j$$

where $\{z_1, \dots, z_N\}$ is the set of elements of the universe W which attain the maximum value of C. If C is not discrete then defuzzified value of a fuzzy set C is defined as

$$z_0 = \frac{\int_G z \, dz}{\int_G dz}$$

where G denotes the set of maximizing element of C.

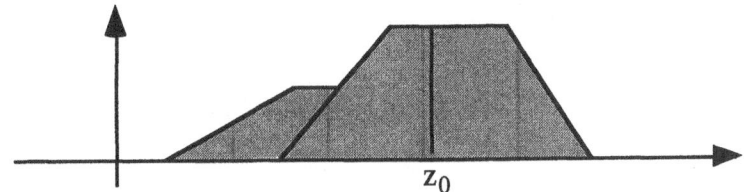

Fig. 1.62. Middle-of-Maxima defuzzification method.

- **Max-Criterion.** This method chooses an arbitrary value, from the set of maximizing elements of C, i.e.

$$z_0 \in \{z \,|\, C(z) = \max_w C(w)\}.$$

- **Height defuzzification** The elements of the universe of discourse W that have membership grades lower than a certain level α are completely discounted and the defuzzified value z_0 is calculated by the application of the Center-of-Area method on those elements of W that have membership grades not less than α:

$$z_0 = \frac{\int_{[C]^\alpha} z C(z) \, dz}{\int_{[C]^\alpha} C(z) \, dz}.$$

where $[C]^\alpha$ denotes the α-level set of C as usually.

Example 1.12.1 *[148] Consider a fuzzy controller steering a car in a way to avoid obstacles. If an obstacle occurs right ahead, the plausible control action depicted in Fig. 1.63 could be interpreted as "turn right or left". Both Center-of-Area and Middle-of-Maxima defuzzification methods results in a control action "drive ahead straightforward" which causes an accident.*

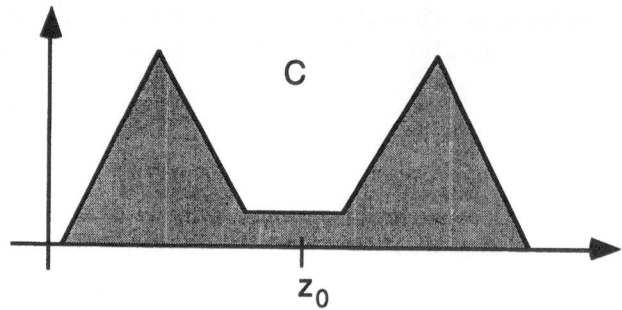

Fig. 1.63. Undisered result by Center-of-Area and Middle-of-Maxima defuzzification methods.

A suitable defuzzification method would have to choose between different control actions (choose one of two triangles in the Figure) and then transform the fuzzy set into a crisp value.

Exercise 1.12.1 Let the overall system output, C, have the following membership function

$$C(x) = \begin{cases} x^2 & \text{if } 0 \leq x \leq 1 \\ 2 - \sqrt{x} & \text{if } 1 \leq x \leq 4 \\ 0 & \text{otherwise} \end{cases}$$

Compute the defuzzified value of C using the Center-of-Area and Height-Defuzzification with $\alpha = 0.7$ methods.

Exercise 1.12.2 Let $C = (a, b, \alpha)$ be a triangular fuzzy number. Compute the defuzzified value of C using the Center-of-Area and Middle-of-Maxima methods.

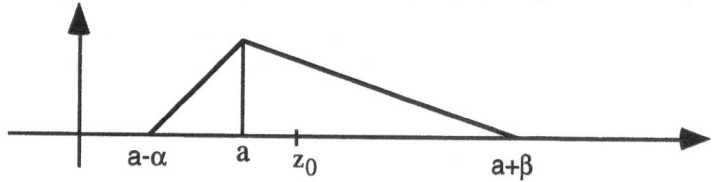

Fig. 1.64. z_0 is the defuzzified value of C.

Exercise 1.12.3 Let $C = (a, b, \alpha, \beta)$ be a trapezoidal fuzzy number. Compute the defuzzified value of C using the Center-of-Area and Middle-of-Maxima methods.

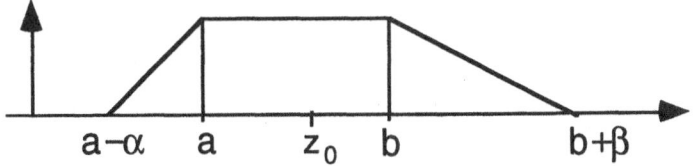

Fig. 1.65. z_0 is the defuzzified value of C.

Exercise 1.12.4 *Let* $C = (a, b, \alpha, \beta)_{LR}$ *be a fuzzy number of type LR. Compute the defuzzified value of* C *using the Center-of-Area and Middle-of-Maxima methods.*

1.13 Inference mechanisms

We present four well-known inference mechanisms in fuzzy logic control systems. For simplicity we assume that we have two fuzzy control rules of the form

$\Re_1$:   if x is A_1 and y is B_1 then z is C_1

also

$\Re_2$:   if x is A_2 and y is B_2 then z is C_2

fact : x is $\bar{x}_0$ and y is $\bar{y}_0$

---

consequence : z is C

- **Mamdani.** The fuzzy implication is modelled by Mamdani's minimum operator and the sentence connective *also* is interpreted as oring the propositions and defined by max operator.
  The firing levels of the rules, denoted by α_i, $i = 1, 2$, are computed by

$$\alpha_1 = A_1(x_0) \wedge B_1(y_0), \quad \alpha_2 = A_2(x_0) \wedge B_2(y_0)$$

The individual rule outputs are obtained by

$$C_1'(w) = (\alpha_1 \wedge C_1(w)), \quad C_2'(w) = (\alpha_2 \wedge C_2(w))$$

Then the overall system output is computed by oring the individual rule outputs

$$C(w) = C_1'(w) \vee C_2'(w) = (\alpha_1 \wedge C_1(w)) \vee (\alpha_2 \wedge C_2(w))$$

Finally, to obtain a deterministic control action, we employ any defuzzification strategy.

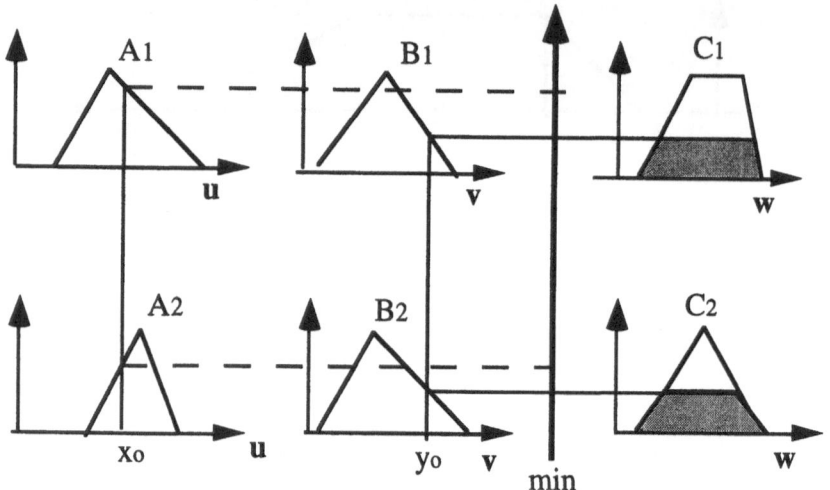

Fig. 1.66. Making inferences with Mamdani's implication operator.

- **Tsukamoto.** All linguistic terms are supposed to have monotonic membership functions.

  The firing levels of the rules, denoted by α_i, $i = 1, 2$, are computed by

$$\alpha_1 = A_1(x_0) \wedge B_1(y_0), \quad \alpha_2 = A_2(x_0) \wedge B_2(y_0)$$

In this mode of reasoning the individual crisp control actions z_1 and z_2 are computed from the equations

$$\alpha_1 = C_1(z_1), \quad \alpha_2 = C_2(z_2)$$

and the overall crisp control action is expressed as

$$z_0 = \frac{\alpha_1 z_1 + \alpha_2 z_2}{\alpha_1 + \alpha_2}$$

i.e. z_0 is computed by the discrete Center-of-Gravity method.

If we have n rules in our rule-base then the crisp control action is computed as

$$z_0 = \sum_{i=1}^{n} \alpha_i z_i \Big/ \sum_{i=1}^{n} \alpha_i,$$

where α_i is the firing level and z_i is the (crisp) output of the i-th rule, $i = 1, \ldots, n$

Example 1.13.1 *We illustrate Tsukamoto's reasoning method by the following simple example*

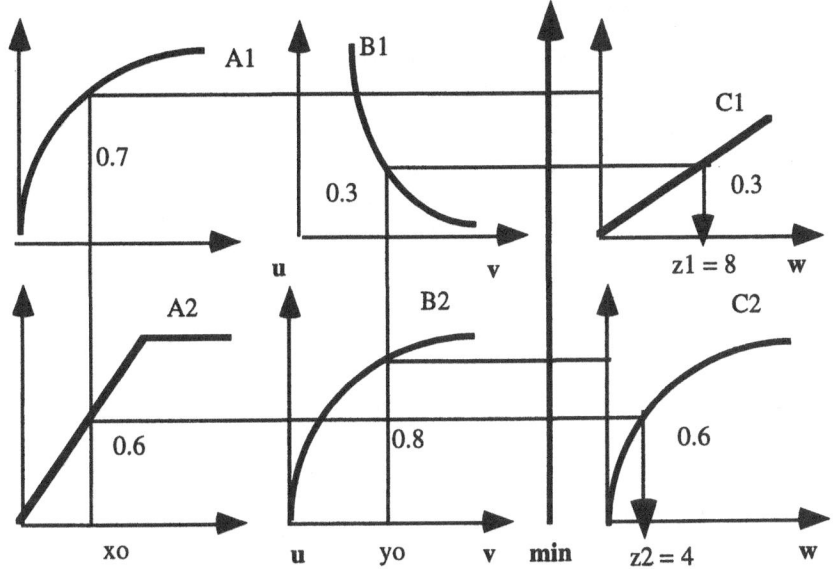

Fig. 1.67. Tsukamoto's inference mechanism.

$\Re_1$:   *if x is A_1 and y is B_1 then z is C_1*

also

$\Re_2$:   *if x is A_2 and y is B_2 then z is C_2*

fact :   *x is $\bar{x}_0$ and y is $\bar{y}_0$*

consequence :   *z is C*

Then according to Fig.1.48 we see that

$$A_1(x_0) = 0.7, \quad B_1(y_0) = 0.3$$

therefore, the firing level of the first rule is

$$\alpha_1 = \min\{A_1(x_0), B_1(y_0)\} = \min\{0.7, 0.3\} = 0.3$$

and from

$$A_2(x_0) = 0.6, \quad B_2(y_0) = 0.8$$

it follows that the firing level of the second rule is

$$\alpha_2 = \min\{A_2(x_0), B_2(y_0)\} = \min\{0.6, 0.8\} = 0.6$$

the individual rule outputs $z_1 = 8$ and $z_2 = 4$ are derived from the equations

$$C_1(z_1) = 0.3, \quad C_2(z_2) = 0.6$$

and the crisp control action is

$$z_0 = (8 \times 0.3 + 4 \times 0.6)/(0.3 + 0.6) = 6.$$

- **Sugeno.** Sugeno and Takagi use the following architecture [208]

$\Re_1$:   if x is A_1 and y is B_1 then $z_1 = a_1 x + b_1 y$

also

$\Re_2$:   if x is A_2 and y is B_2 then $z_2 = a_2 x + b_2 y$

fact : x is $\bar{x}_0$ and y is $\bar{y}_0$

consequence : z_0

The firing levels of the rules are computed by

$$\alpha_1 = A_1(x_0) \wedge B_1(y_0), \quad \alpha_2 = A_2(x_0) \wedge B_2(y_0)$$

then the individual rule outputs are derived from the relationships

$$z_1^* = a_1 x_0 + b_1 y_0, \quad z_2^* = a_2 x_0 + b_2 y_0$$

and the crisp control action is expressed as

$$z_0 = \frac{\alpha_1 z_1^* + \alpha_2 z_2^*}{\alpha_1 + \alpha_2}$$

If we have n rules in our rule-base then the crisp control action is computed as

$$z_0 = \sum_{i=1}^{n} \alpha_i z_i^* \Big/ \sum_{i=1}^{n} \alpha_i,$$

where α_i denotes the firing level of the i-th rule, $i = 1, \dots, n$

Example 1.13.2 *We illustrate Sugeno's reasoning method by the following simple example*

$\Re_1$:   *if x is BIG and y is SMALL then $z_1 = x + y$*

also

$\Re_2$:   *if x is MEDIUM and y is BIG then $z_2 = 2x - y$*

fact : x_0 is 3 and y_0 is 2

conseq : z_0

Then according to Fig.1.49 we see that

$$\mu_{BIG}(x_0) = \mu_{BIG}(3) = 0.8, \quad \mu_{SMALL}(y_0) = \mu_{SMALL}(2) = 0.2$$

therefore, the firing level of the first rule is

$$\alpha_1 = \min\{\mu_{BIG}(x_0), \mu_{SMALL}(y_0)\} = \min\{0.8, 0.2\} = 0.2$$

and from

$$\mu_{MEDIUM}(x_0) = \mu_{MEDIUM}(3) = 0.6, \ \mu_{BIG}(y_0) = \mu_{BIG}(2) = 0.9$$

it follows that the firing level of the second rule is

$$\alpha_2 = \min\{\mu_{MEDIUM}(x_0), \mu_{BIG}(y_0)\} = \min\{0.6, 0.9\} = 0.6.$$

the individual rule outputs are computed as

$$z_1^* = x_0 + y_0 = 3 + 2 = 5, \ z_2^* = 2x_0 - y_0 = 2 \times 3 - 2 = 4$$

so the crisp control action is

$$z_0 = \frac{5 \times 0.2 + 4 \times 0.6}{0.2 + 0.6} = 4.25.$$

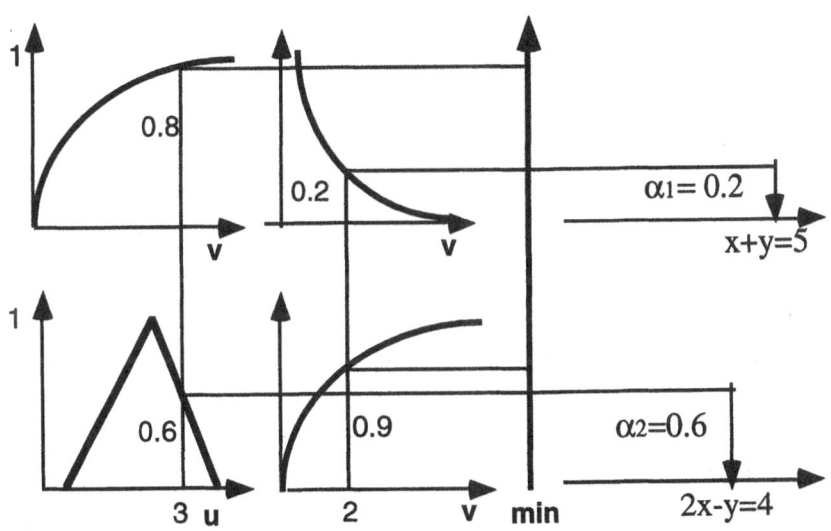

Fig. 1.68. Sugeno's inference mechanism.

- **Larsen.** The fuzzy implication is modelled by Larsen's prduct operator and the sentence connective *also* is interpreted as oring the propositions and defined by max operator. Let us denote α_i the firing level of the i-th rule, $i = 1, 2$

$$\alpha_1 = A_1(x_0) \wedge B_1(y_0), \ \alpha_2 = A_2(x_0) \wedge B_2(y_0).$$

Then membership function of the inferred consequence C is pointwise given by

$$C(w) = (\alpha_1 C_1(w)) \vee (\alpha_2 C_2(w)).$$

To obtain a deterministic control action, we employ any defuzzification strategy.

If we have n rules in our rule-base then the consequence C is computed as

$$C(w) = \bigvee_{i=1}^{n} (\alpha_i C_1(w))$$

where α_i denotes the firing level of the i-th rule, $i = 1, \ldots, n$

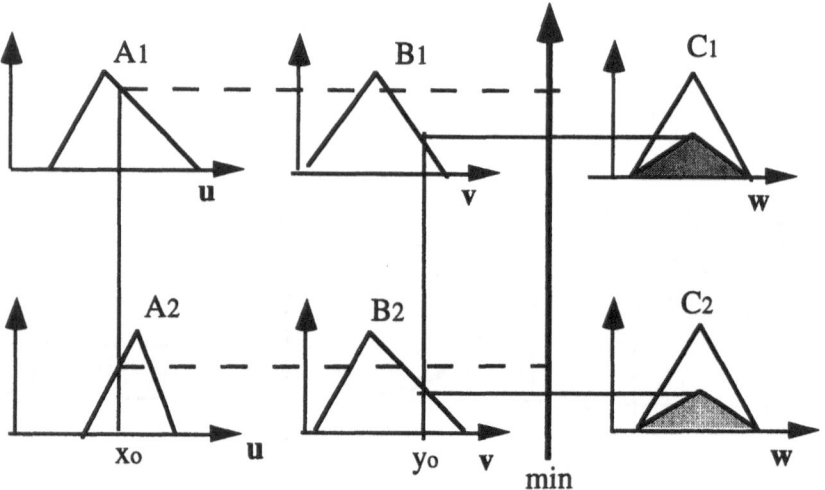

Fig. 1.69. Making inferences with Larsen's product operation rule.

1.14 Construction of data base and rule base of FLC

The knowledge base of an fuzzy logic controller is compromised of two components, namely, a *data base* and a *fuzzy rule base*. The concepts associated with a data base are used to characterize fuzzy control rules and fuzzy data manipulation in an FLC. These concepts are subjectively defined and based on experience and engineering judgment. It should be noted that the *correct choice* of the membership functions of a linguistic term set plays an essential role in the success of an application.

Drawing heavily on [178] we discuss some of the important aspects relating to the construction of the data base and rule base in an FLC.

- **Data base strategy.**

  The data base strategy is concerned with the supports on which primary fuzzy sets are defined. The union of these supports should cover the related universe of discourse in relation to some level set ε. This property of an FLC is called ε-completeness. In general, we choose the level ε at the crossover point, implying that we have a strong belief in the positive sense of the fuzzy control rules which are associated with the FLC.

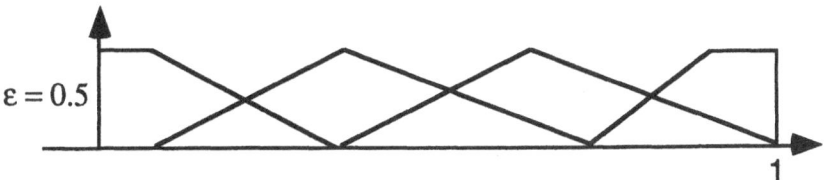

Fig. 1.70. ε-complet fuzzy partition of $[0,1]$ with $\varepsilon = 0.5$.

In this sense, a dominant rule always exists and is associated with the degree of belief greater than 0.5. In the extreme case, two dominant rules are activated with equal belief 0.5.

- **Discretization/normalization of universes of discourse.**

  Discretization of universes of discourse is frequently referred to as quantization. In effect, quantization discretizes a universe into a certain number of segments (quantization levels). Each segment is labeled as a generic element, and forms a discrete universe. A fuzzy set is then defined by assigning grade of membership values to each generic element of the new discrete universe.

  In the case of an FLC with continuous universes, the number of quantization levels should be large enough to provide an adequate approximation and yet be small to save memory storage. The choice of quantization levels has an essential influence on how fine a control can be obtained.

  For example, if a universe is quantized for every five units of measurement instead of ten units, then the controller is twice as sensitive to the observed variables.

  A look-up table based on discrete universes, which defines the output of a controller for all possible combinations of the input signals, can be implemented by off-line processing in order to shorten the running time of the controller.

  However, these findings have purely empirical nature and so far no formal analysis tools exist for studying how the quantization affects controller performance. This explains the preference for continuous domains, since quantization is a source of instability and oscillation problems.

- **Fuzzy partition of the input and output spaces.**

Range	NB	NM	NS	ZE	PS	PM	PB
$x \leq -3$	1.0	0.3	0.0	0.0	0.0	0.0	0.0
$-3 < x \leq -1.6$	0.7	0.7	0.0	0.0	0.0	0.0	0.0
$-1.6 < x \leq -0.8$	0.3	1.0	0.3	0.0	0.0	0.0	0.0
$-0.8 < x \leq -0.4$	0.0	0.7	0.7	0.0	0.0	0.0	0.0
$-0.4 < x \leq -0.2$	0.0	0.3	0.1	0.3	0.0	0.0	0.0
$-0.2 < x \leq -0.1$	0.0	0.0	0.7	0.7	0.0	0.0	0.0
$-0.1 < x \leq 0.1$	0.0	0.0	0.3	0.1	0.3	0.0	0.0
$0.1 < x \leq 0.2$	0.0	0.0	0.0	0.7	0.7	0.0	0.0
$0.2 < x \leq 0.4$	0.0	0.0	0.0	0.3	0.1	0.3	0.0
$0.4 < x \leq 0.8$	0.0	0.0	0.0	0.0	0.7	0.7	0.0
$0.8 < x \leq 1.6$	0.0	0.0	0.0	0.0	0.3	1.0	0.3
$1.6 < x \leq 3.0$	0.0	0.0	0.0	0.0	0.0	0.7	0.7
$3.0 \leq x$	0.0	0.0	0.0	0.0	0.0	0.3	1.0

Table 1.3. Quantization.

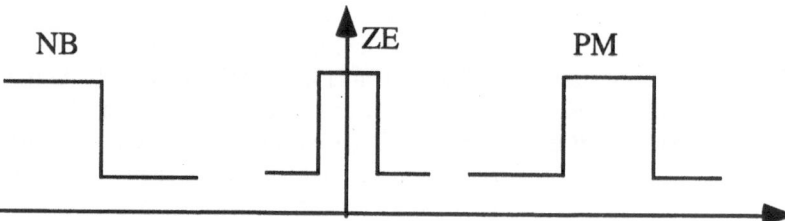

Fig. 1.71. Discretization of the universe of discourses.

A linguistic variable in the antecedent of a fuzzy control rule forms a fuzzy input space with respect to a certain universe of discourse, while that in the consequent of the rule forms a fuzzy output space. In general, a linguistic variable is associated with a term set, with each term in the term set defined on the same universe of discourse. A fuzzy partition, then, determines how many terms should exist in a term set. The primary fuzzy sets usually have a meaning, such as NB, NM, NS, ZE, PS, PM and PB.

Since a normalized universe implies the knowledge of the input/output space via appropriate scale mappings, a well-formed term set can be achieved as shown. If this is not the case, or a nonnormalized universe is used, the terms could be asymmetrical and unevenly distributed in the universe. Furthermore, the cardinality of a term set in a fuzzy input space determines the maximum number of fuzzy control rules that we can construct.

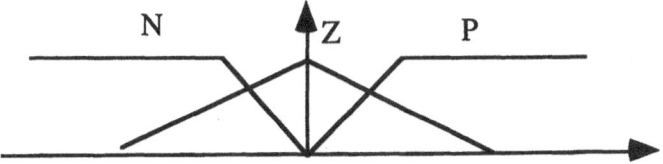

Fig. 1.72. A coarse fuzzy partition of the input space.

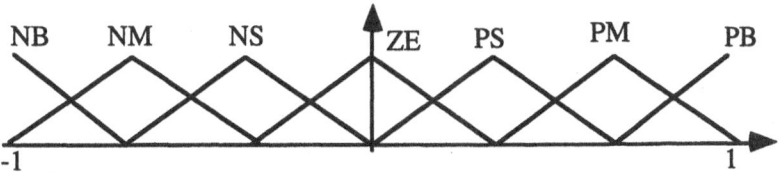

Fig. 1.73. A finer fuzzy partition of $[-1, 1]$.

In the case of two-input-one-output fuzzy systems, if the linguistic variables x and y can take 7 different values, respectively, the maximum rule number is 7×7. It should be noted that the fuzzy partition of the fuzzy input/output space is not deterministic and has no unique solution. A heuristic cut and trial procedure is usually needed to find the optimal fuzzy partition.

- **Completeness.**
  Intuitively, a fuzzy control algorithm should always be able to infer a proper control action for every state of process. This property is called "completeness". The completeness of an FLC relates to its data base, rule base, or both.

- **Choice of the membership function of a primary fuzzy sets.**
  There are two methods used for defining fuzzy sets, depending on whether the universe of discourse is discrete or continuous: *functional* and *numerical.*
  Functional A functional definition expresses the membership function of a fuzzy set in a functional form, typically a bell-shaped function, triangle-shaped function, trapezoid-shaped function, etc.
  Such functions are used in FLC because they lead themselves to manipulation through the use of fuzzy arithmetic. The functional definition can readily be adapted to a change in the normalization of a universe. Either a numerical definition or functional definition may be used to assign the grades of membership is based on the subjective criteria of the decision.
  Numerical In this case, the grade of membership function of a fuzzy set is represented as a vector of numbers whose dimention depends on the degree of discretization.
  In this case, the membership function of each primary fuzzy set has the form

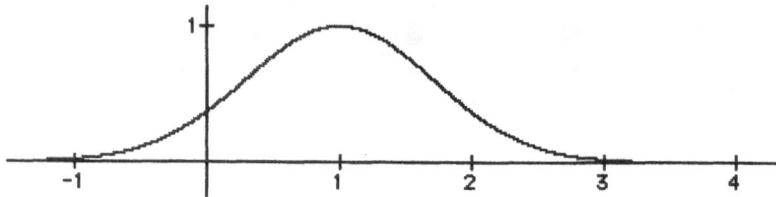

Fig. 1.74. Bell-shaped membership function.

$$A(x) \in \{0.3, 0.7, 1.0\}.$$

- **Rule base.**
  A fuzzy system is characterized by a set of linguistic statements based on expert knowledge. The expert knowledge is usually in the form of "IF-THEN" rules, which are easily implemented by fuzzy conditional statements in fuzzy logic. The collection of fuzzy control rules that are expressed as fuzzy conditional statements forms the rule base or the rule set of an FLC.
- **Choice of process state (input) variables and control (output) variables of fuzzy control rules.**
  Fuzzy control rules are more conveniently formulated in linguistic rather than numerical terms. Typically, the linguistic variables in an FLC are the *state, state error, state error derivative, state error integral,* etc.
- **Source and derivation of fuzzy control rules.**
  There are four modes of derivation of fuzzy control rules.
  - *Expert Experience and Control Engineering Knowledge*
    Fuzzy control rules have the form of fuzzy conditional statements that relate the state variables in the antecedent and process control variables in the consequents. In this connection, it should be noted that in our daily life most of the information on which our decisions are based is linguistic rather than numerical in nature. Seen in this perspective, fuzzy control rules provide a natural framework for the characterization of human behavior and decisions analysis. Many experts have found that fuzzy control rules provide a convenient way to express their domain knowledge.
  - *Operator's Control Actions*
    In many industrial man-machine control system, the input-output relations are not known with sufficient precision to make it possible to employ classical control theory for modeling and simulation.
    And yet skilled human operators can control such systems quite successfully without having any quantitative models in mind. In effect, a human operator employs-consciously or subconsciously - a set of fuzzy IF-THEN rules to control the process.
    As was pointed out by Sugeno, to automate such processes, it is expedient to express the operator's control rules as fuzzy IF-THEN rules

employing linguistic variables. In practice, such rules can be deduced from the observation of human controller's actions in terms of the input-output operating data.

– *Fuzzy Model of a Process*

In the linguistic approach, the linguistic description of the dynamic characteristics of a controlled process may be viewed as a fuzzy model of the process.Based on the fuzzy model, we can generate a set of fuzzy control rules for attaining optimal performance of a dynamic system.

The set of fuzzy control rules forms the rule base of an FLC.

Although this approach is somewhat more complicated, it yields better performance and reliability, and provides a FLC.

– *Learning*

Many fuzzy logic controllers have been built to emulate human decision-making behavior, but few are focused on human learning, namely, the ability to create fuzzy control rules and to modify them based on experience. A very interesting example of a fuzzy rule based system which has a learning capability is Sugeno's fuzzy car. Sugeno's fuzzy car can be trained to park by itself.

• **Types of fuzzy control rules.**
• **Consistency, interactivity, completeness of fuzzy control rules.**

Decision making logic: Definition of a fuzzy implication, Interpretation of the sentence connective *and*, Interpretation of the sentence connective *also*, Definitions of a compositional operator, Inference mechanism.

1.15 The ball and beam problem

We illustrate the applicability of fuzzy logic control systems by the *ball and beam* problem.

The ball and beam system can be found in many undergraduate control laboratories. The beam is made to rotate in a vertical plane by applying a torque at the center of rotation and the ball is free to roll along the beam. We require that the ball remain in contact with the beam. Let $x = (r, \dot{r}, \theta, \dot{\theta})^T$ be the state of the system, and $y = r$ be the output of the system. Then the system can be represented by the state-space model

$$\begin{bmatrix} \dot{x}_1 \\ \dot{x}_2 \\ \dot{x}_3 \\ \dot{x}_4 \end{bmatrix} = \begin{bmatrix} x_2 \\ B(x_1 x_4^2 - G \sin x_3) \\ x_4 \\ 0 \end{bmatrix} + \begin{bmatrix} 0 \\ 0 \\ 0 \\ 1 \end{bmatrix} u$$

$$y = x_1$$

where the control u is the acceleration of θ. The purpose of control is to determine $u(x)$ such that the closed-loop system output y will converge to zero from certain initial conditions.

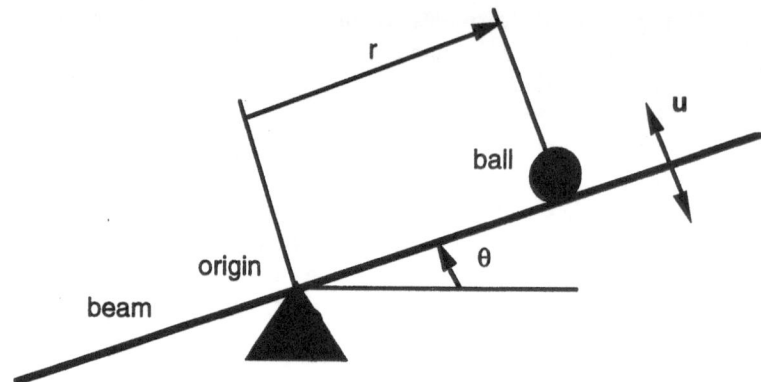

Fig. 1.75. The beam and ball problem.

The input-output linearization algorithm determines the control law $u(x)$ as follows: For state x compute

$$u(x) = -\alpha_3\phi_4(x) - \alpha_2\phi_3(x) - \alpha_1\phi_2(x) - \alpha_0\phi_1(x)$$

where $\phi_1 = x_1, \phi_2 = x_2$

$$\phi_3(x) = -BG\sin x_3, \phi_4(x) = -BGx_4\cos x_3$$

and the α_i are chosen so that

$$s^4 + \alpha_3 s^3 + \alpha_2 s^2 + \alpha_1 s + \alpha_0$$

is a *Hurwitz* polynomial.

Compute $a(x) = -BG\cos x_3$ and $b(x) = BGx_4^2\sin x_3$; then $u(x) = (v(x) - b(x))/a(x)$.

Wang and Mendel [213] use the following four common-sense linguistic control rules for the beam and ball problem:

$\Re_1$:   **if** x_1 is "positive" and x_2 is "near zero" and x_3 is "positive" and x_4 is "near zero" **then** "u is negative"

$\Re_2$:   **if** x_1 is "positive" and x_2 is "near zero" and x_3 is "negative" and x_4 is "near zero" **then** "u is positive big"

$\Re_3$:   **if** x_1 is "negative" and x_2 is "near zero" and x_3 is "positive" and x_4 is "near zero" **then** "u is negative big"

$\Re_4$:   **if** x_1 is "negative" and x_2 is "near zero" and x_3 is "negative" and x_4 is "near zero" **then** "u is positive"

where all fuzzy numbers have Gaussian membership function, e.g. the value "near zero" of the linguistic variable x_2 is defined by $\exp(-x^2/2)$.

Using the Stone-Weierstrass theorem Wang [214] showed that fuzzy logic control systems of the form

$$\Re_i : \text{ if } x \text{ is } A_i \text{ and } y \text{ is } B_i \text{ then } z \text{ is } C_i, \ i = 1, \dots, n$$

with

- Gaussian membership functions

$$A_i(u) = \exp\left[-\frac{1}{2}\left(\frac{u - \alpha_{i1}}{\beta_{i1}}\right)^2\right],$$

$$B_i(v) = \exp\left[-\frac{1}{2}\left(\frac{v - \alpha_{i2}}{\beta_{i2}}\right)^2\right],$$

$$C_i(w) = \exp\left[-\frac{1}{2}\left(\frac{w - \alpha_{i3}}{\beta_{i3}}\right)^2\right],$$

- Singleton fuzzifier

$$fuzzifier(x) := \bar{x}, \quad fuzzifier(y) := \bar{y},$$

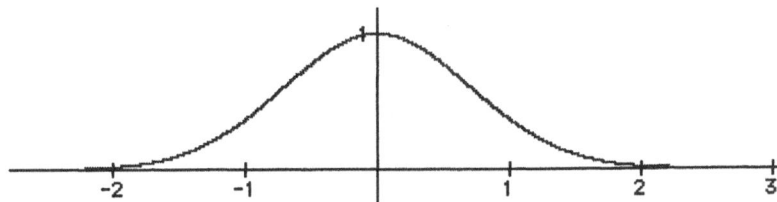

Fig. 1.76. Gaussian membership function for "near zero"

- Product fuzzy conjunction

$$\big[A_i(u) \text{ and } B_i(v)\big] = A_i(u)B_i(v)$$

- Product fuzzy implication (Larsen implication)

$$\big[A_i(u) \text{ and } B_i(v)\big] \to C_i(w) = A_i(u)B_i(v)C_i(w)$$

- Centroid defuzzification method [179]

$$z = \frac{\sum_{i=1}^{n} \alpha_{i3} A_i(x)B_i(y)}{\sum_{i=1}^{n} A_i(x)B_i(y)}$$

where α_{i3} is the center of C_i.

are universal approximators, i.e. they can approximate any continuous function on a compact set to arbitrary accuracy. Namely, he proved the following theorem

Theorem 1.15.1 *For a given real-valued continuous function g on the compact set U and arbitrary $\epsilon > 0$, there exists a fuzzy logic control system with output function f such that*

$$\sup_{x \in U} \|g(x) - f(x)\| \le \epsilon.$$

Castro in 1995 [50] showed that Mamdani's fuzzy logic controllers

$$\Re_i : \text{ if } x \text{ is } A_i \text{ and } y \text{ is } B_i \text{ then } z \text{ is } C_i, \ i = 1, \dots, n$$

with

- Symmetric triangular membership functions

$$A_i(u) = \begin{cases} 1 - \dfrac{|a_i - u|}{\alpha_i} & \text{if } |a_i - u| \le \alpha_i \\ 0 & \text{otherwise} \end{cases}$$

$$B_i(v) = \begin{cases} 1 - \dfrac{|b_i - v|}{\beta_i} & \text{if } |b_i - v| \le \beta_i \\ 0 & \text{otherwise} \end{cases}$$

$$C_i(w) = \begin{cases} 1 - \dfrac{|c_i - w|}{\gamma_i} & \text{if } |c_i - w| \le \gamma_i \\ 0 & \text{otherwise} \end{cases}$$

- Singleton fuzzifier

$$fuzzifier(x_0) := \bar{x}_0$$

- Minimum norm fuzzy conjunction

$$\left[A_i(u) \text{ and } B_i(v) \right] = \min\{A_i(u), B_i(v)\}$$

- Minimum-norm fuzzy implication

$$\left[A_i(u) \text{ and } B_i(v) \right] \to C_i(w) = \min\{A_i(u), B_i(v), C_i(w)\}$$

- Maximum t-conorm rule aggregation

$$\mathbf{Agg}\,(\Re_1, \Re_2, \cdots, \Re_n) = \max\{\Re_1, \Re_2, \cdots, \Re_n\},$$

- Centroid defuzzification method

$$z = \frac{\sum_{i=1}^{n} c_i \min\{A_i(x), B_i(y)\}}{\sum_{i=1}^{n} \min\{A_i(x)B_i(y)\}}$$

where c_i is the center of C_i.

are also universal approximators.

1.16 Aggregation in fuzzy system modeling

Many applications of fuzzy set theory involve the use of a fuzzy rule base to model complex and perhaps ill-defined systems. These applications include fuzzy logic control, fuzzy expert systems and fuzzy systems modeling. Typical of these situations are set of n rules of the form

$$\Re_1 : \text{if } x \text{ is } A_1 \text{ then } y \text{ is } C_1$$

also

$$\Re_2 : \text{if } x \text{ is } A_2 \text{ then } y \text{ is } C_2$$

also

. . . .

also

$$\Re_n : \text{if } x \text{ is } A_n \text{ then } y \text{ is } C_n$$

The fuzzy inference process consists of the following four step algorithm [225]:

- Determination of the relevance or matching of each rule to the current input value.
- Determination of the output of each rule as fuzzy subset of the output space. We shall denote these individual rule outputs as R_j.
- Aggregation of the individual rule outputs to obtain the overall fuzzy system output as fuzzy subset of the output space. We shall denote this overall output as R.
- Selection of some action based upon the output set.

Our purpose here is to investigate the requirements for the operations that can be used to implement this reasoning process. We are particularly concerned with the third step, the rule output aggregation.

Let us look at the process for combining the individual rule outputs. A basic assumption we shall make is that the operation is **pointwise** and **likewise**. By pointwise we mean that for every y, $R(y)$ just depends upon $R_j(y)$, $j = 1, \dots, n$. By likewise we mean that the process used to combine the R_j is the same for all of the y.

Let us denote the pointwise process we use to combine the individual rule outputs as

$$F(y) = \mathbf{Agg}(R_1(y), \dots, R_n(y))$$

In the above **Agg** is called the aggregation operator and the $R_j(y)$ are the arguments. More generally, we can consider this as an operator

$$a = \mathbf{Agg}(a_1, \dots, a_n)$$

where the a_i and a are values from the membership grade space, normally the unit interval.

Let us look at the minimal requirements associated with **Agg**. We first note that the combination of of the individual rule outputs should be independent of the choice of indexing of the rules. This implies that a required property that we must associate with th **Agg** operator is that of commutativity, the indexing of the arguments does not matter. We note that the commutativity property allows to represent the arguments of the **Agg** operator, as an unordered collection of possible duplicate values; such an object is a **bag**.

For an individual rule output, R_j, the membership grade $R_j(y)$ indicates the degree or sterength to which this rule suggests that y is the appropriate solution. In particular if for a pair of elements y' and y'' it is the case that

$$R_i(y') \geq R_i(y''),$$

then we are saying that rule j is preferring y' as the system output over y''. From this we can reasonably conclude that if all rules prefer y' over y'' as output then the overall system output should prefer y' over y''. This observation requires us to impose a monotonicity condition on the **Agg** operation. In particular if

$$R_j(y') \geq R_j(y''),$$

for all j, then

$$R(y') \geq R(y'').$$

There appears one other condition we need to impose upon the aggregation operator. Assume that there exists some rule whose firing level is zero. The implication of this is that the rule provides no information regarding what should be the output of the system. It should not affect the final R. The first observation we can make is that whatever output this rule provides should not make make any distinction between the potential outputs. Thus, we see that the aggregation operator needs an identy element.

In summary, we see that the aggregation operator, **Agg** must satisfy three conditions: *commutativity, monotonicity, must contain a fixed identity.* These conditions are based on the three requirements: that the indexing of the rules be unimportant, a positive association between individual rule output and total system output, and non-firing rules play no role in the decision process.

These operators are called **MICA** (Monotonic Identity Commutative Aggregation) operators (see Yager [225]). **MICA** *operators are the most general class for aggregation in fuzzy modeling.* They include t-norms, t-conorms, averaging and compensatory operators.

Assume X is a set of elements. A bag drawn from X is any collection of elements which is contained in X. A bag is different from a subset in that it allows multiple copies of the same element. A bag is similar to a set in that the ordering of the elements in the bag does not matter. If A is a bag consisiting of a, b, c, d we denote this as $A = < a, b, c, d >$. Assume A and B are two bags. We denote the sum of the bags

$$C = A \oplus B$$

where C is the bag consisting of the members of both A and B.

Example 1.16.1 *Let $A = < a, b, c, d >$ and $B = < b, c, c >$ then*

$$A \oplus B = < a, b, c, d, b, c, c >$$

In the following we let $\mathrm{Bag}(X)$ indicate the set of all bags of the set X.

Definition 1.16.1 *A function*

$$F : \mathrm{Bag}(X) \to X$$

is called a bag mapping from $\mathrm{Bag}(X)$ into the set X.

An important property of bag mappings are that they are commutative in the sense that the ordering of the elements does not matter.

Definition 1.16.2 *Assume $A = < a_1, \ldots, a_n >$ and $B = < b_1, \ldots, b_n >$ are two bags of the same cardinality n. If the elements in A and B can be indexed in such way that $a_i \geq b_i$ for all i then we shall denote this $A \geq B$.*

Definition 1.16.3 *(MICA operator [225]) A bag mapping*

$$M : \mathrm{Bag}([0, 1]) \to [0, 1]$$

is called MICA operator if it has the following two properties

- *If $A \geq B$ then $M(A) \geq M(B)$ (monotonicity)*
- *For every bag A there exists an element, $u \in [0, 1]$, called the identity of A such that if $C = A \oplus < u >$ then $M(C) = M(A)$ (identity)*

Thus the MICA operator is endowed with two properties in addition to the inherent commutativity of the bag operator, *monotonicity and identity.*

- The requirement of monotonicity appears natural for an aggregation operator in that it provides some connection between the arguments and the aggregated value.
- The property of identity allows us to have the facility for aggregating data which does not affect the overall result. This becomes useful for enabling us to include importances among other characteristics.

Fuzzy set theory provides a host of attractive aggregation connectives for integrating membership values representing uncertain information. These connectives can be categorized into the following three classes *union, intersection* and *compensation* connectives.

Union produces a high output whenever any one of the input values representing degrees of satisfaction of different features or criteria is high. Intersection connectives produce a high output only when all of the inputs have high values. Compensative connectives have the property that a higher degree of satisfaction of one of the criteria can compensate for a lower degree of satisfaction of another criteria to a certain extent.

In the sense, union connectives provide full compensation and intersection connectives provide no compensation.

1.17 Averaging operators

In a decision process the idea of *trade-offs* corresponds to viewing the global evaluation of an action as lying between the *worst* and the *best* local ratings. This occurs in the presence of conflicting goals, when a compensation between the corresponding compabilities is allowed.

Averaging operators realize trade-offs between objectives, by allowing a positive compensation between ratings.

Definition 1.17.1 *(averaging operator) An averaging (or mean) operator M is a function*

$$M : [0,1] \times [0,1] \to [0,1]$$

satisfying the following properties

- $M(x,x) = x, \ \forall x \in [0,1], \ (idempotency)$
- $M(x,y) = M(y,x), \ \forall x,y \in [0,1], \ (commutativity)$
- $M(0,0) = 0, \ M(1,1) = 1, \ (extremal \ conditions)$
- $M(x,y) \leq M(x',y') \ if \ x \leq x' \ and \ y \leq y' \ (monotonicity)$
- M *is continuous*

Lemma 1.17.1 *If M is an averaging operator then*

$$\min\{x,y\} \leq M(x,y) \leq \max\{x,y\}, \ \forall x,y \in [0,1]$$

Proof. From idempotency and monotonicity of M it follows that

$$\min\{x,y\} = M(\min\{x,y\}, \min\{x,y\}) \leq M(x,y)$$

and

$$M(x,y) \leq M(\max\{x,y\}, \max\{x,y\}) = \max\{x,y\}.$$

Which ends the proof.

The interesting properties averagings are the following [60]:

Property 1.17.1 *A strictly increasing averaging operator cannot be associative.*

Property 1.17.2 *The only associative averaging operators are defined by*

$$M(x, y, \alpha) = med(x, y, \alpha) = \begin{cases} y & \text{if } x \leq y \leq \alpha \\ \alpha & \text{if } x \leq \alpha \leq y \\ x & \text{if } \alpha \leq x \leq y \end{cases}$$

where $\alpha \in (0, 1)$.

An important family of averaging operators is formed by quasi-arithmetic means

$$M(a_1, \ldots, a_n) = f^{-1}\left(\frac{1}{n}\sum_{i=1}^{n} f(a_i)\right)$$

This family has been characterized by Kolmogorov as being the class of all decomposable continuous averaging operators.

Example 1.17.1 *For example, the quasi-arithmetic mean of a_1 and a_2 is defined by*

$$M(a_1, a_2) = f^{-1}\left[\frac{f(a_1) + f(a_2)}{2}\right].$$

The next table shows the most often used mean operators.

Name	Definition
harmonic mean	$\dfrac{2xy}{x+y}$
geometric mean	$\sqrt{xy}$
arithmetic mean	$\dfrac{x+y}{2}$
dual of geometric mean	$1 - \sqrt{(1-x)(1-y)}$
dual of harmonic mean	$\dfrac{x + y - 2xy}{2 - x - y}$
median	$med(x, y, \alpha)$, $\alpha \in (0, 1)$
generalized p-mean	$\left(\dfrac{x^p + y^p}{2}\right)^{1/p}$, $p \geq 1$

Table 1.4. Mean operators.

The process of information aggregation appears in many applications related to the development of intelligent systems. One sees aggregation in neural networks, fuzzy logic controllers, vision systems, expert systems and multi-criteria decision aids. In [222] Yager introduced a new aggregation technique based on the ordered weighted averaging (OWA) operators.

Definition 1.17.2 *An OWA operator of dimension n is a mapping*

$$F\colon I\!R^n \to I\!R,$$

that has an associated n vector $W = (w_1, w_2, \ldots, w_n)^T$ *such as* $w_i \in [0,1]$, $1 \leq i \leq n$,

$$\sum_{i=1}^n w_i = 1.$$

Furthermore

$$F(a_1, \ldots, a_n) = \sum_{j=1}^n w_j b_j$$

where b_j *is the j-th largest element of the bag* $< a_1, \ldots, a_n >$.

Example 1.17.2 *Assume* $W = (0.4, 0.3, 0.2, 0.1)^T$ *then*

$$F(0.7, 1, 0.2, 0.6) = 0.4 \times 1 + 0.3 \times 0.7 + 0.2 \times 0.6 + 0.1 \times 0.2 = 0.75.$$

A fundamental aspect of this operator is the re-ordering step, in particular an aggregate a_i is not associated with a particular weight w_i but rather a weight is associated with a particular ordered position of aggregate.

When we view the OWA weights as a column vector we shall find it convenient to refer to the weights with the low indices as weights at the top and those with the higher indices with weights at the bottom.

It is noted that different OWA operators are distinguished by their weighting function. In [222] Yager pointed out three important special cases of OWA aggregations:

- F^*: In this case $W = W^* = (1, 0 \ldots, 0)^T$ and

$$F^*(a_1, \ldots, a_n) = \max\{a_1, \ldots, a_n\}.$$

- F_*: In this case $W = W_* = (0, 0 \ldots, 1)^T$ and

$$F_*(a_1, \ldots, a_n) = \min\{a_1, \ldots, a_n\}.$$

- F_A: In this case $W = W_A = (1/n, \ldots, 1/n)^T$ and

$$F_A(a_1, \ldots, a_n) = \frac{1}{n} \sum_{i=1}^n a_i,$$

A number of important properties can be associated with the OWA operators. We shall now discuss some of these.

For any OWA operator F

$$F_*(a_1, \ldots, a_n) \leq F(a_1, \ldots, a_n) \leq F^*(a_1, \ldots, a_n).$$

Thus the upper an lower star OWA operator are its boundaries. From the above it becomes clear that for any F

$$\min\{a_1, \dots, a_n\} \leq F(a_1, \dots, a_n) \leq \max\{a_1, \dots, a_n\}.$$

The OWA operator can be seen to be *commutative*. Let $\{a_1, \dots, a_n\}$ be a bag of aggregates and let $\{d_1, \dots, d_n\}$ be any *permutation* of the a_i. Then for any OWA operator

$$F(a_1, \dots, a_n) = F(d_1, \dots, d_n).$$

A third characteristic associated with these operators is *monotonicity*. Assume a_i and c_i are a collection of aggregates, $i = 1, \dots, n$ such that for each i, $a_i \geq c_i$. Then

$$F(a_1, \dots, a_n) \geq F(c_1, c_2, \dots, c_n)$$

where F is some fixed weight OWA operator.

Another characteristic associated with these operators is *idempotency*. If $a_i = a$ for all i then for any OWA operator

$$F(a_1, \dots, a_n) = a.$$

From the above we can see the OWA operators have the basic properties associated with an *averaging operator*.

Example 1.17.3 *A window type OWA operator takes the average of the m arguments about the center. For this class of operators we have*

$$w_i = \begin{cases} 0 & \text{if } i < k \\ \dfrac{1}{m} & \text{if } k \leq i < k+m \\ 0 & \text{if } i \geq k+m \end{cases} \tag{1.16}$$

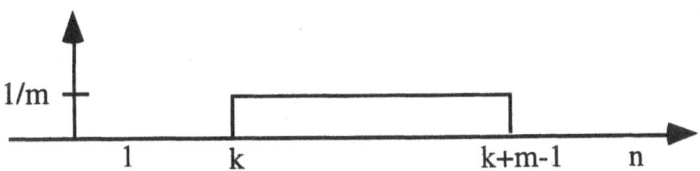

Fig. 1.77. Window type OWA operator.

In order to classify OWA operators in regard to their location between *and* and *or*, a measure of orness, associated with any vector W is introduce by Yager [222] as follows

$$\text{orness}(W) = \frac{1}{n-1} \sum_{i=1}^{n} (n-i)w_i$$

It is easy to see that for any W the orness(W) is always in the unit interval. Furthermore, note that the nearer W is to an *or*, the closer its measure is to one; while the nearer it is to an *and*, the closer is to zero.

Lemma 1.17.2 *Let us consider the the vectors* $W^* = (1, 0 \ldots, 0)^T$, $W_* = (0, 0 \ldots, 1)^T$ *and* $W_A = (1/n, \ldots, 1/n)^T$. *Then it can easily be shown that*

- orness(W^*) = 1
- orness(W_*) = 0
- orness(W_A) = 0.5

A measure of andness is defined as

$$\text{andness}(W) = 1 - \text{orness}(W)$$

Generally, an OWA opeartor with much of nonzero weights near the top will be an *orlike* operator,

$$\text{orness}(W) \geq 0.5$$

and when much of the weights are nonzero near the bottom, the OWA operator will be *andlike*

$$\text{andness}(W) \geq 0.5.$$

Example 1.17.4 *Let* $W = (0.8, 0.2, 0.0)^T$. *Then*

$$\text{orness}(W) = \frac{1}{3}(2 \times 0.8 + 0.2) = 0.6$$

and

$$\text{andness}(W) = 1 - \text{orness}(W) = 1 - 0.6 = 0.4.$$

This means that the OWA operator, defined by

$$F(a_1, a_2, a_3) = 0.8b_1 + 0.2b_2 + 0.0b_3 = 0.8b_1 + 0.2b_2$$

where b_j *is the j-th largest element of the bag* $< a_1, a_2, a_3 >$, *is an orlike aggregation.*

The following theorem shows that as we move weight up the vector we increase the orness, while moving weight down causes us to decrease *orness*(W).

Theorem 1.17.1 *[223] Assume W and W' are two n-dimensional OWA vectors such that*

$$W = (w_1, \ldots, w_n)^T,$$

and

$$W' = (w_1, \ldots, w_j + \epsilon, \ldots, w_k - \epsilon, \ldots, w_n)^T$$

where $\epsilon > 0$, $j < k$. *Then orness*(W') > *orness*(W).

Proof. From the definition of the measure of *orness* we get

$$\text{orness}(W') = \frac{1}{n-1}\sum_{i=1}^{n}(n-i)w'_i$$

$$= \frac{1}{n-1}\sum_{i=1}^{n}(n-1)w_i + (n-j)\epsilon - (n-k)\epsilon,$$

$$\text{orness}(W') = \text{orness}(W) + \frac{1}{n-1}\epsilon(k-j).$$

Since $k > j$ we get

$$\text{orness}(W') > \text{orness}(W)$$

This ends the proof.

In [222] Yager defined the measure of dispersion (or entropy) of an OWA vector by

$$disp(W) = -\sum_i w_i \ln w_i.$$

We can see when using the OWA operator as an averaging operator $Disp(W)$ measures the degree to which we use all the aggregates equally.

If F is an OWA aggregation with weights w_i the *dual* of F denoted $\hat{F}$, is an OWA aggregation of the same dimention where with weights $\hat{w}_i$

$$\hat{w}_i = w_{n-i+1}.$$

We can easily see that if F and $\hat{F}$ are duals then

$$disp(\hat{F}) = disp(F)$$

$$\text{orness}(\hat{F}) = 1 - \text{orness}(F) = \text{andness}(F)$$

Thus is F is orlike its dual is andlike.

Example 1.17.5 *Let* $W = (0.3, 0.2, 0.1, 0.4)^T$. *Then*

$$\hat{W} = (0.4, 0.1, 0.2, 0.3)^T.$$

and

$$\text{orness}(F) = 1/3(3 \times 0.3 + 2 \times 0.2 + 0.1) \approx 0.466$$

$$\text{orness}(\hat{F}) = 1/3(3 \times 0.4 + 2 \times 0.1 + 0.2) \approx 0.533$$

An important application of the OWA operators is in the area of quantifier guided aggregations [222]. Assume

$$\{A_1, \ldots, A_n\}$$

is a collection of criteria. Let x be an object such that for any criterion A_i, $A_i(x) \in [0,1]$ indicates the degree to which this criterion is satisfied by x. If we want to find out the degree to which x satisfies "*all* the criteria" denoting this by $D(x)$, we get following Bellman and Zadeh [2].

$$D(x) = \min\{A_1(x), \dots, A_n(x)\}$$

In this case we are essentially requiring x to satisfy A_1 and A_2 and ... and A_n.

If we desire to find out the degree to which x satisfies "*at least one* of the criteria", denoting this $E(x)$, we get

$$E(x) = \max\{A_1(x), \dots, A_n(x)\}$$

In this case we are requiring x to satisfy A_1 or A_2 or ... or A_n.

In many applications rather than desiring that a solution satisfies one of these extreme situations, "all" or "at least one", we may require that x satisfies *most* or *at least half* of the criteria. Drawing upon Zadeh's concept [241] of linguistic quantifiers we can accomplish these kinds of quantifier guided aggregations.

Definition 1.17.3 *A quantifier Q is called*

- regular monotonically non-decreasing *if*

$$Q(0) = 0, \quad Q(1) = 1, \quad \textit{if } r_1 > r_2 \textit{ then } Q(r_1) \geq Q(r_2).$$

- regular monotonically non-increasing *if*

$$Q(0) = 1, \quad Q(1) = 0, \quad \textit{if } r_1 < r_2 \textit{ then } Q(r_1) \geq Q(r_2).$$

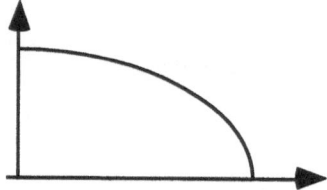 

Fig. 1.78. Monoton linguistic quantifiers.

- regular unimodal *if*

$$Q(r) = \begin{cases} 0 & \textit{if } r = 0 \\ \text{monotone increasing} & \textit{if } 0 \leq r \leq a \\ 1 & \textit{if } a \leq r \leq b,\ 0 < a < b < 1 \\ \text{monotone decreasing} & \textit{if } b \leq r \leq 1 \\ 0 & \textit{if } r = 1 \end{cases}$$

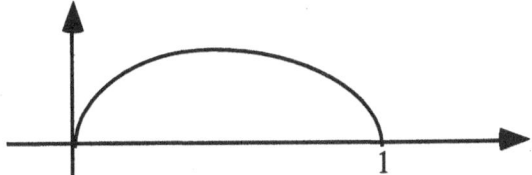

Fig. 1.79. Unimodal linguistic quantifier.

With $a_i = A_i(x)$ the overall valuation of x is $F_Q(a_1, \ldots , a_n)$ where F_Q is an OWA operator. The weights associated with this quantified guided aggregation are obtained as follows

$$w_i = Q\left(\frac{i}{n}\right) - Q\left(\frac{i-1}{n}\right), \quad i = 1, \ldots , n. \tag{1.17}$$

The next figure graphically shows the operation involved in determining the OWA weights directly from the quantifier guiding the aggregation.

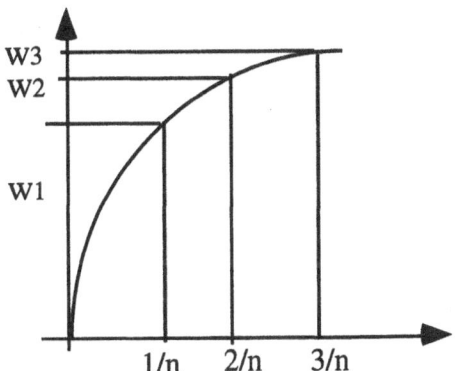

Fig. 1.80. Determining weights from a quantifier.

Theorem 1.17.2 *If we construct w_i via the method (1.17) we always get*

$$\sum_{i=1}^{n} w_i = 1$$

and $w_i \in [0, 1]$ for any function

$$Q \colon [0, 1] \rightarrow [0, 1],$$

satisfying the conditions of a regular nondecreasing quantifier.

Proof. We first see that from the non-decreasing property

$$Q\left(\frac{i}{n}\right) \geq Q\left(\frac{i-1}{n}\right),$$

hence $w_i \geq 0$ and since $Q(r) \leq 1$ then $w_i \leq 1$. Furthermore we see

$$\sum_{i=1}^{n} w_i = \sum_{i=1}^{n}\left[Q\left(\frac{i}{n}\right) - Q\left(\frac{i}{n-1}\right)\right] - Q\left(\frac{n}{n}\right) - Q\left(\frac{0}{n}\right)$$
$$= 1 - 0 = 1.$$

Which ends the proof.

We call any function satisfying the conditions of a regular non-decreasing quantifier an *acceptable OWA weight generating function.*

Let us look at the weights generated from some basic types of quantifiers. The quantifier, *for all* Q_*, is defined such that

$$Q_*(r) = \begin{cases} 0 \text{ for } r < 1, \\ 1 \text{ for } r = 1. \end{cases}$$

Using our method for generating weights

$$w_i = Q_*\left(\frac{i}{n}\right) - Q_*\left(\frac{i-1}{n}\right)$$

we get

$$w_i = \begin{cases} 0 \text{ for } i < n, \\ 1 \text{ for } i = n. \end{cases}$$

This is exactly what we previously denoted as W_*.

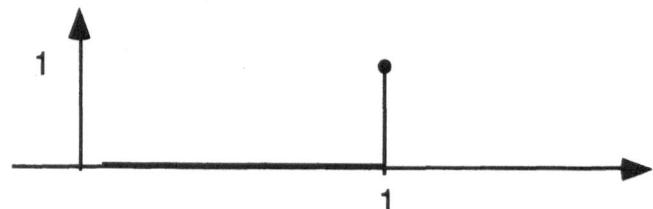

Fig. 1.81. The quantifier *all.*

For the quantifier *there exists* we have

$$Q^*(r) = \begin{cases} 0 \text{ for } r = 0, \\ 1 \text{ for } r > 0. \end{cases}$$

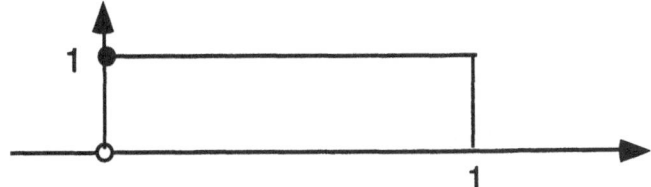

Fig. 1.82. The quantifier *there exists*.

In this case we get

$$w_1 = 1, \quad w_i = 0, \text{ for } i \neq 1.$$

This is exactly what we denoted as W^*.

Consider next the quantifier defined by

$$Q(r) = r.$$

This is *an identity* or *linear type* quantifier.

In this case we get

$$w_i = Q\left(\frac{i}{n}\right) - Q\left(\frac{i-1}{n}\right) = \frac{i}{n} - \frac{i-1}{n} = \frac{1}{n}.$$

This gives us the pure averaging OWA aggregation operator.

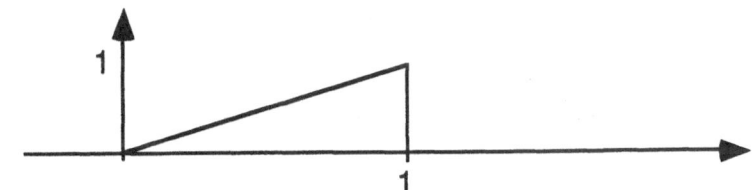

Fig. 1.83. The *identity* quantifier.

Recapitulating using the approach suggested by Yager if we desire to calculate

$$F_Q(a_1, \ldots, a_n)$$

for Q being a regular non-decreasing quantifier we proceed as follows:

1. Calculate

$$w_i = Q\left(\frac{i}{n}\right) - Q\left(\frac{i-1}{n}\right),$$

2. Calculate

$$F_Q(a_i, \ldots, a_n) = \sum_{i=1}^{n} w_i b_i,$$

where b_i is the i-th largest of the $\{a_j\}$.

Example 1.17.6 *The weights of the window-type OWA operator given by equation (1.16) can be derived from the quantifier*

$$Q(r) = \begin{cases} 0 & \text{if } r \le (k-1)/n, \\ 1 - \dfrac{(k-1+m) - nr}{m} & \text{if } \dfrac{k-1}{n} \le r \le \dfrac{k-1+m}{n}, \\ 1 & \text{if } (k-1+m)/n \le r \le 1. \end{cases}$$

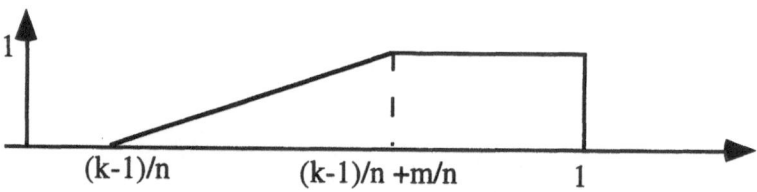

Fig. 1.84. Quantifier for a window-type OWA operator.

Exercise 1.17.1 *Let $W = (0.4, 0.2, 0.1, 0.1, 0.2)^T$. Calculate $disp(W)$.*

Exercise 1.17.2 *Let $W = (0.3, 0.3, 0.1, 0.1, 0.2)^T$. Calculate $orness(F)$, where the OWA operator F is derived from W.*

Exercise 1.17.3 *Prove that $0 \le disp(W) \le \ln(n)$ for any n-dimensional weight vector W.*

Exercise 1.17.4 *Let $Q(x) = x^2$ be a linguistic quintifier. Assume the weights of an OWA operator F are derived from Q. Calculate the value $F(a_1, a_2, a_3, a_4)$ for $a_1 = a_2 = 0.6$, $a_3 = 0.4$ and $a_4 = 0.2$. What is the orness measure of F?*

Exercise 1.17.5 *Let $Q(x) = \sqrt{x}$ be a linguistic quintifier. Assume the weights of an OWA operator F are derived from Q. Calculate the value $F(a_1, a_2, a_3, a_4)$ for $a_1 = a_2 = 0.6$, $a_3 = 0.4$ and $a_4 = 0.2$. What is the orness measure of F?*

1.18 Fuzzy screening systems

In *screening* problems one usually starts with a large subset, X, of possible alternative solutions. Each alternative is essentially represented by a minimal amount of information supporting its appropriateness as the best solution. This minimal amount of information provided by each alternative is used to help select a subset A of X to be further investigated.

Two prototypical examples of this kind of problem can be recalled.

- **Job selection problem.** Here a large number of candidates, X, submit a resume, minimal information, to a job announcement. Based upon these resumes a small subset of X, Y, are called in for interviews. These interviews, which provide more detailed information, are the basis of selecting winning candidate from Y.

- **Proposal selection problem.** Here a large class of candidates, X, submit preliminary proposals, minimal information. Based upon these preliminary proposals a small subset of X, Y, are requested to submit full detailed proposals. These detailed proposals are the basis of selecting winning candidate from Y.

In the above examples the process of selecting the subset A, required to provide further information, is called a *screening process*. In [224] Yager suggests a technique, called *fuzzy screening system*, for managing this screening process.

This kinds of screening problems described above besides being characterized as decision making with minimal information general involve multiple participants in the selection process. The people whose opinion must be considered in the selection process are called *experts*. Thus screening problems are a class of multiple expert decision problems. In addition each individual expert's decision is based upon the use of multiple criteria. So we have ME-MCDM (Multi Expert-Multi Criteria Decision Making) problem with minimal information.

The fact that we have minimal information associated with each of the alternatives complicates the problem because it limits the operations which can be performed in the aggregation processes needed to combine the multi-experts as well as multi-criteria. The Arrow impossibility theorem [1] is a reflection of this difficulty.

Yager [224] suggests an approach to the screening problem which allows for the requisite aggregations but which respects the lack of detail provided by the information associated with each alternative. The technique only requires that preference information be expressed in by elements draw from a scale that essentially only requires a linear ordering. This property allows the experts to provide information about satisfactions in the form of a linguistic values such as *high, medium, low*. This ability to perform the necessary operations will only requiring imprecise linguistic preference valuations will enable the experts to comfortably use the kinds of minimally informative sources of information about the objects described above. The fuzzy screening system is a two stage process.

- In the first stage, experts are asked to provide an evaluation of the alternatives. This evaluation consists of a rating for each alternative on each of the criteria.
- In the second stage, the methodology introduced in [222] is used to aggregate the individual experts evaluations to obtain an overall linguistic value for each object.

The problem consists of three components.

- The first component is a collection

$$X = \{X_1, \ldots, X_p\}$$

of alternative solutions from amongst which we desire to select some subset to be investigated further.
- The second component is a group

$$A = \{A_1, \ldots, A_r\}$$

of experts whose opinion solicited in screening the alternatives.
- The third component is a collection

$$C = \{C_1, \ldots, C_n\}$$

of criteria which are considered relevant in the choice of the objects to be further considered.

For each alternative each expert is required to provided his opinion. In particular for each alternative an expert is asked to evaluate how well that alternative satisfies each of the criteria in the set C. These evaluations of alternative satisfaction to criteria will be given in terms of elements from the following scale S:

Excellent (EX)	S_7
Very High (VH)	S_6
High (H)	S_5
Medium (M)	S_4
Low	S_3
Very Low	S_2
None	S_1

The use of such a scale provides a natural ordering, $S_i > S_j$ if $i > j$ and the maximum and minimum of any two scores re defined by

$$\max(S_i, S_j) = S_i \text{ if } S_i \geq S_j, \quad \min(S_i, S_j) = S_j \text{ if } S_j \leq S_i$$

We shall denote the max by $\vee$ and the min by $\wedge$. Thus for an alternative an expert provides a collection of n values

$$\{\mathcal{P}_1, \ldots, \mathcal{P}_n\}$$

where $\mathcal{P}_j$ is the rating of the alternative on the j-th criteria by the expert. Each $\mathcal{P}_j$ is an element in the set of allowable scores S. Assuming $n = 6$, a typical scoring for an alternative from one expert would be:

(medium, low, excellent, very high, excellent, none).

Independent of this evaluation of alternative satisfaction to criteria each expert must assign a measure of importance to each of the criteria. An expert uses the same scale, S, to provide the importance associated with the criteria.

The next step in the process is to find the overall valuation for a alternative by a given expert.

In order to accomplish this overall evaluation, we use a methodology suggested by Yager [220]. A crucial aspect of this approach is the taking of the negation of the importances as

$$Neg(S_i) = S_{q-i+1}$$

For the scale that we are using, we see that the negation operation provides the following:

$$Neg(EX) = N$$
$$Neg(VH) = VL$$
$$Neg(H) \ \ = L$$
$$Neg(M) \ \ = M$$
$$Neg(L) \ \ \ = H$$
$$Neg(VL) = VH$$
$$Neg(N) \ \ \ = EX$$

Then the unit score of each alternative by each expert, denoted by $\mathcal{U}$, is calculated as follows

$$\mathcal{U} = \min_j \{Neg(I_j) \vee \mathcal{P}_j)\} \tag{1.18}$$

where I_j denotes the importance of the j-th critera.

We note that (1.18) essentially is an *anding* of the criteria satisfactions modified by the importance of the criteria. The formula (1.18) can be seen as a measure of the degree to which an alternative satisfies the following proposition:

All important criteria are satisfied.

Example 1.18.1 *Consider some alternative with the following scores on five criteria*

Criteria:	C_1	C_2	C_3	C_4	C_5
Importance:	VH	VH	M	L	VL
Score:	M	L	EX	VH	EX

In this case we have

$$\mathcal{U} = \min\{Neg(VH) \vee M, Neg(VH) \vee L, Neg(M) \vee EX,$$
$$Neg(L) \vee VH, Neg(VL) \vee EX\}$$
$$= \min\{VL \vee M, VL \vee L, M \vee EX, H \vee VH, VH \vee EX\}$$
$$= \min\{M, L, EX, VH, EX\}$$
$$= L.$$

The essential reason for the low performance of this object is that it performed low on the second criteria which has a very high importance. Linguistically, Eq. 1.18 is saying that

If a criterion is important then an alternative should score well on it.

As a result of the first stage, we have for each alternative a collection of evaluations

$$\{X_{i1}, X_{i2}, \ldots, X_{ir}\}$$

where X_{ik} is the unit evaluation of the i-th alternative by the k-th expert.

In the second stage the technique for combining the expert's evaluation to obtain an overall evaluation for each alternative is based upon the OWA operators.

The first step in this process is for the decision maker to provide an aggregation function which we shall denote as Q. This function can be seen as a generalization of the idea of how many experts it feels need to agree on an alternative for it to be acceptable to pass the screening process. In particular for each number i, $1 \leq i \leq r$, the decision maker must provide a value $Q(i)$ indicating how satisfied it would be in passing an alternative that i of the experts where satisfied with. The values for $Q(i)$ should be drawn from the scale S described above.

It should be noted that Q should have certain characteristics to make it rational:

- As more experts agree the decision maker's satisfaction or confidence should increase: $Q(i) \geq Q(j)$ if $i > j$.
- If all the experts are satisfied then his satisfaction should be the highest possible $Q(r) = \text{'excellent'}$.
- If the decision maker requires all experts to support a alternative then we get $Q(r) = \text{'excellent'}$ and $Q(i) = \text{'none'}$ for $i < r$.
- If the support of just one expert is enough to make a alternative worthy of consideration then $Q(i) = \text{'excellent'}$ for all i.

- If at least m experts' support is needed for consideration then $Q(i) = \,'none'$ if $i < m$ and $Q(i) = \,'excellent'$ if $i \geq m$.

In order to define function Q, Yager [224] introduced the operation $\text{Int}[a]$ as returning the integer value that is closest to the number a. In the following, we shall let q be the number of points on the scale and r be the number of experts participating. This function which emulates the average is denoted as $Q_A(k)$ and is defined by

$$Q_A(k) = S_{b(k)}$$

where

$$b(k) = \text{Int}\left[1 + \left(k \times \frac{q-1}{r}\right)\right]$$

for all $k = 0, 1, \dots, r$. It can easily be checked that the equality

$$Q_A(0) = S_1, \quad Q_A(r) = S_q$$

holds for any values of q and r.

As an example of this function if $r = 3$ and $q = 7$ then

$$b(k) = \text{Int}\left[1 + \left(k \times \frac{6}{3}\right)\right] = \text{Int}[1 + 2k],$$

and

$$Q_A(0) = S_1, \ Q_A(1) = S_3, \ Q_A(2) = S_5, \ Q_A(3) = S_7.$$

If $r = 5$ and $q = 7$ then

$$b(k) = \text{Int}[1 + k \times 1.2]$$

and

$$Q_A(0) = S_1,$$
$$Q_A(1) = S_2,$$
$$Q_A(2) = S_3,$$
$$Q_A(3) = S_5,$$
$$Q_A(4) = S_6,$$
$$Q_A(5) = S_7,$$

Having appropriately selected Q we are now in the position to use the OWA method for aggregating the expert opinions. Assume we have r experts, each of which has a unit evaluation for the i-th projected denoted X_{ik}.

The first step in the OWA procedure is to order the X_{ik}'s in descending order, thus we shall denote B_j as the j-th highest score among the experts unit scores for the project. To find the overall evaluation for the ith project, denoted X_i, we calculate

$$X_i = \max_j \{Q(j) \wedge B_j\}.$$

In order to appreciate the workings for this formulation we must realize that

- B_j can be seen as the worst of the j-th top scores.
- $Q(j) \wedge B_j$ can be seen as an indication of how important the decision maker feels that the support of at least j experts is.
- The term $Q(j) \wedge B_j$ can be seen as a weighting of an objects j best scores, B_j, and the decision maker requirement that j people support the project, $Q(j)$.
- The max operator plays a role akin to the summation in the usual numeric averaging procedure.

Example 1.18.2 *Assume we have four experts each providing a unit evaluation for project i obtained by the methodology discussed in the previous section.*

$$X_{i1} = M$$
$$X_{i2} = H$$
$$X_{i3} = H$$
$$X_{i4} = VH$$

Reording these scores we get

$$B_1 = VH$$
$$B_2 = H$$
$$B_3 = H$$
$$B_4 = M$$

Furthermore, we shall assume that our decision maker chooses as its aggregation function the average like function Q_A. Then with $r = 4$ and scale cardinality $q = 7$, we obtain:

$$Q_A(1) = L \qquad (S_3)$$
$$Q_A(2) = M \qquad (S_4)$$
$$Q_A(3) = VH \qquad (S_6)$$
$$Q_A(4) = EX \qquad (S_7)$$

We calculate the overall evaluation as

$$X_i = \max\{L \wedge VH, M \wedge H, VH \wedge H, EX \wedge M\}$$
$$X_i = \max\{L, M, H, M\}$$
$$X_i = H$$

Thus the overall evaluation of this alternative is high.

Using the methodology suggested by Yager thus far we obtain for each alternative an overall rating X_i. These ratings allow us to obtain a evaluation of all the alternative without resorting to a numeric scale. The decision maker is now in the position to make its selection of alternatives that are be passed through the screening process. A level S^* from the scale S is selected and all those alternatives that have an overall evaluation of S^* or better are passed to the next step in the decision process.

Exercise 1.18.1 *Consider some alternative with the following scores on five criteria*

Criteria:	C_1	C_2	C_3	C_4	C_5	C_6
Importance:	H	VH	M	L	VL	M
Score:	L	VH	EX	VH	EX	M

Calculate the unit score of this alternative.

1.19 Applications of fuzzy systems

For the past few years, particularly in Japan, USA and Germany, approximately 1,000 commercial and industrial fuzzy systems have been successfully developed. The number of industrial and commercial applications worldwide appears likely to increase significantly in the near future.

The first application of fuzzy logic is due to *Mamdani* of the University of London, U.K., who in 1974 designed an experimental fuzzy control for a steam engine. In 1980, a Danish company (F.L. Smidth & Co. A/S) used fuzzy theory in cement kiln control. Three years later, *Fuji Electric Co., Ltd.* (Japan) implemented fuzzy control of chemical injection for water purification plants.

The first fuzzy controller was exhibited at Second IFSA Congress in 1987. This controller originated from Omron Corp., a Japanese company which began research in fuzzy logic in 1984 and has since applied for over 700 patents. Also in 1987, the *Sendai Subway Automatic Train Operations Controller*, designed by the Hitachi team, started operating in Sendai, Japan. The fuzzy logic in this subway system makes the journey more comfortable with smooth braking and acceleration. In 1989, Omron Corp. demonstrated fuzzy workstations at the Business Show in Harumi, Japan. Such a workstation is just a RISC–based computer, equipped with a fuzzy inference board. This fuzzy inference board is used to store and retrieve fuzzy information, and to make fuzzy inferences.

The application of fuzzy theory in consumer products started in 1990 in Japan. An example is the "fuzzy washing machine", which automatically judges the material, the volume and the dirtiness of the laundry and chooses the optimum washing program and water flow. Another example is the fuzzy logic found in the electronic fuel injection controls and automatic cruise control systems of cars, making complex controls more efficient and easier to use.

Product	Company
Washing Machine	AEG, Sharp, Goldstar
Rice Cooker	Goldstar
Cooker/Fryer	Tefal
Microwave Oven	Sharp
Electric Shaver	Sharp
Refrigerator	Whirlpool
Battery Charger	Bosch
Vacuum Cleaner	Philips, Siemens
Camcorders	Canon, Sanyo, JVC
Transmission Control	GM(Saturn), Honda, Mazda
Climate Control	Ford
Temp control	NASA in space shuttle
Credit Card	GE Corporation

Table 1.5. Industrial applications of fuzzy logic controllers.

Fuzzy logic is also being used in vacuum cleaners, camcorders, television sets etc. In 1993, Sony introduced the Sony PalmTop, which uses a fuzzy logic decision tree algorithm to perform handwritten (using a computer lightpen) Kanji character recognition. For instance, if one would write *253*, then the Sony Palmtop can distinguish the number *5* from the letter *S*.

There are many products based on Fuzzy Logic in the market today. Most of the consumer products in SEA/Japan advertise Fuzzy Logic based products for consumers. We are beginning to see many automotive applications based on Fuzzy logic. Here are few examples seen in the market. By no means this list includes all possible fuzzy logic based products in the market.

The most successful domain has been in fuzzy control of various physical or chemical characteristics such as temperature, electric current, flow of liquid/gas, motion of machines, etc.

Also, fuzzy systems can be obtained by applying the principles of fuzzy sets and logic to other areas, for example, fuzzy knowledge-based systems such as fuzzy expert systems which may use fuzzy IF-THEN rules; "fuzzy software engineering" which may incorporate fuzziness in programs and data; fuzzy databases which store and retrieve fuzzy information: fuzzy pattern recognition which deals with fuzzy visual or audio signals; applications to medicine, economics, and management problems which involve fuzzy information processing.

When fuzzy systems are applied to appropriate problems, particularly the type of problems described previously, their typical characteristics are faster and smoother response than with conventional systems. This translates to efficient and more comfortable operations for such tasks as controlling tem-

Year		Number
1986	...	8
1987	...	15
1988	...	50
1989	...	100
1990	...	150
1991	...	300
1992	...	800
1993	...	1500
⋮		
1998	...	2500

Table 1.6. The growing number of fuzzy logic applications.

perature, cruising speed, for example. Furthermore, this will save energy, reduce maintenance costs, and prolong machine life. In fuzzy systems describing the control rules is usually simpler and easier, often requiring fewer rules, and thus the systems execute faster than conventional systems. Fuzzy systems often achieve tractability, robustness, and overall low cost. In turn, all these contribute to better performance. In short, conventional methods are good for simple problems, while fuzzy systems are suitable for complex problems or applications that involve human descriptive or intuitive thinking.

However we have to note some problems and limitations of fuzzy systems which include [191]

- *Stability*: a major issue for fuzzy control.
  There is no theoretical guarantee that a general fuzzy system does not go chaotic and remains stable, although such a possibility appears to be extremely slim from the extensive experience.
- *Learning capability*: Fuzzy systems lack capabilities of learning and have no memory as stated previously.
  This is why hybrid systems, particularly neuro-fuzzy systems, are becoming more and more popular for certain applications.
- Determining or tuning good membership functions and fuzzy rules are not always easy.
  Even after extensive testing, it is difficult to say how many membership functions are really required. Questions such as *why a particular fuzzy expert system needs so many rules* or *when can a developer stop adding more rules* are not easy to answer.
- There exists a general misconception of the term "fuzzy" as meaning imprecise or imperfect.

Many professionals think that fuzzy logic represents some magic without firm mathematical foundation.

- Verification and validation of a fuzzy expert system generally requires extensive testing with hardware in the loop.

Such luxury may not be affordable by all developers.

The basic steps for developing a fuzzy system are the following

- Determine whether a fuzzy system is a right choice for the problem. If the knowledge about the system behavior is described in approximate form or heuristic rules, then fuzzy is suitable. Fuzzy logic can also be useful in understanding and simplifying the processing when the system behavior requires a complicated mathematical model.
- Identify inputs and outputs and their ranges. Range of sensor measurements typically corresponds to the range of input variable, and the range of control actions provides the range of output variable.
- Define a primary membership function for each input and output parameter. The number of membership functions required is a choice of the developer and depends on the system behavior.
- Construct a rule base. It is up to the designer to determine how many rules are necessary.
- Verify that rule base output within its range for some sample inputs, and further validate that this output is correct and proper according to the rule base for the given set of inputs.

Several studies show that fuzzy logic is applicable in *Management Science* (see e.g. [13]).

Bibliography

1. K.J. Arrow, *Social Choice and Individual Values* (John Wiley & Sons, New York, 1951).
2. R.A.Bellman and L.A.Zadeh, Decision-making in a fuzzy environment, *Management Sciences*, Ser. B 17(1970) 141-164.
3. J. F. Brule, Fuzzy systems - a tutorial, (http://www.austinlinks.com/Fuzzy/-tutorial.html, 1985).
4. D. Butnariu and E.P. Klement, *Triangular Norm-Based Measures and Games with Fuzzy Coalitions* (Kluwer, Dordrecht, 1993).
5. D. Butnariu, E.P. Klement and S. Zafrany, On triangular norm-based propositional fuzzy logics, *Fuzzy Sets and Systems*, 69(1995) 241-255.
6. E. Canestrelli and S. Giove, Optimizing a quadratic function with fuzzy linear coefficients, *Control and Cybernetics*, 20(1991) 25-36.
7. E. Canestrelli and S. Giove, Bidimensional approach to fuzzy linear goal programming, in: M. Delgado, J. Kacprzyk, J.L. Verdegay and M.A. Vila eds., *Fuzzy Optimization* (Physical Verlag, Heildelberg, 1994) 234-245.
8. E. Canestrelli, S. Giove and R. Fullér, Sensitivity analysis in possibilistic quadratic programming, *Fuzzy Sets and Systems*, 82(1996) 51-56.
9. B. Cao, New model with T-fuzzy variations in linear programming, *Fuzzy Sets and Systems*, 78(1996) 289-292.
10. C. Carlsson, A. Törn and M. Zeleny eds., *Multiple Criteria Decision Making: Selected Case Studies*, McGraw Hill, New York 1981.
11. C. Carlsson, Tackling an MCDM-problem with the help of some results from fuzzy sets theory, *European Journal of Operational Research*, 3(1982) 270-281.
12. C. Carlsson, An approach to handle fuzzy problem structures, *Cybernet. and Systems*, 14(1983) 33-54.
13. C. Carlsson, On the relevance of fuzzy sets in management science methodology, *TIMS/Studies in the Management Sciences*, 20(1984) 11-28.
14. C. Carlsson, Fuzzy multiple criteria for decision support systems, in: M.M. Gupta, A. Kandel and J.B. Kiszka eds., *Approximate Reasoning in Expert Systems* (North-Holland, Amsterdam, 1985) 48-60.
15. C. Carlsson, and P.Korhonen, A parametric approach to fuzzy linear programming, *Fuzzy Sets and Systems*, 20(1986) 17-30.
16. C. Carlsson, Approximate Reasoning for solving fuzzy MCDM problems, *Cybernetics and Systems: An International Journal* 18(1987) 35-48.
17. C. Carlsson, Approximate reasoning through fuzzy MCDM-models, *Operation Research'87* (North-Holland, Amsterdam, 1988) 817-828.
18. C. Carlsson, On interdependent fuzzy multiple criteria, in: R. Trappl ed., *Cybernetics and Systems'90* (World Scientific, Singapore, 1990) 139-146.
19. C. Carlsson, On optimization with interdependent multiple criteria, in: R. Lowen and M. Roubens eds., *Proc. of Fourth IFSA Congress, Vol. Computer, Management and Systems Science*, Brussels,1991 19-22.

20. C. Carlsson, On optimization with interdependent multiple criteria, in: R. Lowen and M. Roubens eds., *Fuzzy Logic: State of the Art*, Kluwer Academic Publishers, Dordrecht, 1992 415-422.

21. C. Carlsson, Cognitive Maps and Hyperknowledge. A Blueprint for Active Decision Support Systems, in: *Cognitive Maps and Strategic Thinking*, C. Carlsson ed. May 1995, Åbo, Finland, (Meddelanden Från Ekonomisk-Statsvetenskapliga Fakulteten Vid Åbo Akademi, IAMSR, Ser. A:442) 27-59.

22. C. Carlsson, D. Ehrenberg, P. Eklund, M. Fedrizzi, P. Gustafsson, P. Lindholm, G. Merkuryeva, T. Riissanen and A. Ventre, Consensus in distributed soft environments, *European Journal of Operational Research*, 61(1992) 165-185

23. C. Carlsson and R. Fullér, Fuzzy if-then rules for modeling interdependencies in FMOP problems, in: *Proceedings of EUFIT'94 Conference*, September 20-23, 1994 Aachen, Germany (Verlag der Augustinus Buchhandlung, Aachen, 1994) 1504-1508.

24. C. Carlsson and R. Fullér, Interdependence in fuzzy multiple objective programming, *Fuzzy Sets and Systems*, 65(1994) 19-29.

25. C. Carlsson and R. Fullér, Fuzzy reasoning for solving fuzzy multiple objective linear programs, in: R.Trappl ed., *Cybernetics and Systems '94, Proceedings of the Twelfth European Meeting on Cybernetics and Systems Research* (World Scientific Publisher, London, 1994) 295-301.

26. C. Carlsson and R. Fullér, Multiple Criteria Decision Making: The Case for Interdependence, *Computers & Operations Research* 22(1995) 251-260.

27. C. Carlsson and R. Fullér, On linear interdependences in MOP, in: *Proceedings of CIFT'95 Workshop*, June 8-10, 1995, Trento, Italy, University of Trento, 1995 48-52.

28. C. Carlsson and R. Fullér, Active DSS and approximate reasoning, in: *Proceedings of EUFIT'95 Conference*, August 28-31, 1995, Aachen, Germany, Verlag Mainz, Aachen, 1995 1209-1215.

29. C. Carlsson and R. Fullér, On fuzzy screening system, in: *Proceedings of the Third European Congress on Intelligent Techniques and Soft Computing (EUFIT'95)*, August 28-31, 1995 Aachen, Germany, Verlag Mainz, Aachen, [ISBN 3-930911-67-1], 1995 1261-1264.

30. C. Carlsson and R. Fullér, Fuzzy multiple criteria decision making: Recent developments, *Fuzzy Sets and Systems*, 78(1996) 139-153.

31. C. Carlsson and R. Fullér, Additive interdependences in MOP, in: M.Brännback and M.Kuula eds., *Proceedings of the First Finnish Noon-to-noon seminar on Decision Analysis*, Åbo, December 11-12, 1995, Meddelanden Från Ekonomisk-Statsvetenskapliga Fakulteten vid Åbo Akademi, Ser: A:459, Åbo Akademis tryckeri, Åbo, 1996 77-92.

32. C. Carlsson and R. Fullér, Compound interdependences in MOP, in: *Proceedings of the Fourth European Congress on Intelligent Techniques and Soft Computing (EUFIT'96)*, September 2-5, 1996, Aachen, Germany, Verlag Mainz, Aachen, 1996 1317-1322.

33. C. Carlsson and R. Fullér, Problem-solving with multiple interdependent criteria: Better solutions to complex problems, in: D.Ruan, P.D'hondt, P.Govaerts and E.E.Kerre eds., *Proceedings of the Second International FLINS Workshop on Intelligent Systems and Soft Computing for Nuclear Science and Industry*, September 25-27, 1996, Mol, Belgium, World Scientific Publisher, 1996 89-97.

34. C. Carlsson and R. Fullér, Adaptive Fuzzy Cognitive Maps for Hyperknowledge Representation in Strategy Formation Process, in: *Proceedings of International Panel Conference on Soft and Intelligent Computing*, Budapest, October 7-10, 1996, Technical University of Budapest, 1996 43-50.

35. C. Carlsson and R. Fullér, A neuro-fuzzy system for portfolio evaluation, in: R.Trappl ed., *Cybernetics and Systems '96, Proceedings of the Thirteenth European Meeting on Cybernetics and Systems Research*, Vienna, April 9-12, 1996, Austrian Society for Cybernetic Studies, Vienna, 1996 296-299.

36. C. Carlsson, R. Fullér and S.Fullér, Possibility and necessity in weighted aggregation, in: R.R. Yager and J. Kacprzyk eds., *The ordered weighted averaging operators: Theory, Methodology, and Applications*, Kluwer Academic Publishers, Boston, 1997 18-28.

37. C. Carlsson, R. Fullér and S.Fullér, OWA operators for doctoral student selection problem, in: R.R. Yager and J. Kacprzyk eds., *The ordered weighted averaging operators: Theory, Methodology, and Applications*, Kluwer Academic Publishers, Boston, 1997 167-178.

38. C. Carlsson and R. Fullér, Problem solving with multiple interdependent criteria, in: J. Kacprzyk, H.Nurmi and M.Fedrizzi eds., *Consensus under Fuzziness*, The Kluwer International Series in Intelligent Technologies, Vol. 10, Kluwer Academic Publishers, Boston, 1997 231-246.

39. C. Carlsson and R. Fullér, OWA operators for decision support, in: *Proceedings of the Fifth European Congress on Intelligent Techniques and Soft Computing (EUFIT'97)*, September 8-11, 1997, Aachen, Germany, Verlag Mainz, Aachen, Vol. II, 1997 1539-1544.

40. C. Carlsson and R. Fullér, Soft computing techniques for portfolio evaluation, in: A. Zempléni ed., *Statistics at Universities: Its Impact for Society, Tempus (No. 9521) Workshop*, Budapest, May 22-23, 1997, Eötvös University Press, Budapest, Hungary, 1997 47-54.

41. C. Carlsson and R. Fullér, A novel approach to linguistic importance weighted aggregations, in: C. Carlsson and I.Eriksson eds., *Global & Multiple Criteria Optimization and Information Systems Quality*, Åbo Akademis tryckeri, Åbo, 1998 143-153.

42. C. Carlsson and R. Fullér, A new look at linguistic importance weighted aggregations, *Cybernetics and Systems '98, Proceedings of the Fourteenth European Meeting on Cybernetics and Systems Research*, Austrian Society for Cybernetic Studies, Vienna, 1998 169-174

43. C. Carlsson and R. Fullér, Benchmarking in linguistic importance weighted aggregations, *Fuzzy Sets and Systems*, 1999 (to appear).

44. C. Carlsson and R. Fullér, Optimization with linguistic values, *TUCS Technical Reports*, Turku Centre for Computer Science, No. 157/1998. [ISBN 952-12-0138-X, ISSN 1239-1891].

45. C. Carlsson and R. Fullér, Multiobjective optimization with linguistic variables, in: *Proceedings of the Sixth European Congress on Intelligent Techniques and Soft Computing (EUFIT'98)*, Aachen, September 7-10, 1998, Verlag Mainz, Aachen, [ISBN 3-89653-500-5], Vol. II, 1998 1038-1042.

46. C. Carlsson and R. Fullér, Optimization under fuzzy if-then rules, *Fuzzy Sets and Systems*, 1999, (to appear).

47. C. Carlsson and R. Fullér, On mean value and variance of fuzzy numbers, *Research Reports, Institute for Advanced Management Systems Research, Åbo Akademi University*, July, 1999, No. 1999/8. [ISBN 952-12-0500-8, ISSN 1235-9505]

48. C. Carlsson and R. Fullér, On mean value and variance of fuzzy numbers, *Fuzzy Sets and Systems*, (submitted).

49. C. Carlsson and P. Walden, Active DSS and Hyperknowledge: Creating Strategic Visions, in: *Proceedings of EUFIT'95 Conference*, August 28-31, 1995, Aachen, Germany, (Verlag Mainz, Aachen, 1995) 1216-1222.

50. J.L. Castro, Fuzzy logic contollers are universal approximators, *IEEE Transactions on Syst. Man Cybernet.*, 25(1995) 629-635.

51. S.M. Chen, A weighted fuzzy reasoning akgorithm for medical diagnosis, *Decision Support Systems*, 11(1994) 37-43.

52. E. Cox, The Fuzzy system Handbook. *A Practitioner's Guide to Building, Using, and Maintaining Fuzzy Systems* (Academic Press, New York, 1994).

53. M. Delgado, E. Trillas, J.L. Verdegay and M.A. Vila, The generalized "modus ponens" with linguistic labels, in: *Proceedings of the Second International Conference on Fuzzy Logics andd Neural Network*, IIzuka, Japan, 1990 725-729.

54. M.Delgado, J. Kacprzyk, J.L.Verdegay and M.A.Vila eds., *Fuzzy Optimization* (Physical Verlag, Heildelberg, 1994).

55. J. Dombi, A general class of fuzzy operators, the DeMorgan class of fuzzy operators and fuziness measures induced by fuzzy operators, *Fuzzy Sets and Systems*, 8(1982) 149-163.

56. J. Dombi, Membership function as an evaluation, *Fuzzy Sets and Systems*, 35(1990) 1-21.

57. D. Driankov, H. Hellendoorn and M. Reinfrank, *An Introduction to Fuzzy Control* (Springer Verlag, Berlin, 1993).

58. D. Dubois, R. Martin-Clouaire and H. Prade, Practical computation in fuzzy logic, in: M.M. Gupta and T. Yamakawa eds., *Fuzzy Computing* (Elsevier Science Publishing, Amsterdam, 1988) 11-34.

59. D. Dubois and H. Prade, *Fuzzy Sets and Systems: Theory and Applications* (Academic Press, London, 1980).

60. D. Dubois and H. Prade, Criteria aggregation and ranking of alternatives in the framework of fuzzy set theory *TIMS/Studies in the Management Sciences*, 20(1984) 209-240.

61. D.Dubois and H.Prade, *Possibility Theory* (Plenum Press, New York,1988).

62. D.Dubois, H.Prade and R.R Yager eds., *Readings in Fuzzy Sets for Intelligent Systems* (Morgan & Kaufmann, San Mateo, CA, 1993).

63. G. W. Evans ed., *Applications of Fuzzy Set Methodologies in Industrial Engineering* (Elsevier, Amsterdam, 1989).

64. M. Fedrizzi, J. Kacprzyk and S. Zadrozny, An interactive multi-user decision support system for consensus reaching processes using fuzzy logic with linguistic quantifiers, *Decision Support Systems*, 4(1988) 313-327.

65. M. Fedrizzi and L. Mich, Decision using production rules, in: *Proc. of Annual Conference of the Operational Research Society of Italy*, September 18-10, Riva del Garda. Italy, 1991 118-121.

66. M. Fedrizzi and R.Fullér, On stability in group decision support systems under fuzzy production rules, in: R.Trappl ed., *Proceedings of the Eleventh European Meeting on Cybernetics and Systems Research* (World Scientific Publisher, London, 1992) 471-478.

67. M. Fedrizzi and R.Fullér, Stability in possibilistic linear programming problems with continuous fuzzy number parameters, *Fuzzy Sets and Systems*, 47(1992) 187-191.

68. M. Fedrizzi, M, Fedrizzi and W. Ostasiewicz, Towards fuzzy modeling in economics, *Fuzzy Sets and Systems* (54)(1993) 259-268.

69. J.C. Fodor and M. Roubens, *Fuzzy Preference Modelling and Multicriteria Decision Aid* (Kluwer Academic Publisher, Dordrecht, 1994).

70. M.J. Frank, On the simultaneous associativity of $F(x, y)$ and $x + y - F(x, y)$, *Aequat. Math.*, 19(1979) 194-226.

71. R. Fullér and T. Keresztfalvi, On Generalization of Nguyen's theorem, *Fuzzy Sets and Systems*, 41(1991) 371-374.

72. R. Fullér, On Hamacher-sum of triangular fuzzy numbers, *Fuzzy Sets and Systems*, 42(1991) 205-212.

73. R. Fullér, Well-posed fuzzy extensions of ill-posed linear equality systems, *Fuzzy Systems and Mathematics*, 5(1991) 43-48.

74. R. Fullér and B. Werners, The compositional rule of inference: introduction, theoretical considerations, and exact calculation formulas, Working Paper, RWTH Aachen, institut für Wirtschaftswissenschaften, No.1991/7.

75. R. Fullér, On law of large numbers for L-R fuzzy numbers, in: R. Lowen and M. Roubens eds., *Proceedings of the Fourth IFSA Congress, Volume: Mathematics*, Brussels, 1991 74-77.

76. R. Fullér and H.-J. Zimmermann, On Zadeh's compositional rule of inference, in: R.Lowen and M.Roubens eds., *Proceedings of the Fourth IFSA Congress, Volume: Artifical intelligence*, Brussels, 1991 41-44.

77. R. Fullér and H.-J. Zimmermann, On computation of the compositional rule of inference under triangular norms, *Fuzzy Sets and Systems*, 51(1992) 267-275.

78. R. Fullér and T. Keresztfalvi, t-Norm-based addition of fuzzy intervals, *Fuzzy Sets and Systems*, 51(1992) 155-159.

79. R. Fullér and B.Werners, The compositional rule of inference with several relations, *Tatra Mountains Mathematical Publications*, 1(1992) 39-44.

80. R. Fullér and H.-J.Zimmermann, Fuzzy reasoning for solving fuzzy mathematical programming problems, Working Paper, RWTH Aachen, institut für Wirtschaftswissenschaften, No.1992/01.

81. R. Fullér, A law of large numbers for fuzzy numbers, *Fuzzy Sets and Systems*, 45(1992) 299-303.

82. R. Fullér and H.-J. Zimmermann, On Zadeh's compositional rule of inference, In: R.Lowen and M.Roubens eds., *Fuzzy Logic: State of the Art*, Theory and Decision Library, Series D (Kluwer Academic Publisher, Dordrecht, 1993) 193-200.

83. R. Fullér and H.-J. Zimmermann, Fuzzy reasoning for solving fuzzy mathematical programming problems, *Fuzzy Sets and Systems* 60(1993) 121-133.

84. R. Fullér and E. Triesch, A note on law of large numbers for fuzzy variables, *Fuzzy Sets and Systems*, 55(1993).

85. R. Fullér and M. Fedrizzi, On stability in multiobjective possibilistic linear programs, *European Journal of Operational Reseach*, 74(1994) 179-187.

86. R. Fullér, L. Gaio, L.Mich and A.Zorat, OCA functions for consensus reaching in group decisions in fuzzy environment, in: *Proceedings of the 3rd International Conference on Fuzzy Logic, Neural Nets and Soft Computing*, Iizuka, Japan, August 1-7, 1994, Iizuka, Japan, 1994, Fuzzy Logic Systems institute, 1994 101-102.

87. R. Fullér and S.Giove, A neuro-fuzzy approach to FMOLP problems, in: *Proceedings of CIFT'94*, June 1-3, 1994, Trento, Italy, University of Trento, 1994 97-101.

88. R. Fullér, *Neural Fuzzy Systems*, Åbo Akademis tryckeri, Åbo, ESF Series, A:443, 1995, 249 pages.

89. R. Fullér, Hyperknowledge representation: challenges and promises, in: P. Walden, M. Brännback, B.Back and H. Vanharanta eds., *The Art and Science of Decision-Making*, Åbo Akademi University Press, Åbo, 1996 61-89.

90. R. Fullér, OWA operators for decision making, in: C. Carlsson ed., *Exploring the Limits of Support Systems*, TUCS General Publications, No. 3, Turku Centre for Computer Science, Åbo, 1996 85-104.

91. R. Fullér, *Fuzzy Reasoning and Fuzzy Optimization*, TUCS General Publications, No. 9, Turku Centre for Computer Science, Åbo, 1998, 270 pages.

92. R. Goetschel and W. Voxman, Elementary fuzzy calculus, *Fuzzy Sets and Systems*, 18(1986) 31-43.
93. M.M. Gupta and D.H. Rao, On the principles of fuzzy neural networks, *Fuzzy Sets and Systems*, 61(1994) 1-18.
94. H. Hamacher, H.Leberling and H.-J.Zimmermann, Sensitivity analysis in fuzzy linear programming, *Fuzzy Sets and Systems*, 1(1978) 269-281.
95. H. Hamacher, Über logische Aggregationen nicht binär explizierter Entscheidung-kriterien (Rita G.Fischer Verlag, Frankfurt, 1978).
96. H. Hellendoorn, Closure properties of the compositional rule of inference, *Fuzzy Sets and Systems*, 35(1990) 163-183.
97. F. Herrera, M. Kovács, and J. L. Verdegay, An optimum concept for fuzzified linear programming problems: a parametric approach, *Tatra Mountains Mathematical Publications*, 1(1992) 57–64.
98. F. Herrera, M. Kovács, and J. L. Verdegay, Fuzzy linear programming problems with homogeneous linear fuzzy functions, in: *Proc. of IPMU'92*, Universitat de les Illes Balears, 1992 361-364.
99. F. Herrera, M. Kovács, and J. L. Verdegay. Optimality for fuzzified mathematical programming problems: a parametric approach, *Fuzzy Sets and Systems*, 54(1993) 279-285.
100. F. Herrera, J. L. Verdegay, and M. Kovács, A parametric approach for (g,p)-fuzzified linear programming problems, *Journal of Fuzzy Mathematics*, 1(1993) 699-713.
101. F. Herrera, M. Kovács, and J. L. Verdegay, Homogeneous linear fuzzy functions and ranking methods in fuzzy linear programming problems, *Int. J. on Uncertainty, Fuzziness and Knowledge-based Systems*, 1(1994) 25-35.
102. F. Herrera, E. Herrera-Viedma and J. L. Verdegay, Aggregating Linguistic Preferences: Properties of the LOWA Operator, in: *Proceedings of the 6th IFSA World Congress*, Sao Paulo (Brasil), Vol. II, 1995 153-157.
103. F. Herrera, E. Herrera-Viedma, J.L. Verdegay, Direct approach processes in group decision making using linguistic OWA operators, *Fuzzy Sets and Systems*, 79(1996) 175-190.
104. F. Herrera and J.L. Verdegay, Fuzzy boolean *Fuzzy Sets and Systems*, 81(1996) 57-76.
105. F. Herrera, E. Herrera-Viedma, J.L. Verdegay, A model of consensus in group decision making under linguistic assessments, *Fuzzy Sets and Systems*, 78(1996) 73-87.
106. F. Herrera and E. Herrera-Viedma, Aggregation Operators for Linguistic Weighted Information, *IEEE Transactions on Systems, Man and Cybernetics - Part A: Systems and Humans*, (27)1997 646-656.
107. F. Herrera, E. Herrera-Viedma, J.L. Verdegay, Linguistic Measures Basedon Fuzzy Coincidence for Reaching Consensus in Group Decision Making, *International Journal of Approximate Reasoning*, 16(1997) 309-334.
108. E.Herrera and E. Herrera-Viedma, On the linguistic OWA operator and extensions, in: R.R.Yager and J.Kacprzyk eds., *The ordered weighted averaging operators: Theory, Methodology, and Applications*, Kluwer Academic Publishers, Boston, 1997 60-72.
109. F. Herrera, E. Herrera-Viedma and J.L. Verdegay, Applications of the Linguistic OWA Operator in Group Decision Making, in: R.R.Yager and J.Kacprzyk eds., *The ordered weighted averaging operators: Theory, Methodology, and Applications*, Kluwer Academic Publishers, Boston, 1997 207-218.
110. F. Herrera, E. Herrera-Viedma, J.L. Verdegay, A rational consensus model in group decision making using linguistic assessments, *Fuzzy Sets and Systems*, 88(1997) 31-49.

111. F. Herrera, E. Herrera-Viedma and J.L.Verdegay, Choice processes for non-homogeneous group decision making in linguistic setting, *International Journal for Fuzzy Sets and Systems*, 94(1998) 287-308.

112. D.H. Hong and S.Y.Hwang, On the convergence of T-sum of L-R fuzzy numbers, *Fuzzy Sets and Systems*, 63(1994) 175-180.

113. D.H. Hong, A note on product-sum of L-R fuzzy numbers, *Fuzzy Sets and Systems*, 66(1994) 381-382.

114. D.H. Hong and S.Y.Hwang, On the compositional rule of inference under triangular norms, *Fuzzy Sets and Systems*, 66(1994) 25-38.

115. D.H. Hong A note on the law of large numbers for fuzzy numbers, *Fuzzy Sets and Systems*, 64(1994) 59-61.

116. D.H. Hong A note on the law of large numbers for fuzzy numbers, *Fuzzy Sets and Systems*, 68(1994) 243.

117. D.H. Hong, A note on t-norm-based addition of fuzzy intervals, *Fuzzy Sets and Systems*, 75(1995) 73-76.

118. D.H. Hong and Y.M.Kim, A law of large numbers for fuzzy numbers in a Banach space, *Fuzzy Sets and Systems*, 77(1996) 349-354.

119. D.H. Hong and C. Hwang, Upper bound of T-sum of LR-fuzzy numbers, in: *Proceedings of IPMU'96 Conference* (July 1-5, 1996, Granada, Spain), 1996 343-346.

120. D.H. Hong, A convergence theorem for arrays of L-R fuzzy numbers, *Information Sciences*, 88(1996) 169-175.

121. D.H. Hong and S.Y.Hwang, The convergence of T-product of fuzzy numbers, *Fuzzy Sets and Systems*, 85(1997) 373-378.

122. D.H. Hong and C. Hwang, A T-sum bound of LR-fuzzy numbers, *Fuzzy Sets and Systems*, 91(1997) 239-252.

123. S. Horikowa, T. Furuhashi and Y. Uchikawa, On identification of structures in premises of a fuzzy model using a fuzzy neural network, in: *Proc. IEEE International Conference on Fuzzy Systems*, San Francisco, 1993 661-666.

124. H. Hsi-Mei and C. Chen-Tung, Aggregation of fuzzy opinions under group decision making, *Fuzzy Sets and Systems*, 79(1996) 279-285.

125. C. Huey-Kuo and C, Huey-Wen,Solving multiobjective linear programming problems - a generic approach, *Fuzzy Sets and Systems*, 82(1996) 35-38.

126. Suwarna Hulsurkar, M.P. Biswal and S.B. Sinha, Fuzzy programming approach to multi-objective stochastic linear programming problems, *Fuzzy Sets and Systems*, 88(1997) 173-181.

127. M.L. Hussein and M.A. Abo-Sinna, A fuzzy dynamic approach to the multicriterion resource allocation problem, *Fuzzy Sets and Systems*, 69(1995) 115-124.

128. M.L. Hussein and M. Abdel Aaty Maaty, The stability notions for fuzzy nonlinear programming problem, *Fuzzy Sets and Systems*, 85(1997) 319-323

129. M. L. Hussein, Complete solutions of multiple objective transportation problems with possibilistic coefficients, *Fuzzy Sets and Systems*, 93(1998) 293-299.

130. C.L. Hwang and A.S.M. Masud, *Multiobjective Decision Making - Methods and Applications, A State-of-the-Art Survey* (Springer-Verlag, New-York, 1979).

131. C.L. Hwang and K. Yoon, *Multiple Attribute Decision Making - Methods and Applications, A State-of-the-Art Survey* (Springer-Verlag, New-York, 1981).

132. C.L. Hwang and M.J. Lin, *Group Decision Making Under Multiple Criteria* (Springer-Verlag, New-York, 1987).

133. S.Y.Hwang and D.H.Hong, The convergence of T-sum of fuzzy numbers on Banach spaces, *Applied Mathematics Letters* 10(1997) 129-134.

134. M.Inuiguchi, H.Ichihashi and H. Tanaka, Fuzzy Programming: A Survey of Recent Developments, in: Slowinski and Teghem eds., *Stochastic versus Fuzzy*

Approaches to Multiobjective Mathematical Programming under Uncertainty,
Kluwer Academic Publishers, Dordrecht 1990, pp 45-68

135. M. Inuiguchi and M. Sakawa, A possibilistic linear program is equivalent to a stochastic linear program in a special case, *Fuzzy Sets and Systems,* 76(1995) 309-317.

136. M. Inuiguchi and M. Sakawa, Possible and necessary efficiency in possibilistic multiobjective linear programming problems and possible efficiency test *Fuzzy Sets and Systems,* 78(1996) 231-241.

137. M. Inuiguchi, Fuzzy linear programming: what, why and how? *Tatra Mountains Math. Publ.,* 13(1997) 123-167.

138. J.-S. Roger Jang, ANFIS: Adaptive-network-based fuzzy inference system, *IEEE Trans. Syst., Man, and Cybernetics,* 23(1993) 665-685.

139. L.C.Jang and J.S.Kwon, A note on law of large numbers for fuzzy numbers in a Banach space, *Fuzzy Sets and Systems,* 98(1998) 77-81.

140. S.Jenei, Continuity in approximate reasoning, *Annales Univ. Sci. Budapest, Sect. Comp.,* 15(1995) 233-242.

141. B.Julien, An extension to possibilistic linear programming, *Fuzzy Sets and Systems,* 64(1994) 195-206.

142. J. Kacprzyk and R.R. Yager, "Softer" optimization and control models via fuzzy linguistic quantifiers, *Information Sciences,* 34(1984) 157-178.

143. J. Kacprzyk and R.R. Yager, *Management Decision Support Systems Using Fuzzy Sets and Possibility Theory,* Springer Verlag, Berlin 1985.

144. J. Kacprzyk, Group decision making with a fuzzy linguistic majority, *Fuzzy Sets and Systems,* 18(1986) 105-118.

145. J. Kacprzyk and S.A. Orlovski eds., *Optimization Models Using Fuzzy Sets and Possibility Theory* (D.Reidel, Boston,1987).

146. J. Kacprzyk and R.R. Yager, Using fuzzy logic with linguistic quantifiers in multiobjective decision-making and optimization: A step towards more human-consistent models, in: R.Slowinski and J.Teghem eds., *Stochastic versus Fuzzy Approaches to Multiobjective Mathematical Programming under Uncertainty,* Kluwer Academic Publishers, Dordrecht, 1990 331-350.

147. J. Kacprzyk and M. Fedrizzi, *Multiperson decision-making Using Fuzzy Sets and Possibility Theory* (Kluwer Academic Publisher, Dordrecht, 1990).

148. J. Kacprzyk and A.O. Esogbue, Fuzzy dynamic programming: Main developments and applications, *Fuzzy Sets and Systems,* 81(1996) 31-45.

149. J. Kacprzyk and R.R. Yager eds., *The ordered weighted averaging operators: Theory, Methodology, and Applications,* Kluwer Academic Publishers, Boston, 1997.

150. O. Kaleva, Fuzzy differential equations, *Fuzzy Sets and Systems,* 24(1987) 301-317.

151. M.A.E. Kassem, Interactive stability of multiobjective nonlinear programming problems with fuzzy parameters in the constraints, *Fuzzy Sets and Systems,* 73(1995) 235-243.

152. M.A.El-Hady Kassem and E.I. Ammar, Stability of multiobjective nonlinear programming problems with fuzzy parameters in the constraints, *Fuzzy Sets and Systems,* 74(1995) 343-351.

153. M.A.El-Hady Kassem and E.I. Ammar, A parametric study of multiobjective NLP problems with fuzzy parameters in the objective functions *Fuzzy Sets and Systems,* 80(1996) 187-196.

154. M.F.Kawaguchi and T.Da-te, A calculation method for solving fuzzy arithmetic equations with triangular norms, in: *Proceedings of Second IEEE international Conference on Fuzzy Systems,* 1993 470-476.

155. M.F.Kawaguchi and T.Da-te, Some algebraic properties of weakly non-interactive fuzzy numbers, *Fuzzy Sets and Systems*, 68(1994) 281-291.

156. T. Keresztfalvi and H. Rommelfanger, Fuzzy linear programming with *t*-norm based extended addition, *Operations Research Proceedings 1991* (Springer-Verlag, Berlin, Heidelberg, 1992) 492–499.

157. T. Keresztfalvi and M. Kovács, g,p-fuzzification of arithmetic operations, *Tatra Mountains Mathematical Publications*, 1(1992) 65-71.

158. P.E. Klement and R. Mesiar, Triangular norms, *Tatra Mountains Mathematical Publications*, 13(1997) 169-193.

159. G.J.Klir and B.Yuan, Fuzzy Sets and Fuzzy Logic: Theory and Applications, Prentice Hall, 1995.

160. L.T. Kóczy and K. Hirota, Ordering, distance and Closeness of Fuzzy Sets, *Fuzzy Sets and Systems*, 59(1993) 281-293.

161. L.T. Kóczy, A fast algorithm for fuzzy inference by compact rules, in: L.A. Zadeh and J. Kacprzyk eds., *Fuzzy Logic for the Management of Uncertainty* (J. Wiley, New York, 1993) 297-317.

162. S. Korner, *Laws of Thought*, Encyclopedia of Philosophy, Vol. 4, (MacMillan, New York 1967) 414-417.

163. B.Kosko, *Neural networks and fuzzy systems*, Prentice-Hall, New Jersey, 1992.

164. B. Kosko, Fuzzy systems as universal approximators, in: *Proc. IEEE 1992 Int. Conference Fuzzy Systems*, San Diego, 1992 1153-1162.

165. M.Kovács, Fuzzification of ill-posed linear systems, in: D. Greenspan and P.Rózsa, Eds., Colloquia mathematica Societitas János Bolyai 50, Numerical Methods, North-Holland, Amsterdam, 1988, 521-532.

166. M.Kovács, F.P.Vasiljev and R. Fullér, On stability in fuzzified linear equality systems, *Proceedings of the Moscow State University*, Ser. 15, 1(1989), 5-9 (in Russian), translation in *Moscow Univ. Comput. Math. Cybernet.*, 1(1989), 4-9.

167. M.Kovács and R.Fullér, On fuzzy extended systems of linear equalities and inequalities, in: A.A.Tihonov and A.A.Samarskij eds., *Current Problems in Applied Mathematics*, Moscow State University, Moscow, [ISBN 5-211-00342-X], 1989 73-80 (in Russian).

168. M. Kovács and L. H. Tran, Algebraic structure of centered M-fuzzy numbers, *Fuzzy Sets and Systems*, 39(1991) 91-99.

169. M. Kovács, Linear programming with centered fuzzy numbers, *Annales Univ. Sci. Budapest, Sectio Comp.*, 12(1991) 159-165.

170. M. Kovács, An optimum concept for fuzzified mathematical programming problems, in: M. Fedrizzi, J. Kacprzyk, and M. Roubens, eds., *Interactive Fuzzy Optimization*, Lecture Notes Econ. Math. Systems, Vol. 368, Springer, Berlin, 1991 36-44.

171. M. Kovács, Fuzzy linear model fitting to fuzzy observations, in: M. Fedrizzi and J. Kacprzyk, eds., *Fuzzy Regression Analysis*, Studies in Fuzziness, Omnitech Press, Warsaw, 1991 116-123.

172. M. Kovács, Fuzzy linear programming problems with min- and max-extended algebraic operations on centered fuzzy numbers, in: R. Lowen and M. Roubens, eds., *Proceedings of the Fourth IFSA Congress*, Vol. *Computer, Management & Systems Science*, Brussels, 1991 125-128.

173. M. Kovács, A stable embedding of ill-posed linear systems into fuzzy systems, *Fuzzy Sets and Systems*, 45(1992) 305–312.

174. M. Kovács, A concept of optimality for fuzzified linear programming based on penalty function, in: V. Novák at.al., eds., *Fuzzy Approach to Reasoning and Decision Making*, Kluwer, Dordecht, 1992 133-139.

175. M. Kovács, Fuzzy linear programming problems with min- and max-extended algebraic operations on centered fuzzy numbers. in: R. Lowen and M. Roubens, eds., *Fuzzy Logic: State of Arts*, Kluwer, 1993 265-275.

176. M. Kovács, Fuzzy linear programming with centered fuzzy numbers, in: M. Delgado et.al., eds., *Fuzzy Optimization: Recent Advances*, Omnitech Physica Verlag, Heidelberg, 1994 135-147.

177. J.R.Layne, K.M.Passino and S.Yurkovich, Fuzzy learning control for antiskid braking system, *IEEE Transactions on Contr. Syst. Tech.*, 1(993) 122-129.

178. C.-C. Lee, Fuzzy logic in control systems: Fuzzy logic controller - Part I, *IEEE Transactions on Syst., Man, Cybern.*, 20(1990) 419-435.

179. C.-C. Lee, Fuzzy logic in control systems: Fuzzy logic controller - Part II, *IEEE Transactions on Syst., Man, Cybern.*, 20(1990) 404-418.

180. C. Lejewski, *Jan Lukasiewicz*, Encyclopedia of Philosophy, Vol. 5, (MacMillan, New York 1967) 104-107.

181. Hong Xing Li and Vincent C. Yen, *Fuzzy sets and fuzzy decision-making*, (CRC Press, Boca Raton, FL, 1995).

182. E.H. Mamdani and S. Assilian, An experiment in linquistic synthesis with a fuzzy logic controller. *International Journal of Man-Machine Studies* 7(1975) 1-13.

183. J.K. Mattila, On some logical points of fuzzy conditional decision making, *Fuzzy Sets and Systems*, 20(1986) 137-145.

184. Jorma K. Mattila, *Text Book of Fuzzy Logic*, Art House, Helsinki, [ISBN 951-884-152-7], 1998.

185. G.F. Mauer, A fuzzy logic controller for an ABS braking system, *IEEE Transactions on Fuzzy Systems*, 3(1995) 381-388.

186. R. Mesiar, A note to the T-sum of L-R fuzzy numbers, *Fuzzy Sets and Systems*, 79(1996) 259-261.

187. R. Mesiar, Shape preserving additions of fuzzy intervals, *Fuzzy Sets and Systems*, 86(1997) 73-78.

188. R. Mesiar, Triangular-norm-based addition of fuzzy intervals, *Fuzzy Sets and Systems*, 91(1997) 231-237.

189. L. Mich, M. Fedrizzi and L. Gaio, Approximate Reasoning in the Modelling of Consensus in Group Decisions, in: E.P. Klement and W. Slany eds., *Fuzzy Logic in Artifiacial intelligence*, Lectures Notes in Artifiacial intelligence, Vol. 695, Springer-Verlag, Berlin, 1993 91-102.

190. D. McNeil and P, Freiberger, *Fuzzy Logic* (Simon and Schuster, New York, 1993).

191. T. Munkata and Y.Jani, Fuzzy systems: An overview, *Communications of ACM*, 37(1994) 69-76.

192. C. V. Negoita, *Fuzzy Systems* (Abacus Press, Turnbridge-Wells, 1981).

193. H.T. Nguyen, A note on the extension principle for fuzzy sets, *Journal of Mathematical Analysis and Applications*, 64(1978) 369-380.

194. S.A. Orlovsky, *Calculus of Decomposable Properties. Fuzzy Sets and Decisions* (Allerton Press, 1994).

195. A.R. Ralescu, A note on rule representation in expert systems, *Information Sciences*, 38(1986) 193-203.

196. R.A. Riberio, Fuzzy multiple attribute decision making: A review and new preference elicitation techniques, *Fuzzy Sets and Systems*, 78(1996) 155-181.

197. M. Rommelfanger and R. Hanuscheck and J. Wolf, Linear Programming with Fuzzy Objectives, *Fuzzy Sets and Systems*, 29(1989) 31-48.

198. H. Rommelfanger, FULPAL: An interactive method for solving (multiobjective) fuzzy linear programming problems, in: R. Słowiński and J. Teghem Jr.

eds., *Stochastic versus Fuzzy Approaches to Multiobjective Mathematical Programming under Uncertainty*, Kluwer Academic Publishers, Dordrecht, 1990 279-299.

199. H.Rommelfanger, FULP - A PC-supported procedure for solving multicriteria linear programming problems with fuzzy data, in: M. Fedrizzi, J. Kacprzyk and M.Roubens eds., *Interactive Fuzzy Optimization*, Springer-Verlag, Berlin, 1991 154-167.

200. H.Rommelfanger and T.Keresztfalvi, Multicriteria fuzzy optimization based on Yager's parametrized t-norm, *Foundations of Computing and Decision Sciences*, 16(1991) 99-110.

201. H. Rommelfanger, *Fuzzy Decision Support-Systeme*, Springer-Verlag, Heidelberg, 1994 (Second Edition).

202. H. Rommelfanger, Fuzzy linear programming and applications, *European Journal of Operational Research*, 92(1996) 512-527.

203. B.Schweizer and A.Sklar, Associative functions and abstract semigroups, *Publ. Math. Debrecen*, 10(1963) 69-81.

204. R. Słowiński, A multicriteria fuzzy linear programming method for water supply system development planning, *Fuzzy Sets and Systems*, 19(1986) 217-237.

205. R. Słowiński and J. Teghem Jr. Fuzzy versus Stochastic Approaches to Multicriteria Linear Programming under Uncertainty, *Naval Research Logistics*, 35(1988) 673-695.

206. R. Słowiński and J. Teghem Jr. *Stochastic versus Fuzzy Approaches to Multiobjective Mathematical Programming under Uncertainty*, Kluwer Academic Publishers, Dordrecht 1990.

207. T. Sudkamp, Similarity, interpolation, and fuzzy rule construction, *Fuzzy Sets and Systems*, 58(1993) 73-86.

208. T.Takagi and M.Sugeno, Fuzzy identification of systems and its applications to modeling and control, *IEEE Trans. Syst. Man Cybernet.*, 1985, 116-132.

209. M. Sugeno, *Industrial Applications of Fuzzy Control* (North Holland, Amsterdam, 1992).

210. T. Tilli, *Fuzzy Logik: Grundlagen, Anwendungen, Hard- und Software* (Franzis-Verlag, München, 1992).

211. T. Tilli, *Automatisierung mit Fuzzy Logik* (Franzis-Verlag, München, 1992).

212. I.B. Turksen, Fuzzy normal forms, *Fuzzy Sets and Systems*, 69(1995) 319-346.

213. L.-X. Wang and J.M. Mendel, Fuzzy basis functions, universal approximation, and orthogonal least-squares learning, *IEEE Transactions on Neural Networks*, 3(1992) 807-814.

214. L.-X. Wang, Fuzzy systems are universal approximators, in: *Proc. IEEE 1992 Int. Conference Fuzzy Systems*, San Diego, 1992 1163-1170.

215. S. Weber, A general concept of fuzzy connectives, negations, and implications based on t-norms and t-conorms, *Fuzzy Sets and Systems*, 11(9183) 115-134.

216. B.Werners, Interaktive Entscheidungsunterstützung durch ein flexibles mathematisches Programmierungssystem, Minerva, Publikation, München, 1984.

217. B. Werners, Interactive Multiple Objective Programming Subject to Flexible Constraints, *European Journal of Operational Research*, 31(1987) 324-349.

218. B.Werners and H.-J. Zimmermann, Evaluation and selection of alternatives considering multiple criteria,in: A.S. Jovanovic, K.F. Kussmaul, A.C.Lucia and P.P.Bonissone eds., *Proceedings of an International Course on Expert Systems in Structural Safety Assessment* (Stuttgart, October 2-4, 1989) Springer-Verlag, Heilderberg, 1989 167-183.

219. R.R. Yager, Fuzzy decision making using unequal objectives, *Fuzzy Sets and Systems*,1(1978) 87-95.

220. R.R. Yager, A new methodology for ordinal multiple aspect decisions based on fuzzy sets, *Decision Sciences* 12(1981) 589-600.
221. R.R. Yager ed., *Fuzzy Sets and Applications. Selected Papers by L.A.Zadeh* (John Wiley & Sons, New York, 1987).
222. R.R.Yager, Ordered weighted averaging aggregation operators in multi-criteria decision making, *IEEE Trans. on Systems, Man and Cybernetics*, 18(1988) 183-190.
223. R.R.Yager, Families of OWA operators, *Fuzzy Sets and Systems*, 59(1993) 125-148.
224. R.R.Yager, Fuzzy Screening Systems, in: R.Lowen and M.Roubens eds., *Fuzzy Logic: State of the Art* (Kluwer, Dordrecht, 1993) 251-261.
225. R.R.Yager, Aggregation operators and fuzzy systems modeling, *Fuzzy Sets and Systems*, 67(1994) 129-145.
226. R.R.Yager and D.Filev, *Essentials of Fuzzy Modeling and Control* (Wiley, New York, 1994).
227. R.R. Yager, Fuzzy sets as a tool for modeling, in: J. van Leeuwen ed., *Computer Science Today: Recent Trends and Development*, Springer-Verlag, Berlin, 1995 536-548.
228. R.R. Yager, Constrained OWA aggregation, *Fuzzy Sets and Systems*, 81(1996) 89-101.
229. R.R.Yager, Including importances in OWA aggregations using fuzzy systems modeling, Technical Report, #MII-1625, Mashine Intelligence Institute, Iona College, New York, 1996.
230. R.R.Yager, On the inclusion of importances in OWA aggregations, in: R.R.Yager and J.Kacprzyk eds., *The ordered weighted averaging operators: Theory, Methodology, and Applications*, Kluwer Academic Publishers, Boston, 1997 41-59.
231. T. Yamakawa and K. Sasaki, Fuzzy memory device, in: *Proceedings of 2nd IFSA Congress*, Tokyo, Japan, 1987 551-555.
232. T. Yamakawa, Fuzzy controller hardware system, in: *Proceedings of 2nd IFSA Congress*, Tokyo, Japan, 1987.
233. T. Yamakawa, Fuzzy microprocessors - rule chip and defuzzifier chip, in: *International Workshop on Fuzzy System Applications*, Iizuka, Japan, 1988 51-52.
234. J. Yen, R. Langari and L.A. Zadeh eds., *Industrial Applications of Fuzzy Logic and Intelligent Systems* (IEEE Press, New York, 1995).
235. L.A. Zadeh, Fuzzy Sets, *Information and Control*, 8(1965) 338-353.
236. L.A. Zadeh, Towards a theory of fuzzy systems, in: R.E. Kalman and N. DeClaris eds., *Aspects of Network and System Theory* (Hort, Rinehart and Winston, New York, 1971) 469-490.
237. L.A. Zadeh, Outline of a new approach to the analysis of complex systems and decision processes, *IEEE Transanctins on Systems, Man and Cybernetics*, 3(1973) 28-44.
238. L.A. Zadeh, Concept of a linguistic variable and its application to approximate reasoning, **I, II, III**, *Information Sciences*, 8(1975) 199-249, 301-357; 9(1975) 43-80.
239. L.A. Zadeh, Fuzzy sets as a basis for a theory of possibility, *Fuzzy Sets and Systems*, 1(1978) 3-28.
240. L.A. Zadeh, A theory of approximate reasoning, In: J.Hayes, D.Michie and L.I.Mikulich eds., *Machine Intelligence, Vol.9* (Halstead Press, New York, 1979) 149-194.
241. L.A. Zadeh, A computational theory of dispositions, *Int. Journal of Intelligent Systems*, 2(1987) 39-63.

242. L.A. Zadeh, Knowledge representation in fuzzy logic, In: R.R.Yager and L.A. Zadeh eds., *An introduction to fuzzy logic applications in intelligent systems* (Kluwer Academic Publisher, Boston, 1992) 2-25.

243. H.-J. Zimmermann, Description and optimization of fuzzy systems, *International Journal of General Systems*, 2(1975) 209-215.

244. H.-J. Zimmermann, Description and optimization of fuzzy system, *International Journal of General Systems*, 2(1976) 209-215.

245. H.-J. Zimmermann, Fuzzy programming and linear programming with several objective functions, *Fuzzy Sets and Systems*, 1(1978) 45-55.

246. H.-J. Zimmermann and P. Zysno, Latent connectives in human decision making, *Fuzzy Sets and Systems*, 4(1980) 37-51.

247. H.-J. Zimmermann, Applications of fuzzy set theory to mathematical programming, *Information Sciences*, 36(1985) 29-58.

248. H.-J. Zimmermann, *Fuzzy Set Theory and Its Applications*, Dordrecht, Boston 1985.

249. H.J.Zimmermann, Fuzzy set theory and mathematical programming, in: A.Jones et al. eds., Fuzzy Sets Theory and Applications, 1986, D.Reidel Publishing Company, Dordrecht, 99-114.

250. H.-J. Zimmermann, *Fuzzy Sets, decision-making and Expert Systems*, Kluwer Academic Publisher, Boston 1987.

251. H.-J. Zimmermann and B.Werners, Uncertainty representation in knowledge-based systems, in: A.S. Jovanovic, K.F. Kussmal, A.C. Lucia and P.P. Bonissone eds., *Proc. of an International Course on Expert Systems in Structural Safety Assessment* Stuttgart, October 2-4, 1989, (Springer-Verlag, Berlin, Heidelberg, 1989) 151-166.

252. H.-J. Zimmermann, decision-making in ill-structured environments and with multiple criteria, in: Bana e Costa ed., *Readings in Multiple Criteria Decision Aid* Springer Verlag, 1990 119-151.

253. H.-J. Zimmermann, Cognitive sciences, decision technology, and fuzzy sets, *Information Sciences*, 57-58(1991) 287-295.

254. H.-J. Zimmermann, Fuzzy Mathematical Programming, in: Stuart C. Shapiro ed., *Encyclopedia of Artificial Intelligence*, John Wiley & Sons, Inc., Vol. 1, 1992 521-528.

255. H.-J. Zimmermann, Methods and applications of fuzzy mathematical programming, in: R.R. Yager and L.A.Zadeh eds., *An Introduction to Fuzzy Logic Applications in Intelligent Systems*, Kluwer Academic Publisher, Boston, 1992 97-120.

256. H.-J. Zimmermann, Fuzzy Decisions: Approaches, Problems and Applications, in: *Publications Series of the Japanese-German Center*, Berlin, Series 3, Vol. 8, 1994 115-136.

257. H.-J. Zimmermann, Fuzzy Mathematical Programming, in: Tomas Gal and Harvey J. Greenberg eds., *Advances in Sensitivity Analysis and Parametric Programming*, Kluwer Academic Publishers, 1997 1-40.

258. H.-J. Zimmermann, A Fresh Perspective on Uncertainty Modeling: Uncertainty vs. Uncertainty Modeling. in: Bilal M. Ayyub and Madan M.Gupta eds., *Uncertainty Analysis in Engineering and Sciences: Fuzzy Logic, Statistics, and Neural Network Approach*, International Series in Intelligent Technologies, Kluwer Academic Publishers, 1997 353-364.

259. H.-J. Zimmermann, Fuzzy logic on the frontiers of decision analysis and expert systems, in: *Proceedings of First International Workshop on Preferences and Decisions*, Trento, June 5-7, 1997, University of Trento, 1997 97-103.

2. Artificial neural networks

2.1 The perceptron learning rule

Artificial neural systems can be considered as simplified mathematical models of brain-like systems and they function as parallel distributed computing networks. However, in contrast to conventional computers, which are programmed to perform specific task, most neural networks must be taught, or trained. They can learn new associations, new functional dependencies and new patterns. Although computers outperform both biological and artificial neural systems for tasks based on precise and fast arithmetic operations, artificial neural systems represent the promising new generation of information processing networks.

The study of brain-style computation has its roots over 50 years ago in the work of McCulloch and Pitts (1943) [19] and slightly later in Hebb's famous *Organization of Behavior* (1949) [11]. The early work in artificial intelligence was torn between those who believed that intelligent systems could best be built on computers modeled after brains, and those like Minsky and Papert (1969) [20] who believed that intelligence was fundamentally symbol processing of the kind readily modeled on the *von Neumann* computer. For a variety of reasons, the symbol-processing approach became the dominant theme in *Artifcial Intelligence* in the 1970s. However, the 1980s showed a rebirth in interest in neural computing:

1982 Hopfield [14] provided the mathematical foundation for understanding the dynamics of an important class of networks.

1984 Kohonen [16] developed unsupervised learning networks for feature mapping into regular arrays of neurons.

1986 Rumelhart and McClelland [22] introduced the backpropagation learning algorithm for complex, multilayer networks.

Beginning in 1986-87, many neural networks research programs were initiated. The list of applications that can be solved by neural networks has expanded from small test-size examples to large practical tasks. Very-large-scale integrated neural network chips have been fabricated.

In the long term, we could expect that artificial neural systems will be used in applications involving vision, speech, decision making, and reasoning, but also as signal processors such as filters, detectors, and quality control systems.

Definition 2.1.1 *[32] Artificial neural systems, or neural networks, are physical cellular systems which can acquire, store, and utilize experiental knowledge.*

The knowledge is in the form of stable states or mappings embedded in networks that can be recalled in response to the presentation of cues.

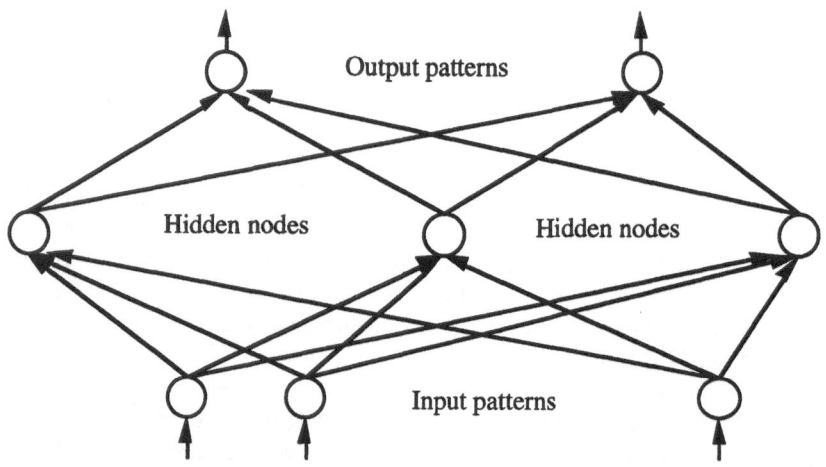

Fig. 2.1. A multi-layer feedforward neural network.

The basic processing elements of neural networks are called *artificial neurons*, or simply *neurons* or *nodes*.

Each processing unit is characterized by an activity level (representing the state of polarization of a neuron), an output value (representing the firing rate of the neuron), a set of input connections, (representing synapses on the cell and its dendrite), a bias value (representing an internal resting level of the neuron), and a set of output connections (representing a neuron's axonal projections). Each of these aspects of the unit are represented mathematically by real numbers. Thus, each connection has an associated weight (synaptic strength) which determines the effect of the incoming input on the activation level of the unit. The weights may be positive (excitatory) or negative (inhibitory).

The signal flow from of neuron inputs, x_j, is considered to be unidirectionalas indicated by arrows, as is a neuron's output signal flow. The neuron output signal is given by the following relationship

$$o = f(\langle w, x \rangle) = f(w^T x) = f\left(\sum_{j=1}^{n} w_j x_j \right),$$

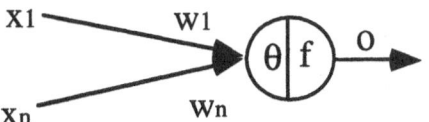

Fig. 2.2. A processing element with single output connection.

where $w = (w_1, \ldots, w_n)^T \in \mathbb{R}^n$ is the weight vector. The function $f(w^T x)$ is often referred to as an *activation (or transfer) function*. Its domain is the set of activation values, *net*, of the neuron model, we thus often use this function as $f(net)$. The variable *net* is defined as a scalar product of the weight and input vectors

$$\text{net} = \langle w, x \rangle = w^T x = w_1 x_1 + \cdots + w_n x_n,$$

and in the simplest case the output value o is computed as

$$o = f(\text{net}) = \begin{cases} 1 \text{ if } w^T x \geq \theta \\ 0 \text{ otherwise,} \end{cases}$$

where θ is called threshold-level and this type of node is called a *linear threshold unit*.

Example 2.1.1 *Suppose we have two Boolean inputs $x_1, x_2 \in \{0,1\}$, one Boolean output $o \in \{0,1\}$ and the training set is given by the following input/output pairs*

	x_1	x_2	$o(x_1, x_2) = x_1 \wedge x_2$
1.	*1*	*1*	*1*
2.	*1*	*0*	*0*
3.	*0*	*1*	*0*
4.	*0*	*0*	*0*

Then the learning problem is to find weight w_1 and w_2 and threshold (or bias) value θ such that the computed output of our network (which is given by the linear threshold function) is equal to the desired output for all examples. A straightforward solution is $w_1 = w_2 = 1/2$, $\theta = 0.6$. Really, from the equation

$$o(x_1, x_2) = \begin{cases} 1 \text{ if } x_1/2 + x_2/2 \geq 0.6 \\ 0 \text{ otherwise} \end{cases}$$

it follows that the output neuron fires if and only if both inputs are on.

Example 2.1.2 *Suppose we have two Boolean inputs $x_1, x_2 \in \{0,1\}$, one Boolean output $o \in \{0,1\}$ and the training set is given by the following input/output pairs*

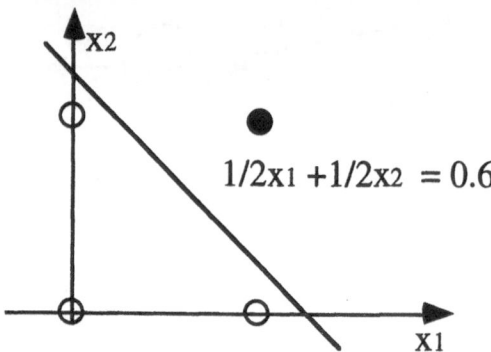

Fig. 2.3. A solution to the learning problem of Boolean *and* function.

	x_1	x_2	$o(x_1, x_2) = x_1 \vee x_2$
1.	*1*	*1*	*1*
2.	*1*	*0*	*1*
3.	*0*	*1*	*1*
4.	*0*	*0*	*0*

Then the learning problem is to find weight w_1 and w_2 and threshold value θ such that the computed output of our network is equal to the desired output for all examples. A straightforward solution is $w_1 = w_2 = 1$, $\theta = 0.8$. Really, from the equation

$$o(x_1, x_2) = \begin{cases} 1 \ if \ x_1 + x_2 \geq 0.8 \\ 0 \ otherwise \end{cases}$$

it follows that the output neuron fires if and only if at least one of the inputs is on.

The removal of the threshold from our network is very easy by increasing the dimension of input patterns. Really, the identity

$$w_1 x_1 + \cdots + w_n x_n > \theta \iff w_1 x_1 + \cdots + w_n x_n - 1 \times \theta > 0$$

means that by adding an extra neuron to the input layer with fixed input value -1 and weight θ the value of the threshold becomes zero. It is why in the following we suppose that the thresholds are always equal to zero.

We define now the scalar product of n-dimensional vectors, which plays a very important role in the theory of neural networks.

Definition 2.1.2 *Let $w = (x_n, \ldots, w_n)^T$ and $x = (x_1, \ldots, x_n)^T$ be two vectors from $\mathbb{R}^n$. The scalar (or inner) product of w and x, denoted by $\langle w, x \rangle$ or $w^T x$, is defined by*

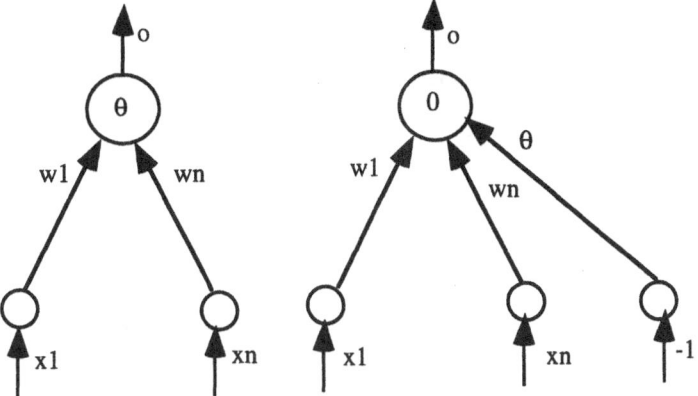

Fig. 2.4. Removing the threshold θ.

$$\langle w, x \rangle = w_1 x_1 + \cdots + w_n x_n = \sum_{j=1}^{n} w_j x_j$$

Other definition of scalar product in two dimensional case is

$$\langle w, x \rangle = \|w\| \|x\| \cos(w, x)$$

where $\|.\|$ denotes the Eucledean norm in the real plane, i.e.

$$\|w\| = \sqrt{w_1^2 + w_2^2}, \ \|x\| = \sqrt{x_1^2 + x_2^2}$$

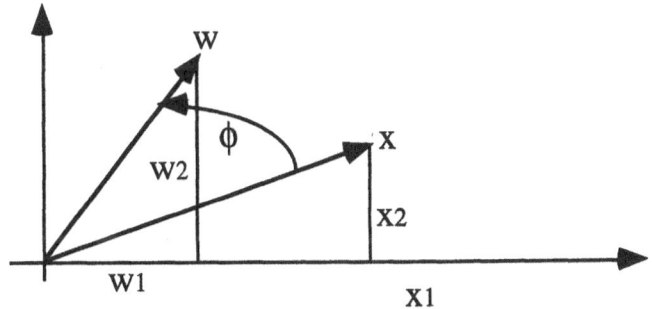

Fig. 2.5. $w = (w_1, w_2)^T$ and $x = (x_1, x_2)^T$.

Lemma 2.1.1 *The following property holds*

$$\langle w, x \rangle = w_1 x_1 + w_2 x_2 = \sqrt{w_1^2 + w_2^2} \sqrt{x_1^2 + x_2^2} \cos(w, x) = \|w\| \|x\| cos(w, x)$$

Proof.

$$\cos(w, x) = \cos((w, \text{1-st axis}) - (x, \text{1-st axis}))$$

$$= \cos((w, \text{1-st axis}) \cos(x, \text{1-st axis})) + \sin(w, \text{1-st axis}) \sin(x, \text{1-st axis})$$

$$= \frac{w_1 x_1}{\sqrt{w_1^2 + w_2^2} \sqrt{x_1^2 + x_2^2}} + \frac{w_2 x_2}{\sqrt{w_1^2 + w_2^2} \sqrt{x_1^2 + x_2^2}}.$$

That is,

$$\|w\| \|x\| \cos(w, x) = \sqrt{w_1^2 + w_2^2} \sqrt{x_1^2 + x_2^2} \cos(w, x) = w_1 x_1 + w_2 x_2.$$

From $\cos(\pi/2) = 0$ it follows that $\langle w, x \rangle = 0$ whenever w and x are perpendicular. If $\|w\| = 1$ (we say that w is normalized) then $|\langle w, x \rangle|$ is nothing else but the projection of x onto the direction of w. Really, if $\|w\| = 1$ then we get

$$\langle w, x \rangle = \|w\| \|x\| \cos(w, x) = \|x\| \cos(w, x)$$

The problem of learning in neural networks is simply the problem of finding a set of connection strengths (weights) which allow the network to carry out the desired computation. The network is provided with a set of example input/output pairs (a training set) and is to modify its connections in order to approximate the function from which the input/output pairs have been drawn. The networks are then tested for ability to generalize.

The error correction learning procedure is simple enough in conception. The procedure is as follows: During training an input is put into the network and flows through the network generating a set of values on the output units. Then, the actual output is compared with the desired target, and a match is computed. If the output and target match, no change is made to the net. However, if the output differs from the target a change must be made to some of the connections.

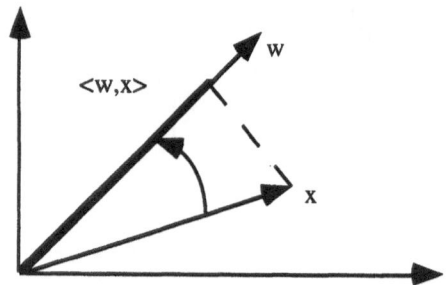

Fig. 2.6. Projection of x onto the direction of w.

The perceptron learning rule, introduced by Rosenblatt [21], is a typical error correction learning algorithm of single-layer feedforward networks with linear threshold activation function.

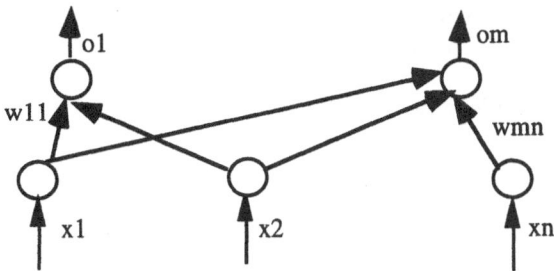

Fig. 2.7. Single-layer feedforward network.

Usually, w_{ij} denotes the weight from the j-th input unit to the i-th output unit and w_i denotes the weight vector of the i-th output node.

We are given a training set of input/output pairs

No.	input values	desired output values
1.	$x^1 = (x_1^1, \ldots x_n^1)$	$y^1 = (y_1^1, \ldots, y_m^1)$
$\vdots$	$\vdots$	$\vdots$
K.	$x^K = (x_1^K, \ldots x_n^K)$	$y^K = (y_1^K, \ldots, y_m^K)$

Our problem is to find weight vectors w_i such that

$$o_i(x^k) = \text{sign}(\langle w_i, x^k \rangle) = y_i^k, \ i = 1, \ldots, m$$

for all training patterns k.

The activation function of the output nodes is linear threshold function of the form

$$o_i(x) = \text{sign}(\langle w_i, x \rangle) = \begin{cases} +1 \text{ if } \langle w_i, x \rangle \geq 0 \\ -1 \text{ if } \langle w_i, x \rangle < 0 \end{cases}$$

and the weight adjustments in the perceptron learning method are performed by

$$w_i := w_i + \eta(y_i^k - \text{sign}(\langle w_i, x^k \rangle))x^k, \ i = 1, \ldots, m$$

$$w_{ij} := w_{ij} + \eta(y_i^k - \text{sign}(\langle w_i, x^k \rangle))x_j^k, \ j = 1, \ldots, n$$

where $\eta > 0$ is the learning rate.

From this equation it follows that if the desired output is equal to the computed output,

$$y_i^k = \text{sign}(\langle w_i, x^k \rangle),$$

then the weight vector of the i-th output node remains unchanged, i.e. w_i is adjusted if and only if the computed output, $o_i(x^k)$, is incorrect. The learning stops when all the weight vectors remain unchanged during a complete training cycle.

Consider now a single-layer network with one output node. Then the input components of the training patterns can be classified into two disjunct classes

$$C_1 = \{x^k | y^k = 1\}, \quad C_2 = \{x^k | y^k = -1\}$$

i.e. x belongs to class C_1 if there exists an input/output pair $(x, 1)$, and x belongs to class C_2 if there exists an input/output pair $(x, -1)$.

Taking into consideration the definition of the activation function it is easy to see that we are searching for a weight vector w such that

$$\langle w, x \rangle \geq 0 \text{ for each } x \in C_1, \text{ and } \langle w, x \rangle < 0 \text{ for each } x \in C_2.$$

If such vector exists then the problem is called *linearly separable*.

Summary 2.1.1 *Perceptron learning algorithm.*

Given are K training pairs arranged in the training set

$$(x^1, y^1), \dots, (x^K, y^K)$$

where $x^k = (x_1^k, \dots, x_n^k)$, $y^k = (y_1^k, \dots, y_m^k)$, $k = 1, \dots, K$.

- **Step 1** $\eta > 0$ is chosen
- **Step 2** Weigts w_i are initialized at small random values, the running error E is set to 0, $k := 1$
- **Step 3** Training starts here. x^k is presented, $x := x^k$, $y := y^k$ and output o is computed

$$o_i(x) = \text{sign}(\langle w_i, x \rangle), \ i = 1, \dots, m$$

- **Step 4** Weights are updated

$$w_i := w_i + \eta(y_i - \text{sign}(\langle w_i, x \rangle))x, \ i = 1, \dots, m$$

- **Step 5** Cumulative cycle error is computed by adding the present error to E

$$E := E + \frac{1}{2}\|y - o\|^2$$

- **Step 6** If $k < K$ then $k := k + 1$ and we continue the training by going back to **Step 3**, otherwise we go to **Step 7**
- **Step 7** The training cycle is completed. For $E = 0$ terminate the training session. If $E > 0$ then E is set to 0, $k := 1$ and we initiate a new training cycle by going to **Step 3**

The following theorem shows that if the problem has solutions then the perceptron learning algorithm will find one of them.

Theorem 2.1.1 *(Convergence theorem) If the problem is linearly separable then the program will go to* **Step 3** *only finetely many times.*

Example 2.1.3 *Illustration of the perceptron learning algorithm.*

Consider the following training set

No.	input values	desired output value
1.	$x^1 = (1, 0, 1)^T$	-1
2.	$x^2 = (0, -1, -1)^T$	1
3.	$x^3 = (-1, -0.5, -1)^T$	1

The learning constant is assumed to be 0.1. The initial weight vector is $w^0 = (1, -1, 0)^T$.

Then the learning according to the perceptron learning rule progresses as follows.

Step 1 Input x^1, desired output is -1:

$$\langle w^0, x^1 \rangle = (1, -1, 0) \begin{pmatrix} 1 \\ 0 \\ 1 \end{pmatrix} = 1$$

Correction in this step is needed since $y_1 = -1 \neq \text{sign}(1)$. We thus obtain the updated vector

$$w^1 = w^0 + 0.1(-1 - 1)x^1$$

Plugging in numerical values we obtain

$$w^1 = \begin{pmatrix} 1 \\ -1 \\ 0 \end{pmatrix} - 0.2 \begin{pmatrix} 1 \\ 0 \\ 1 \end{pmatrix} = \begin{pmatrix} 0.8 \\ -1 \\ -0.2 \end{pmatrix}$$

Step 2 Input is x^2, desired output is 1. For the present w^1 we compute the activation value

$$\langle w^1, x^2 \rangle = (0.8, -1, -0.2) \begin{pmatrix} 0 \\ -1 \\ -1 \end{pmatrix} = 1.2$$

Correction is not performed in this step since $1 = \text{sign}(1.2)$, so we let $w^2 := w^1$.

Step 3 Input is x_3, desired output is 1.

$$< w^2, x^3 >= (0.8, -1, -0.2) \begin{pmatrix} -1 \\ -0.5 \\ -1 \end{pmatrix} = -0.1$$

Correction in this step is needed since $y_3 = 1 \neq \text{sign}(-0.1)$. We thus obtain the updated vector

$$w^3 = w^2 + 0.1(1+1)x^3$$

Plugging in numerical values we obtain

$$w^3 = \begin{pmatrix} 0.8 \\ -1 \\ -0.2 \end{pmatrix} + 0.2 \begin{pmatrix} -1 \\ -0.5 \\ -1 \end{pmatrix} = \begin{pmatrix} 0.6 \\ -1.1 \\ -0.4 \end{pmatrix}$$

Step 4 Input x^1, desired output is -1:

$$\langle w^3, x^1 \rangle = (0.6, -1.1, -0.4) \begin{pmatrix} 1 \\ 0 \\ 1 \end{pmatrix} = 0.2$$

Correction in this step is needed since $y_1 = -1 \neq \text{sign}(0.2)$. We thus obtain the updated vector

$$w^4 = w^3 + 0.1(-1-1)x^1$$

Plugging in numerical values we obtain

$$w^4 = \begin{pmatrix} 0.6 \\ -1.1 \\ -0.4 \end{pmatrix} - 0.2 \begin{pmatrix} 1 \\ 0 \\ 1 \end{pmatrix} = \begin{pmatrix} 0.4 \\ -1.1 \\ -0.6 \end{pmatrix}$$

Step 5 Input is x^2, desired output is 1. For the present w^4 we compute the activation value

$$\langle w^4, x^2 \rangle = (0.4, -1.1, -0.6) \begin{pmatrix} 0 \\ -1 \\ -1 \end{pmatrix} = 1.7$$

Correction is not performed in this step since $1 = \text{sign}(1.7)$, so we let $w^5 := w^4$.

Step 6 Input is x_3, desired output is 1.

$$\langle w^5, x^3 \rangle = (0.4, -1.1, -0.6) \begin{pmatrix} -1 \\ -0.5 \\ -1 \end{pmatrix} = 0.75$$

Correction is not performed in this step since $1 = \text{sign}(0.75)$, so we let $w^6 := w^5$.

This terminates the learning process, because

$$\langle w^6, x_1 \rangle = -0.2 < 0, \quad \langle w^6, x_2 \rangle = 1.7 > 0, \quad \langle w^6, x_3 \rangle = 0.75 > 0.$$

Minsky and Papert [20] provided a very careful analysis of conditions under which the perceptron learning rule is capable of carrying out the required mappings. They showed that the perceptron can not succesfully solve the problem

	x_1	x_1	$o(x_1, x_2)$
1.	1	1	0
2.	1	0	1
3.	0	1	1
4.	0	0	0

This Boolean function is known in the literature as *exclusive or* (XOR). We will refer to the above function as two-dimensional parity function.

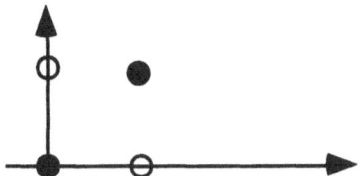

Fig. 2.8. Linearly nonseparable XOR problem.

The n-dimensional parity function is a binary Boolean function, which takes the value 1 if we have odd number of 1-s in the input vector, and zero otherwise. For example, the 3-dimensional parity function is defined as

	x_1	x_1	x_3	$o(x_1, x_2, x_3)$
1.	1	1	1	1
2.	1	1	0	0
3.	1	0	1	0
4.	1	0	0	1
5.	0	0	1	1
6.	0	1	1	0
7.	0	1	0	1
8.	0	0	0	0

2.2 The delta learning rule

The error correction learning procedure is simple enough in conception. The procedure is as follows: During training an input is put into the network and flows through the network generating a set of values on the output units. Then, the actual output is compared with the desired target, and a match is computed. If the output and target match, no change is made to the net.

However, if the output differs from the target a change must be made to some of the connections.

Let's first recall the definition of derivative of single-variable functions.

Definition 2.2.1 *The derivative of f at (an interior point of its domain) x, denoted by $f'(x)$, and defined by*

$$f'(x) = \lim_{x_n \to x} \frac{f(x) - f(x_n)}{x - x_n}$$

Let us consider a differentiable function $f : \mathbb{R} \to \mathbb{R}$. The derivative of f at (an interior point of its domain) x is denoted by $f'(x)$. If $f'(x) > 0$ then we say that f is increasing at x, if $f'(x) < 0$ then we say that f is decreasing at x, if $f'(x) = 0$ then f can have a local maximum, minimum or inflextion point at x.

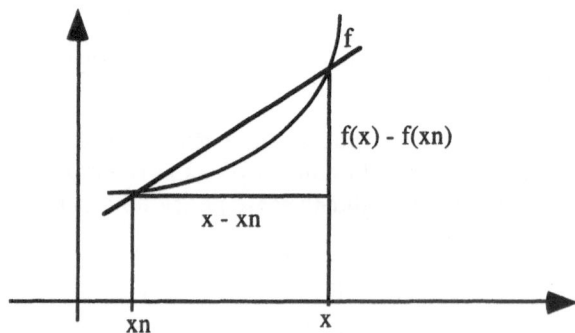

Fig. 2.9. Derivative of function f.

A differentiable function is always increasing in the direction of its derivative, and decreasing in the opposite direction. It means that if we want to find one of the local minima of a function f starting from a point x^0 then we should search for a second candidate in the right-hand side of x^0 if $f'(x^0) < 0$ (when f is decreasing at x^0) and in the left-hand side of x^0 if $f'(x^0) > 0$ (when f increasing at x^0).

The equation for the line crossing the point $(x^0, f(x^0))$ is given by

$$\frac{y - f(x^0)}{x - x^0} = f'(x^0)$$

that is

$$y = f(x^0) + (x - x^0)f'(x^0)$$

The next approximation, denoted by x^1, is a solution to the equation

$$f(x^0) + (x - x^0)f'(x^0) = 0$$

which is,

$$x^1 = x^0 - \frac{f(x^0)}{f'(x^0)}$$

This idea can be applied successively, that is

$$x^{n+1} = x^n - \frac{f(x^n)}{f'(x^n)}.$$

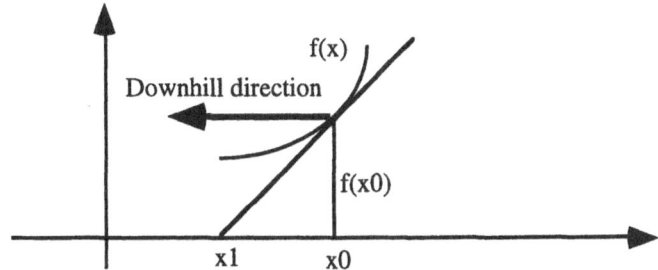

Fig. 2.10. The downhill direction is negative at x_0.

The above procedure is a typical descent method. In a descent method the next iteration w^{n+1} should satisfy the following property

$$f(w^{n+1}) < f(w^n)$$

i.e. the value of f at w^{n+1} is smaller than its previous value at w^n.

In error correction learning procedure, each iteration of a descent method calculates the downhill direction (opposite of the direction of the derivative) at w^n which means that for a sufficiently small $\eta > 0$ the inequality

$$f(w^n - \eta f'(w^n)) < f(w^n)$$

should hold, and we let w^{n+1} be the vector

$$w^{n+1} = w^n - \eta f'(w^n)$$

Let $f: I\!R^n \to I\!R$ be a real-valued function. In a descent method, whatever is the next iteration, w^{n+1}, it should satisfy the property

$$f(w^{n+1}) < f(w^n)$$

i.e. the value of f at w^{n+1} is smaller than its value at previous approximation w^n.

Each iteration of a descent method calculates a downhill direction (opposite of the direction of the derivative) at w^n which means that for a sufficiently small $\eta > 0$ the inequality

$$f(w^n - \eta f'(w^n)) < f(w^n)$$

should hold, and we let w^{n+1} be the vector

$$w^{n+1} = w^n - \eta f'(w^n).$$

Let $f: I\!R^n \to I\!R$ be a real-valued function and let $e \in I\!R^n$ with $\|e\| = 1$ be a given direction. The derivative of f with respect e at w is defined as

$$\partial_e f(w) = \lim_{t \to +0} \frac{f(w + te) - f(w)}{t}$$

If $e = (0, \dots \overset{i\text{-th}}{1} \dots, 0)^T$, i.e. e is the i-th basic direction then instead of $\partial_e f(w)$ we write $\partial_i f(w)$, which is defined by

$$\partial_i f(w) = \lim_{t \to +0} \frac{f(w_1, \dots, w_i + t, \dots w_n) - f(w_1, \dots, w_i, \dots, w_n)}{t}.$$

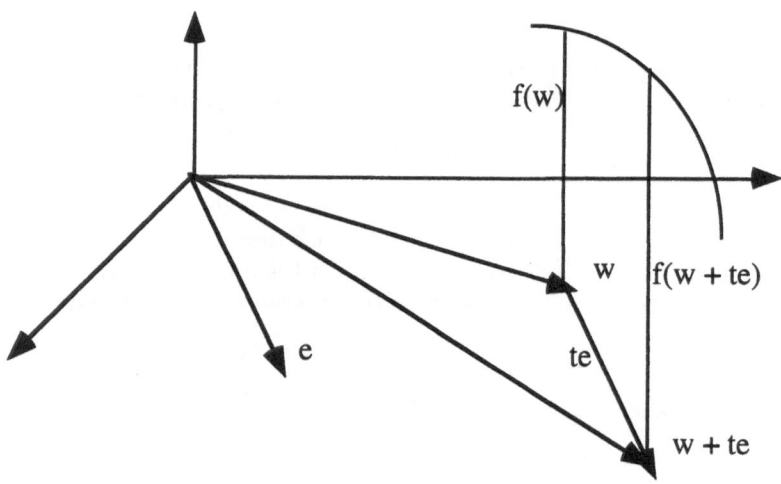

Fig. 2.11. The derivative of f with respect to the direction e.

The gradient of f at w, denoted by $f'(w)$ is defined by

$$f'(w) = (\partial_1 f(w), \dots, \partial_n f(w))^T$$

Example 2.2.1 Let $f(w_1, w_2) = w_1^2 + w_2^2$ then the gradient of f is given by

$$f'(w) = 2w = (2w_1, 2w_2)^T.$$

The gradient vector always points to the uphill direction of f. The downhill (steepest descent) direction of f at w is the opposite of the uphill direction, i.e. the downhill direction is $-f'(w)$, which is

$$(-\partial_1 f(w), \dots, -\partial_n f(w))^T.$$

Definition 2.2.2 *(linear activation function) A linear activation function is a mapping from $f: \mathbb{R} \to \mathbb{R}$ such that*

$$f(t) = t$$

for all $t \in \mathbb{R}$.

Suppose we are given a single-layer network with n input units and m linear output units, i.e. the output of the i-th neuron can be written as

$$o_i = net_i = \langle w_i, x \rangle = w_{i1}x_1 + \cdots + w_{in}x_n, \ i = 1, \dots, m.$$

Assume we have the following training set

$$\{(x^1, y^1), \dots, (x^K, y^K)\}$$

where $x^k = (x_1^k, \dots, x_n^k)$, $y^k = (y_1^k, \dots, y_m^k)$, $k = 1, \dots, K$.

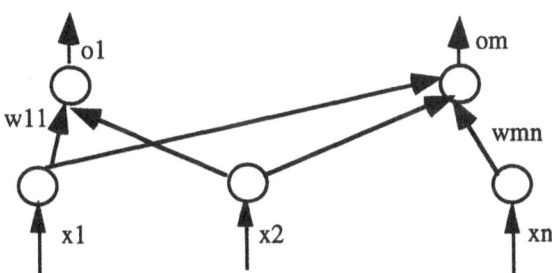

Fig. 2.12. Single-layer feedforward network with m output units

The basic idea of the delta learning rule is to define a measure of the overall performance of the system and then to find a way to optimize that performance. In our network, we can define the performance of the system as

$$E = \sum_{k=1}^{K} E_k = \frac{1}{2} \sum_{k=1}^{K} \|y^k - o^k\|^2$$

That is

$$E = \frac{1}{2} \sum_{k=1}^{K} \sum_{i=1}^{m} (y_i^k - o_i^k)^2 = \frac{1}{2} \sum_{k=1}^{K} \sum_{i=1}^{m} (y_i^k - \langle w_i, x^k \rangle)^2$$

where i indexes the output units; k indexes the input/output pairs to be learned; y_i^k indicates the target for a particular output unit on a particular pattern; $o_i^k := \langle w_i, x^k \rangle$ indicates the actual output for that unit on that pattern; and E is the total error of the system. The goal, then, is to minimize this function. It turns out, if the output functions are differentiable, that this problem has a simple solution: namely, we can assign a particular unit blame in proportion to the degree to which changes in that unit's activity lead to changes in the error. That is, we change the weights of the system in proportion to the derivative of the error with respect to the weights.

The rule for changing weights following presentation of input/output pair (x^k, y^k) is given by the gradient descent method, i.e. we minimize the quadratic error function by using the following iteration process

$$w_{ij} := w_{ij} - \eta \frac{\partial E_k}{\partial w_{ij}}$$

where $\eta > 0$ is the learning rate.

Let us compute now the partial derivative of the error function E_k with respect to w_{ij}

$$\frac{\partial E_k}{\partial w_{ij}} = \frac{\partial E_k}{\partial net_i^k} \frac{\partial net_i^k}{\partial w_{ij}} = -(y_i^k - o_i^k) x_j^k$$

where $net_i^k = w_{i1} x_1^k + \cdots + w_{in} x_n^k$.

That is,

$$w_{ij} := w_{ij} + \eta (y_i^k - o_i^k) x_j^k$$

for $j = 1, \ldots, n$.

Definition 2.2.3 *The error signal term, denoted by δ_i^k and called* delta, *produced by the i-th output neuron is defined as*

$$\delta_i^k = -\frac{\partial E_k}{\partial net_i^k} = (y_i^k - o_i^k)$$

For linear output units δ_i^k is nothing else but the difference between the desired and computed output values of the i-th neuron.

So the delta learning rule can be written as

$$w_{ij} := w_{ij} + \eta \delta_i^k x_j^k$$

for $i = 1, \ldots, m$ and $j = 1, \ldots, n$.

A key advantage of neural network systems is that these simple, yet powerful learning procedures can be defined, allowing the systems to adapt to their environments.

The essential character of such networks is that they map similar input patterns to similar output patterns.

This characteristic is what allows these networks to make reasonable generalizations and perform reasonably on patterns that have never before been presented. The similarity of patterns in a connectionist system is determined by their overlap. The overlap in such networks is determined outside the learning system itself whatever produces the patterns. The standard delta rule essentially implements gradient descent in sum-squared error for linear activation functions.

It should be noted that the delta learning rule was introduced only recently for neural network training by McClelland and Rumelhart [22]. This rule parallels the discrete perceptron training rule. It also can be called the continuous perceptron training rule.

Summary 2.2.1 *The delta learning rule with linear activation functions.*
Given are K training pairs arranged in the training set

$$\{(x^1, y^1), \dots, (x^K, y^K)\}$$

where $x^k = (x_1^k, \dots, x_n^k)$ and $y^k = (y_1^k, \dots, y_m^k)$, $k = 1, \dots, K$.

- **Step 1** $\eta > 0$, $E_{\max} > 0$ are chosen
- **Step 2** Weights w_{ij} are initialized at small random values, $k := 1$, and the running error E is set to 0
- **Step 3** Training starts here. Input x^k is presented, $x := x^k$, $y := y^k$, and output $o = (o_1, \dots, o_m)^T$ is computed

$$o = \langle w_i, x \rangle = w_i^T x$$

for $i = 1, \dots, m$.
- **Step 4** Weights are updated

$$w_{ij} := w_{ij} + \eta(y_i - o_i)x_j$$

- **Step 5** Cumulative cycle error is computed by adding the present error to E

$$E := E + \frac{1}{2}\|y - o\|^2$$

- **Step 6** If $k < K$ then $k := k + 1$ and we continue the training by going back to **Step 3**, otherwise we go to **Step 7**
- **Step 7** The training cycle is completed. For $E < E_{\max}$ terminate the training session. If $E > E_{\max}$ then E is set to 0 and we initiate a new training cycle by going back to **Step 3**

2.3 The delta learning rule with semilinear activation function

In many practical cases instead of linear activation functions we use semilinear ones. The next table shows the most-often used types of activation functions.

Linear	$f(\langle w, x \rangle) = w^T x$
Piecewise linear	$f(\langle w, x \rangle) = \begin{cases} 1 & \text{if } \langle w, x \rangle > 1 \\ \langle w, x \rangle & \text{if } |\langle w, x \rangle| \leq 1 \\ -1 & \text{if } \langle w, x \rangle < -1 \end{cases}$
Hard limiter	$f(\langle w, x \rangle) = \text{sign}(w^T x)$
Unipolar sigmoidal	$f(\langle w, x \rangle) = \dfrac{1}{1 + \exp(-w^T x)}$
Bipolar sigmoidal (1)	$f(\langle w, x \rangle) = \tanh(w^T x)$
Bipolar sigmoidal (2)	$f(\langle w, x \rangle) = \dfrac{2}{1 + \exp(w^T x)} - 1$

Table 2.1. Activation functions.

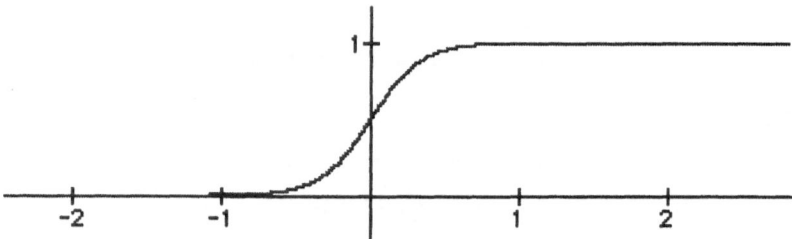

Fig. 2.13. Unipolar activation function.

The derivatives of sigmoidal activation functions are extensively used in learning algorithms.

- If f is a bipolar sigmoidal activation function of the form

$$f(t) = \frac{2}{1 + \exp(-t)} - 1.$$

Then the following equality holds

$$f'(t) = \frac{2 \exp(-t)}{(1 + \exp(-t))^2} = \frac{1}{2}(1 - f^2(t)).$$

- If f is a unipolar sigmoidal activation function of the form

$$f(t) = \frac{1}{1 + \exp(-t)}.$$

Then f' satisfies the following equality

$$f'(t) = f(t)(1 - f(t)).$$

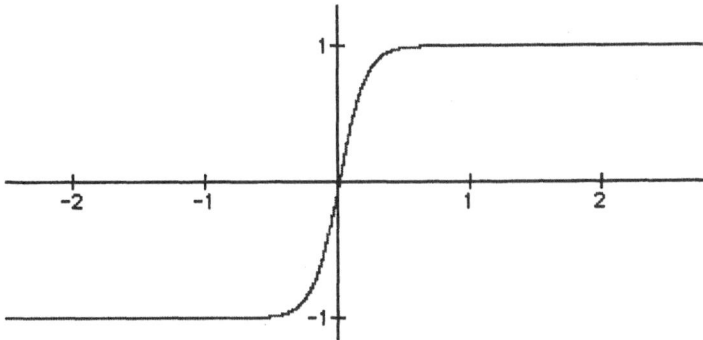

Fig. 2.14. Bipolar activation function.

We shall describe now the delta learning rule with semilinear activation function. For simplicity we explain the learning algorithm in the case of a single-output network.

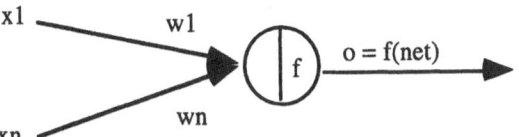

Fig. 2.15. Single neuron network.

The output of the neuron is computed by unipolar sigmoidal activation function

$$o(\langle w, x \rangle) = \frac{1}{1 + \exp(-w^T x)}.$$

Suppose we are given the following training set

No.	input values	desired output value
1.	$x^1 = (x_1^1, \ldots x_n^1)$	y^1
2.	$x^2 = (x_1^2, \ldots x_n^2)$	y^2
$\vdots$	$\vdots$	$\vdots$
K.	$x^K = (x_1^K, \ldots x_n^K)$	y^K

The system first uses the input vector, x^k, to produce its own output vector, o^k, and then compares this with the desired output, y^k. Let

$$E_k = \frac{1}{2}(y^k - o^k)^2 = \frac{1}{2}(y^k - o(\langle w, x^k \rangle))^2 =$$

$$\frac{1}{2}\left(y^k - \frac{1}{1 + \exp{(-w^T x^k)}}\right)^2$$

be our measure of the error on input/output pattern k and let

$$E = \sum_{k=1}^{K} E_k$$

be our overall measure of the error.

The rule for changing weights following presentation of input/output pair k is given by the gradient descent method, i.e. we minimize the quadratic error function by using the following iteration process

$$w := w - \eta E'_k(w).$$

Let us compute now the gradient vector of the error function E_k at point w:

$$E'_k(w) = \frac{d}{dw}\left(\frac{1}{2} \times \left[y^k - \frac{1}{1 + \exp{(-w^T x^k)}}\right]^2\right) =$$

$$\frac{1}{2} \times \frac{d}{dw}\left[y^k - \frac{1}{1 + \exp{(-w^T x^k)}}\right]^2 = -(y^k - o^k)o^k(1 - o^k)x^k$$

where

$$o^k = \frac{1}{1 + \exp{(-w^T x^k)}}.$$

Therefore our learning rule for w is

$$w := w + \eta(y^k - o^k)o^k(1 - o^k)x^k$$

which can be written as

$$w := w + \eta \delta_k o^k(1 - o^k)x^k$$

where

$$\delta_k = (y^k - o^k)o^k(1 - o^k).$$

Summary 2.3.1 *The delta learning rule with unipolar sigmoidal activation function.*

Given are K training pairs arranged in the training set

$$\{(x^1, y^1), \dots, (x^K, y^K)\}$$

where $x^k = (x_1^k, \dots, x_n^k)$ and $y^k \in \mathbb{R}$, $k = 1, \dots, K$.

- **Step 1** $\eta > 0$, $E_{\max} > 0$ are chosen
- **Step 2** Weigts w are initialized at small random values, $k := 1$, and the running error E is set to 0
- **Step 3** Training starts here. Input x^k is presented, $x := x^k$, $y := y^k$, and output o is computed

$$o = o(\langle w, x \rangle) = \frac{1}{1 + \exp(-w^T x)}$$

- **Step 4** Weights are updated

$$w := w + \eta(y - o)o(1 - o)x$$

- **Step 5** Cumulative cycle error is computed by adding the present error to E

$$E := E + \frac{1}{2}(y - o)^2$$

- **Step 6** If $k < K$ then $k := k + 1$ and we continue the training by going back to **Step 3**, otherwise we go to **Step 7**
- **Step 7** The training cycle is completed. For $E < E_{\max}$ terminate the training session. If $E > E_{\max}$ then E is set to 0 and we initiate a new training cycle by going back to **Step 3**

In this case, without hidden units, the error surface is shaped like a bowl with only one minimum, so gradient descent is guaranteed to find the best set of weights. With hidden units, however, it is not so obvious how to compute the derivatives, and the error surface is not concave upwards, so there is the danger of getting stuck in local minima.

We illustrate the delta learning rule with bipolar sigmoidal activation function

$$f(t) = \frac{2}{1 + \exp(-t)} - 1.$$

Example 2.3.1 *The delta learning rule with bipolar sigmoidal activation function.*

Given are K training pairs arranged in the training set

$$\{(x^1, y^1), \ldots, (x^K, y^K)\}$$

where $x^k = (x_1^k, \ldots, x_n^k)$ and $y^k \in \mathbb{R}$, $k = 1, \ldots, K$.

- **Step 1** $\eta > 0$, $E_{\max} > 0$ are chosen
- **Step 2** Weigts w are initialized at small random values, $k := 1$, and the running error E is set to 0
- **Step 3** Training starts here. Input x^k is presented, $x := x^k$, $y := y^k$, and output o is computed

$$o = o(\langle w, x \rangle) = \frac{2}{1 + \exp(-w^T x)} - 1$$

- **Step 4** Weights are updated

$$w := w + \frac{1}{2}\eta(y - o)(1 - o^2)x$$

- **Step 5** Cumulative cycle error is computed by adding the present error to E

$$E := E + \frac{1}{2}(y - o)^2$$

- **Step 6** If $k < K$ then $k := k + 1$ and we continue the training by going back to **Step 3**, otherwise we go to **Step 7**
- **Step 7** The training cycle is completed. For $E < E_{max}$ terminate the training session. If $E > E_{max}$ then E is set to 0 and we initiate a new training cycle by going back to **Step 3**

2.4 The generalized delta learning rule

We now focus on generalizing the delta learning rule for feedforward layered neural networks. The architecture of the two-layer network considered below is shown in Fig. 2.16. It has strictly speaking, two layers of processing neurons. If, however, the layers of nodes are counted, then the network can also be labeled as a three-layer network.

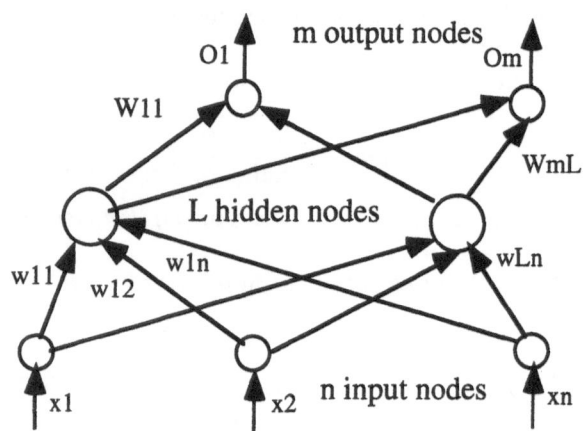

Fig. 2.16. Layered neural network with two continuous perceptron layers.

There is no agreement in the literature as to which approach is to be used to describe network architectures. In this text we will use the term *layer* in reference to the actual number of existing and processing neuron layers. Layers with neurons whose outputs are not directly accesible are called internal or hidden layers. Thus the network of Fig. 2.16 is a two-layer network, which can be called a single hidden-layer network.

The generalized delta rule is the most often used supervised learning algorithm of feedforward multi-layer neural networks. For simplicity we consider only a neural network with one hidden layer and one output node (see Fig. 2.17).

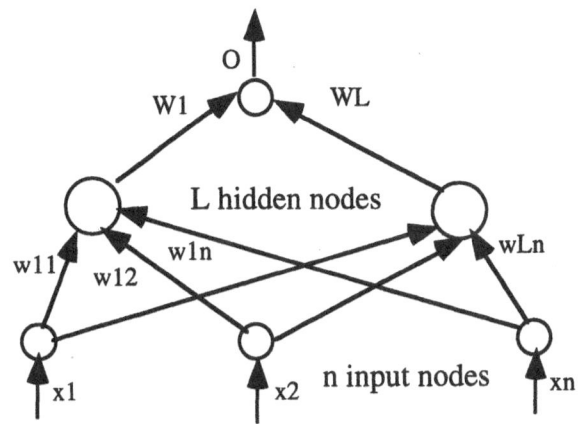

Fig. 2.17. Two-layer neural network with one output node.

The measure of the error on an input/output training pattern (x^k, y^k) is defined by

$$E_k(W, w) = \frac{1}{2}(y^k - O^k)^2$$

where O^k is the computed output and the overall measure of the error is

$$E(W, w) = \sum_{k=1}^{K} E_k(W, w).$$

If an input vector x^k is presented to the network then it generates the following output

$$O^k = \frac{1}{1 + \exp(-W^T o^k)}$$

where o^k is the output vector of the hidden layer

$$o_l^k = \frac{1}{1 + \exp(-w_l^T x^k)}$$

and w_l denotes the weight vector of the l-th hidden neuron, $l = 1, \ldots, L$.

The rule for changing weights following presentation of input/output pair k is given by the gradient descent method, i.e. we minimize the quadratic error function by using the following iteration process

$$W := W - \eta \frac{\partial E_k(W, w)}{\partial W},$$

$$w_l := w_l - \eta \frac{\partial E_k(W, w)}{\partial w_l},$$

for $l = 1, \ldots, L$, and $\eta > 0$ is the learning rate.

By using the chain rule for derivatives of composed functions we get

$$\frac{\partial E_k(W, w)}{\partial W} = \frac{1}{2} \frac{\partial}{\partial W} \left[y^k - \frac{1}{1 + \exp(-W^T o^k)} \right]^2$$

$$= -(y^k - O^k)O^k(1 - O^k)o^k$$

i.e. the rule for changing weights of the output unit is

$$W := W + \eta(y^k - O^k)O^k(1 - O^k)o^k = W + \eta \delta_k o^k$$

that is

$$W_l := W_l + \eta \delta_k o_l^k,$$

for $l = 1, \ldots, L$, and we have used the notation

$$\delta_k = (y^k - O^k)O^k(1 - O^k).$$

Let us now compute the partial derivative of E_k with respect to w_l

$$\frac{\partial E_k(W, w)}{\partial w_l} = -O^k(1 - O^k)W_l o_l^k(1 - o_l^k)x^k$$

i.e. the rule for changing weights of the hidden units is

$$w_l := w_l + \eta \delta_k W_l o_l^k(1 - o_l^k)x^k, \quad l = 1, \ldots, L.$$

that is

$$w_{lj} := w_{lj} + \eta \delta_k W_l o_l^k(1 - o_l^k)x_j^k, \quad j = 1, \ldots, n.$$

Summary 2.4.1 *The generalized delta learning rule (error backpropagation learning)*

We are given the training set

$$\{(x^1, y^1), \ldots, (x^K, y^K)\}$$

where $x^k = (x_1^k, \ldots, x_n^k)$ and $y^k \in \mathbb{R}$, $k = 1, \ldots, K$.

• **Step 1** $\eta > 0$, $E_{\max} > 0$ are chosen

- **Step 2** Weigts w are initialized at small random values, $k := 1$, and the running error E is set to 0
- **Step 3** Training starts here. Input x^k is presented, $x := x^k$, $y := y^k$, and output O is computed

$$O = \frac{1}{1 + \exp(-W^T o)}$$

where o_l is the output vector of the hidden layer

$$o_l = \frac{1}{1 + \exp(-w_l^T x)}$$

- **Step 4** Weights of the output unit are updated

$$W := W + \eta \delta o$$

where $\delta = (y - O)O(1 - O)$.
- **Step 5** Weights of the hidden units are updated

$$w_l = w_l + \eta \delta W_l o_l (1 - o_l) x, \; l = 1, \dots, L$$

- **Step 6** Cumulative cycle error is computed by adding the present error to E

$$E := E + \frac{1}{2}(y - O)^2$$

- **Step 7** If $k < K$ then $k := k + 1$ and we continue the training by going back to **Step 2**, otherwise we go to **Step 8**
- **Step 8** The training cycle is completed. For $E < E_{\max}$ terminate the training session. If $E > E_{\max}$ then $E := 0$, $k := 1$ and we initiate a new training cycle by going back to **Step 3**

Exercise 2.4.1 *Derive the backpropagation learning rule with bipolar sigmoidal activation function*

$$f(t) = \frac{2}{1 + \exp(-t)} - 1.$$

2.5 Effectivity of neural networks

Funahashi [8] showed that infinitely large neural networks with a single hidden layer are capable of approximating all continuous functions. Namely, he proved the following theorem

Theorem 2.5.1 *Let $\phi(x)$ be a nonconstant, bounded and monotone increasing continuous function. Let $K \subset {I\!\!R}^n$ be a compact set and*

$$f \colon K \to {I\!\!R}$$

be a real-valued continuous function on K. Then for arbitrary $\epsilon > 0$, there exists an integer N and real constants w_i, w_{ij} such that

$$\tilde{f}(x_1, \ldots, x_n) = \sum_{i=1}^{N} w_i \phi \left(\sum_{j=1}^{n} w_{ij} x_j \right)$$

satisfies

$$\|f - \tilde{f}\|_\infty = \sup_{x \in K} |f(x) - \tilde{f}(x)| \le \epsilon.$$

In other words, any continuous mapping can be approximated in the sense of uniform topology on K by input-output mappings of two-layers networks whose output functions for the hidden layer are $\phi(x)$ and are linear for the output layer.

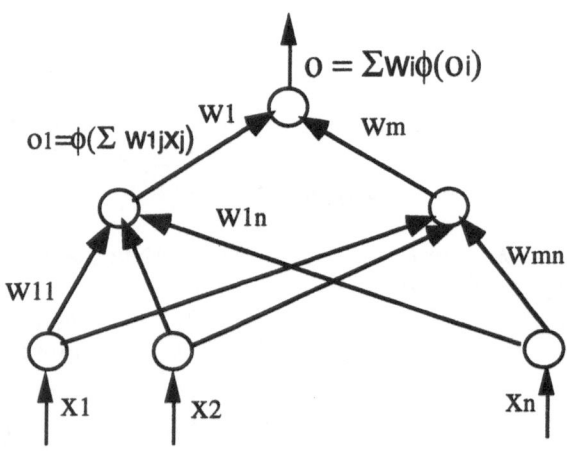

Fig. 2.18. Funahashi's network.

The Stone-Weierstrass theorem from classical real analysis can be used to show certain network architectures possess the universal approximation capability. By employing the Stone-Weierstrass theorem in the designing of our networks, we also guarantee that the networks can compute certain polynomial expressions: if we are given networks exactly computing two functions, f_1 and f_2, then a larger network can exactly compute a polynomial expression of f_1 and f_2.

Theorem 2.5.2 *(Stone-Weierstrass) Let domain K be a compact space of n dimensions, and let $\mathcal{G}$ be a set of continuous real-valued functions on K, satisfying the following criteria:*

1. *The constant function $f(x) = 1$ is in $\mathcal{G}$.*
2. *For any two points $x_1 \neq x_2$ in K, there is an f in $\mathcal{G}$ such that $f(x_1) \neq f(x_2)$.*
3. *If f_1 and f_2 are two functions in $\mathcal{G}$, then fg and $\alpha_1 f_1 + \alpha_2 f_2$ are in $\mathcal{G}$ for any two real numbers α_1 and α_2.*

Then $\mathcal{G}$ is dense in $C(K)$, the set of continuous real-valued functions on K. In other words, for any $\epsilon > 0$ and any function g in $C(K)$, there exists g in $\mathcal{G}$ such that

$$\|f - g\|_\infty = \sup_{x \in K} |f(x) - g(x)| \leq \epsilon.$$

The key to satisfying he Stone-Weierstrass theorem is to find functions that *transform multiplication into addition* so that products can be written as summations. There are at least three generic functions that accomplish this transfomation: exponential functions, partial fractions, and step functions. The following networks satisfy the Stone-Weierstrass theorem.

- **Decaying-exponential networks** Exponential functions are basic to the process of transforming multiplication into addition in several kinds of networks:

$$\exp(x_1) \exp(x_2) = \exp(x_1 + x_2).$$

Let $\mathcal{G}$ be the set of all continuous functions that can be computed by arbitrarily large decaying-exponential networks on domain $K = [0, 1]^n$:

$$\mathcal{G} = \left\{ f(x_1, \dots, x_n) = \sum_{i=1}^N w_i \exp(-\sum_{j=1}^n w_{ij} x_j), w_i, w_{ij} \in \mathbb{R} \right\}.$$

Then $\mathcal{G}$ is dense in $C(K)$
- **Fourier networks**
- **Exponentiated-function networks**
- **Modified logistic networks**
- **Modified sigma-pi and polynomial networks** Let $\mathcal{G}$ be the set of all continuous functions that can be computed by arbitrarily large modified sigma-pi or polynomial networks on domain $K = [0, 1]^n$:

$$\mathcal{G} = \left\{ f(x_1, \dots, x_n) = \sum_{i=1}^N w_i \prod_{j=1}^n x_j^{w_{ij}}, w_i, w_{ij} \in \mathbb{R} \right\}.$$

Then $\mathcal{G}$ is dense in $C(K)$.
- **Step functions and perceptron networks**
- **Partial fraction networks**

2.6 Winner-take-all learning

Unsupervised classification learning is based on clustering of input data. No *a priori* knowledge is assumed to be available regarding an input's membership in a particular class. Rather, gradually detected characteristics and a history of training will be used to assist the network in defining classes and possible boundaries between them.

Clustering is understood to be the grouping of similar objects and separating of dissimilar ones.

We discuss Kohonen's network [16], which classifies input vectors into one of the specified number of m categories, according to the clusters detected in the training set

$$\{x^1, \dots, x^K\}.$$

The learning algorithm treats the set of m weight vectors as variable vectors that need to be learned. Prior to the learning, the normalization of all (randomly chosen) weight vectors is required.

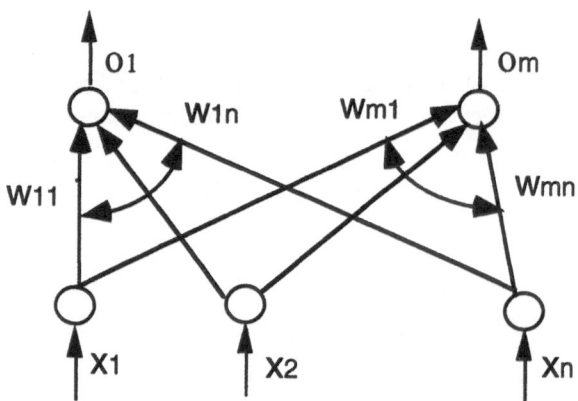

Fig. 2.19. The winner-take-all learning network.

The weight adjustment criterion for this mode of training is the selection of w_r such that

$$\|x - w_r\| = \min_{i=1,\dots,m} \|x - w_i\|$$

The index r denotes the *winning* neuron number corresponding to the vector w_r, which is the closest approximation of the current input x. Using the equality

$$\|x - w_i\|^2 = <x - w_i, x - w_i> = \langle x, x \rangle - 2\langle w_i, x \rangle + \langle w_i, w_i \rangle =$$

$$\|x\|^2 - 2\langle w, x \rangle + \|w_i\|^2 = \|x\|^2 - 2\langle w_i, x \rangle + 1$$

we can infer that searching for the minimum of m distances corresponds to finding the maximum among the m scalar products

$$\langle w_r, x \rangle = \max_{i=1,\ldots,m} \langle w_i, x \rangle$$

Taking into consideration that $\|w_i\| = 1$, $\forall i \in \{1, \ldots, m\}$ the scalar product $\langle w_i, x \rangle$ is nothing else but the projection of x on the direction of w_i. It is clear that the closer the vector w_i to x the bigger the projection of x on w_i.

Note that $< w_r, x >$ is the activation value of the *winning* neuron which has the largest value net$_i$, $i = 1, \ldots, m$.

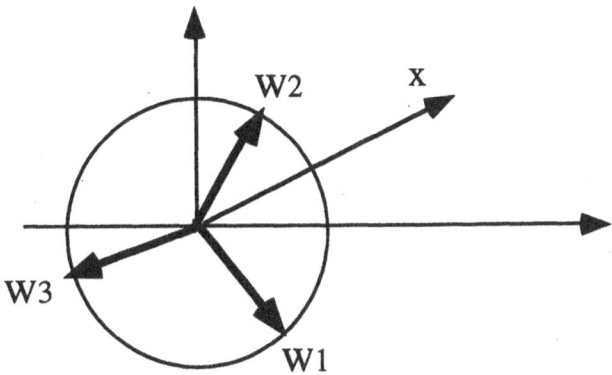

Fig. 2.20. The winner weight is w_2.

When using the scalar product metric of similarity, the synaptic weight vectors should be modified accordingly so that they become more similar to the current input vector.

With the similarity criterion being $\cos(w_i, x)$, the weight vector lengths should be identical for this training approach. However, their directions should be modified.

Intuitively, it is clear that a very long weight vector could lead to a very large output ofits neuron even if there were a large angle between the weight vector and the pattern. This explains the need for weight normalization.

After the winning neuron has been identified and declared a winner, its weight must be adjusted so that the distance $\|x - w_r\|$ is reduced in the current training step.

Thus, $\|x - w_r\|$ must be reduced, preferebly along the gradient direction in the weight space $w_{r1}, \ldots, w_{rn}$,

$$\frac{d\|x - w\|^2}{dw} = \frac{d}{dw}(< x - w, x - w >) =$$

$$\frac{d}{dw}(\langle x, x \rangle - 2\langle w, x \rangle + < w, w >) =$$

$$\frac{d}{dw}(\langle x, x \rangle) - \frac{d}{dw}(2\langle w, x \rangle) + \frac{d}{dw}(< w, w >) =$$

$$-2 \times \frac{d}{dw}(w_1 x_1 + \cdots + w_n x_n) + \frac{d}{dw}(w_1^2 + \cdots + w_n^2) =$$

$$-2 \times \left[\frac{d}{dw_1}(w_1 x_1 + \cdots + w_n x_n), \ldots, \frac{d}{dw_n}(w_1 x_1 + \cdots + w_n x_n) \right]^T +$$

$$\left[\frac{d}{dw_1}(w_1^2 + \cdots + w_n^2), \ldots, \frac{d}{dw_n}(w_1^2 + \cdots + w_n^2) \right]^T =$$

$$-2(x_1, \ldots, x_n)^T + 2(w_1, \ldots, w_n)^T = -2(x - w).$$

It seems reasonable to reward the weights of the winning neuron with an increment of weight in the negative gradient direction, thus in the direction $(x - w_r)$. We thus have

$$w_r := w_r + \eta(x - w_r) \tag{2.1}$$

where η is a small lerning constant selected heuristically, usually between 0.1 and 0.7. The remaining weight vectors are left unaffected.

Summary 2.6.1 *Kohonen's learning algorithm can be summarized in the following three steps*

- **Step 1** $w_r := w_r + \eta(x - w_r)$, $o_r := 1$, (r is the winner neuron)
- **Step 2** $w_r := \dfrac{w_r}{\|w_r\|}$ (normalization)
- **Step 3** $w_i := w_i$, $o_i := 0$, $i \neq r$ (losers are unaffected)

It should be noted that from the identity

$$w_r := w_r + \eta(x - w_r) = (1 - \eta)w_r + \eta x$$

it follows that the updated weight vector is a convex linear combination of the old weight and the pattern vectors.

In the end of the training process the final weight vectors point to the center of gravity of classes.

The network will only be trainable if classes/clusters of patterns are linearly separable from other classes by hyperplanes passing through origin.

To ensure separability of clusters with *a priori* unknown numbers of training clusters, the unsupervised training can be performed with an excessive number of neurons, which provides a certain separability safety margin.

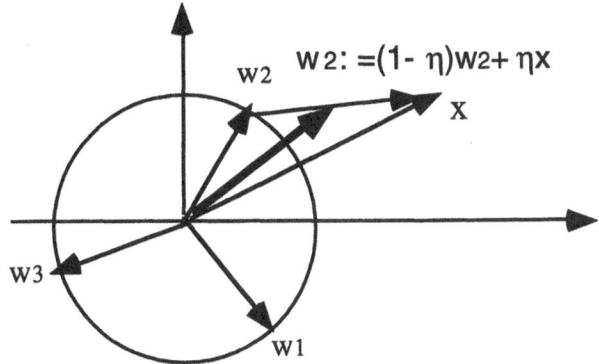

Fig. 2.21. Updating the weight of the winner neuron.

During the training, some neurons are likely not to develop their weights, and if their weights change chaotically, they will not be considered as indicative of clusters.

Therefore such weights can be omitted during the recall phase, since their output does not provide any essential clustering information. The weights of remaining neurons should settle at values that are indicative of clusters.

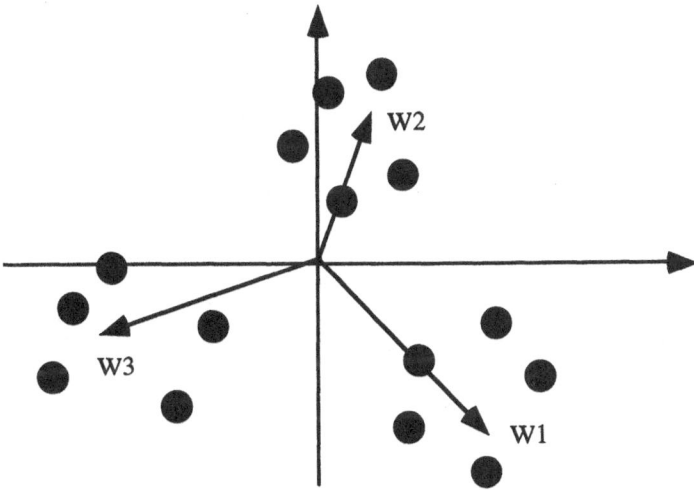

Fig. 2.22. The final weight vectors point to the center of gravity of the classes.

Another learning extension is possible for this network when the proper class for some patterns is known *a priori* [29]. Although this means that the encoding of data into weights is then becoming supervised, this information accelerates the learning process significantly. Weight adjustments are computed in the superviesed mode as in (2.1), i.e.

$$\Delta w_{ij} := \eta(x - w_r) \tag{2.2}$$

and only for correct classifications. For inproper clustering responses of the network, the weight adjustment carries the opposite sign compared to (2.2). That is, $\eta > 0$ for proper node responses, and $\eta < 0$ otherwise, in the supervised learning mode for the Kohonen layer.

Another mofification of the winner-take-all learning rule is that both the winners' and losers' weights are adjusted in proportion to their level of responses. This is called *leakly competative learning* and provides more subtle learning in the case for which clusters may be hard to distinguish.

2.7 Applications of artificial neural networks

There are large classes of problems that appear to be more amenable to solution by neural networks than by other available techniques. These tasks often involve ambiguity, such as that inherent in handwritten character recognition. Problems of this sort are difficult to tackle with conventional methods such as matched filtering or nearest neighbor classification, in part because the metrics used by the brain to compare patterns may not be very closely related to those chosen by an engineer designing a recognition system. Likewise, because reliable rules for recognizing a pattern are usually not at hand, fuzzy logic and expert system designers also face the difficult and sometimes impossible task of finding acceptable descriptions of the complex relations governing class inclusion. In trainable neural network systems, these relations are abstracted directly from training data. Moreover, because neural networks can be constructed with numbers of inputs and outputs ranging into thousands, they can be used to attack problems that require consideration of more input variables than could be feasibly utilized by most other approaches. It should be noted, however, that neural networks will not work well at solving problems for which sufficiently large and general sets of training data are not obtainable. Drawing heavily on [25] we provide a comprehensive list of applications of neural networks in **Industry, Business** and **Science**.

- **The telecommunications industry.**
  Many neural network applications are under development in the telecommunications industry for solving problems ranging from control of a nationwide switching network to management of an entire telephone company. Other aplications at the telephone circuit level turn out to be the most significant commercial applications of neural networks in the world today. Modems, commonly used for computer-to-commputer communications and in every fax machine, have adaptive circuits for telephone line equalization and for echo cancellation.

- **Control of sound and vibration**
  Active control of vibration and noise is accomplished by using an adaptive actuator to generate equal and opposite vibration and noise. This is being used in air-conditioning systems, in automotive systems, and in industrial applications.
- **Particle accelerator beam control.**
  The Stanford linear accelerator Center is now using adaptive techniques to cancel disturbances that diminish the positioning accuracy of opposing beams of positrons and electrons in a particle colloder.
- **Credit card fraud detection.**
  Several banks and credit card companies including *American Express, Mellon Bank, First USA Bank*, and others are currently using neural networks to study patterns of credit card usage and and to detect transactions that are potentially fraudulent.
- **Machine-printed character recognition.**
  Commercial products performing machine-printed character recognition have been introduced by a large number of companies and have been described in the literature.
- **Hand-printed character recognition.**
  Hecht-Nielsen Corp.'s *Quickstrokes Automated Data Entry System* is being used to recognize handwritten forms at Avon's order-processing center and at the state of Wyoming's Department of revenue. In the June 1992 issue of *Systems Integration Business*, Dennis Livingston reports that before implementing the system, Wyoming was losing an estimated $300,000 per year in interest income because so many cheks were being deposited late. Cardiff Software offers a product called Teleform which uses Nestor's hand-printed character recognition system to convert a fax machine into an OCR scanner. Poqet Computer, now a subsidiary of Fujitsu, uses Nestor's NestorWriter neural network software to perform handwriting recognition for the penbased PC it announced in January 1992 [26].
- **Cursive handwriting recognition.**
  Neural networks have proved useful in the development of algorithms for on-line cursive handwriting recognition [23]: A recent startup company in Palo Alto, Lexicus, beginning with this basic technology has developed an impressive PC-based cursive handwriting system.
- **Quality control in manufacturing.**
  Neural networks are being used in a large number of quality control and quality assurance programs throughout industry. Applications include contaminant-level detection from spectroscopy data at chemical plants and loudspeaker defect classification by CTS Electronics.
- **Petroleum exploration.**
  Oil companies including Arco and Texaco are using neural networks to help determine the locations of underground oil and gas deposits.

- **Medical applications.**
  Commercial products by *Neuromedical Systems Inc.* are used for cancer screening and other medical applications [28]. The company markets electrocardiograph and pap smear systems that rely on neural network technology. The pap smear system. *Papnet*, is able to help cytotechnologists spot cancerous cells, drastically reducing false/negative classifications. The system is used by the *U.S. Food and Drug Administration* [7].
- **Financial forecasting and portfolio management.**
  Neural networks are used for financial forecasting at a large number of investment firms and financial entities including Merill Lynch & Co., Salomon Brothers, Shearson Lehman Brothers Inc., Citibank, and the World Bank. Using neural networks trained by genetic algorithms, Citibank's Andrew Colin claims to be able to earn 25 % returns per year investing in the currency markets. A startup company, Promised Land Technologies, offers a $249 software package that is claimed to yield impressive annual returns [27].
- **Loan approval.**
  Chase Manhattan Bank reportedly uses a hybrid system utilizing pattern analysis and neural networks to evaluate corporate loan risk. Robert Marose reports in the May 1990 issue of *AI Expert* that the system, Creditview, helps loan officers estimate the credit worthiness of corporate loan candidates.
- **Marketing analysis.**
  The *Target Marketing System* developed by Churchill System is currently in use by Veratex Corp. to optimize marketing strategy and cut marketing costs by removing unlikely future customers from a list of potential customers [10].
- **Electric arc furnace electrode position control.**
  Electric arc furnaces are used to melt scrap steel. The Intelligent Arc furnace controller systems installed by Neural Applications Corp. are reportedly saving millions of dollars per year per furnace in increased furnace through-put and reduced electrode wear and electricity consumption. The controller is currently being installed at furnaces worldwide.
- **Semiconductor process control.**
  Kopin Corp. has used neural networks to cut dopant concentration and deposition thickness errors in solar cell manufacturing by more than a factor of two.
- **Chemical process control.**
  Pavilion Technologies has developed a neural network process control package, Process Insights, which is helping Eastman Kodak and a number of other companies reduce waste, improve product quality, and increase plant throughput [9]. Neural network models are being used to perform sensitivity studies, determine process set points, detect faults, and predict process performance.

- **Petroleum refinery process control.**
- **Continuous-casting control during steel production**
- **Food and chemical formulation optimization.**
- **Nonlinear Applications on the Horizon.**
  A large number of research programs are developing neural network solutions that are either likely to be used in the near future or, particularly in the case of military applications, that may already be incorporated into products, albeit unadvertised. This category is much larger than the foregoing, so we present here only a few representative examples.
- **Fighter flight and battle pattern quidance..**
- **Optical telescope focusing.**
- **Automobile applications.**
  Ford Motor Co., General Motors, and other automobile manufacturers are currently researching the possibility of widespread use of neural networks in automobiles and in automobile production. Some of the areas that are yielding promising results in the laboratory include engine fault detection and diagnosis, antilock brake control, active-suspension control, and idle-speed control. *General Motors* is having preliminary success using neural networks to model subjective customer ratings of automobiles based on their dynamic characteristics to help engineers tailor vehicles to the market.
- **Electric motor failure prediction.**
  Siemens has reportedly developed a neural network system that can accurately and inexpensively predict failure of large induction motors.
- **Speech recognition.**
  The *Stanford Research Institute* is currently involved in research combining neural networks with hidden *Markov models* and other technologies in a highly successful speaker independent speech recognition system. The technology will most likely be licensed to interested companies once perfected.
- **Biomedical applications.**
  Neural networks are rapidly finding diverse applications in the biomedical sciences. They are being used widely in research on amino acid sequencing in RNA and DNA, ECG and EEG waveform classification, prediction of patients' reactions to drug treatments, prevention of anesthesia-related accidents, arrhythmia recognition for implantable defibrillators patient mortality predictions, quantitative cytology, detection of breast cancer from mammograms, modeling schizophrenia, clinical diagnosis of lower-back pain, enhancement and classification of medical images, lung nodule detection, diagnosis of hepatic masses, prediction of pulmonary embolism likelihood from ventilation-perfusion lung scans, and the study of interstitial lung disease.
- **Drug development.**
  One particularly promising area of medical research involves the use of neural networks in predicting the medicinal properties of substances without expensive, time-consuming, and often inhumane animal testing.

- **Control of copies.**
  The *Ricoh Corp.* has successfully employed neural learning techniques for control of several voltages in copies in order to preserve uniform copy quality despite changes in temperature, humidity, time since last copy, time since change in toner cartridge, and other variables. These variables influence copy quality in highly nonlinear ways, which were learned through training of a backpropagation network.
- **The truck backer-upper.**
  Vehicular control by artificial neural networks is a topic that has generated widespread interest. At *Purdue University*, tests have been performed using neural networks to control a model helicopter.

Perhaps the most important advantage of neural networks is their adaptivity. Neural networks can automatically adjust their parameters (weights) to optimize their behavior as pattern recognizers, decision makers, system controllers, predictors, and so on.

Self-optimization allows the neural network to "design" itself. The system designer first defines the neural network architecture, determines how the network connects to other parts of the system, and chooses a training methodology for the network. The neural network then adapts to the application. Adaptivity allows the neural network to perform well even when the environment or the system being controlled varies over time. There are many control problems that can benefit from continual nonlinear modeling and adaptation. Neural networks, such as those used by Pavilion in chemical process control, and by Neural Application Corp. in arc furnace control, are ideally suited to track problem solutions in changing environments. Additionally, with some "programmability", such as the choices regarding the number of neurons per layer and number of layers, a practitioner can use the same neural network in a wide variety of applications. Engineering time is thus saved.

Another example of the advantages of self-optimization is in the field of *Expert Systems*. In some cases, instead of obtaining a set of rules through interaction between an experienced expert and a knowledge engineer, a neural system can be trained with examples of expert behavior.

Bibliography

1. I. Aleksander and H. Morton, *An Introduction to Neural Computing* (Chapmann and Hal, 1990).
2. J.A. Anderson and E. Rosenfels eds., *Neurocomputing: Foundations of Research* (MIT Press, Cambridge, MA,1988).
3. E. K. Blum and L. K. Li, Approximation theory and feedforward networks, *Neural Networks*, 4(1991) 511-515.
4. P. Cardaliaguet, Approximation of a function and its derivative with a neural network, *Neural Networks*, 5(1992) 207-220.
5. N.E.Cotter, The Stone-Weierstrass theorem and its applications to neural networks, *IEEE Transactions on Neural Networks*, 1(1990) 290-295.
6. J.E. Dayhoff, *Neural Network Architectures: An Introduction* (Van Nostrand Reinhold, New York,1990).
7. A. Fuochi, Neural networks: No zealots yet but progress being made, *Comput. Can.*, (January 20, 1992).
8. K. Funahashi, On the Approximate Realization of Continuous Mappings by Neural Networks, *Neural Networks* 2(1989) 183-192.
9. C. Hall, Neural net technology: Ready for prime time? *IEEE Expert* (December 1992) 2-4.
10. D. Hammerstrom, Neural networks at work. *IEEE Spectr.* (June 1993) 26-32.
11. D.O.Hebb, *The Organization of Behavior* (Wiley, New York, 1949).
12. R.Hecht-Nielsen, Theory of the Backpropagation Neural Network, *Proceedings of International Conference on Neural Networks*, Vol. 1, 1989 593-605.
13. R.Hecht-Nielsen, *Neurocomputing* (Addison-Wesley Publishing Co., Reading, Mass., 1990).
14. J.J.Hopfield, Neural networks and to physical systems with emergent collective computational abilities, *Proc. Natl. Acad. Sci.*, 79(1982) 2554-2558.
15. K. Hornik, M. Stinchcombe and H. White, Universal Approximation of an Unknown Mapping and Its Derivatives Using Multilayer Freedforward Networks, *Neural Networks*, 3(1990) 551-560.
16. T.Kohonen, *Self-organization and Associative Memory*, (Springer-Verlag, New York 1984).
17. S.Y.Kung, *Digital Neural Networks* (Prentice Hall, Englewood Cliffs, New York, 1993).
18. V. Kurkova, Kolmogorov's theorem and multilayer neural networks, *Neural Networks*, 5(1992) 501-506.
19. W.S.McCulloch and W.A.Pitts, A logical calculus of the ideas imminent in nervous activity, *Bull. Math. Biophys.* 5(1943) 115-133.
20. M. Minsky and S. Papert, *Perceptrons* (MIT Press, Cambridge, Mass., 1969).
21. F.Rosenblatt, The perceptron: A probabilistic model for information storage and organization in the brain, *Physic. Rev*, 65(1958) 386-408.

22. D.E.Rumelhart and J.L. McClelland and the PDP Research Group, *Parallel Distributed Processing: Explorations in the Microstructure of Cognition* (MIT Press/Bradford Books, Cambridge, Mass., 1986).
23. D.E. Rumelhart, Theory to practice: A case study - recognizing cursive handwriting. In *Proceedings of the Third NEC Research Symposium. SIAM*, Philadelphia, Pa., 1993.
24. D.E.Rumelhart, B.Widrow and M.A.Lehr, The basic ideas in neural networks, *Communications of ACM*, 37(1994) 87-92.
25. D.E.Rumelhart, B.Widrow and M.A.Lehr, Neural Networks: Applications in **Industry, Business** and **Science**, *Communications of ACM*, 37(1994) 93-105.
26. E.I. Schwartz and J.B. Treece, Smart programs go to work: How applied intelligence software makes decisions for the real world. *Business Week* (Mar. 2, 1992) 97-105.
27. E.I. Schwartz, Where neural networks are already at work: Putting AI to work in the markets, *Business Week* (Nov. 2, 1992) 136-137.
28. J. Shandle, Neural networks are ready for prime time, *Elect. Des.*, (February 18, 1993), 51-58.
29. P.I.Simpson, *Artificial Neural Systems: Foundations, Paradigms, Applications, and Implementation* (Pergamon Press, New York, 1990).
30. P.D.Wasserman, *Advanced Methods in Neural Computing*, Van Nostrand Reinhold, New York 1993.
31. H. White, Connectionist Nonparametric Regression: Multilayer feedforward Networks Can Learn Arbitrary Mappings, *Neural Networks* 3(1990) 535-549.
32. J.M.Zurada, *Introduction to Artificial Neural Systems* (West Publishing Company, New York, 1992).

3. Fuzzy neural networks

3.1 Integration of fuzzy logic and neural networks

Hybrid systems combining fuzzy logic, neural networks, genetic algorithms, and expert systems are proving their effectiveness in a wide variety of real-world problems.

Every intelligent technique has particular computational properties (e.g. ability to learn, explanation of decisions) that make them suited for particular problems and not for others. For example, while neural networks are good at recognizing patterns, they are not good at explaining how they reach their decisions. Fuzzy logic systems, which can reason with imprecise information, are good at explaining their decisions but they cannot automatically acquire the rules they use to make those decisions. These limitations have been a central driving force behind the creation of intelligent hybrid systems where two or more techniques are combined in a manner that overcomes the limitations of individual techniques. Hybrid systems are also important when considering the varied nature of application domains. Many complex domains have many different component problems, each of which may require different types of processing. If there is a complex application which has two distinct sub-problems, say a signal processing task and a serial reasoning task, then a neural network and an expert system respectively can be used for solving these separate tasks. The use of intelligent hybrid systems is growing rapidly with successful applications in many areas including process control, engineering design, financial trading, credit evaluation, medical diagnosis, and cognitive simulation.

> While fuzzy logic provides an inference mechanism under cognitive uncertainty, computational neural networks offer exciting advantages, such as learning, adaptation, fault-tolerance, parallelism and generalization. To enable a system to deal with cognitive uncertainties in a manner more like humans, one may incorporate the concept of fuzzy logic into the neural networks.

The computational process envisioned for fuzzy neural systems is as follows. It starts with the development of a "fuzzy neuron" based on the understanding of biological neuronal morphologies, followed by learning mechanisms. This leads to the following three steps in a fuzzy neural computational process

- development of fuzzy neural models motivated by biological neurons,
- models of synaptic connections which incorporates *fuzziness* into neural network,
- development of learning algorithms (that is the method of adjusting the synaptic weights)

Two possible models of fuzzy neural systems are

- In response to linguistic statements, the fuzzy interface block provides an input vector to a multi-layer neural network. The neural network can be adapted (trained) to yield desired command outputs or decisions.

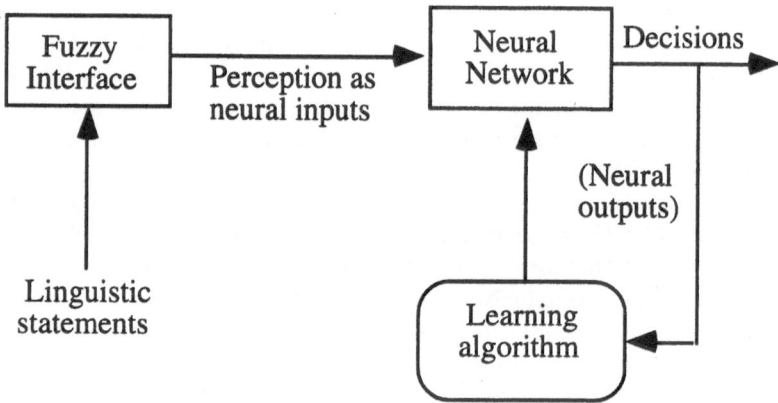

Fig. 3.1. The first model of fuzzy neural system.

- A multi-layered neural network drives the fuzzy inference mechanism.

Neural networks are used to *tune* membership functions of fuzzy systems that are employed as decision-making systems for controlling equipment. Although fuzzy logic can encode expert knowledge directly using rules with linguistic labels, it usually takes a lot of time to design and tune the membership functions which quantitatively define these linquistic labels. Neural network learning techniques can automate this process and substantially reduce development time and cost while improving performance.

In theory, neural networks, and fuzzy systems are equivalent in that they are convertible, yet in practice each has its own advantages and disadvantages. For neural networks, the knowledge is automatically acquired by the backpropagation algorithm, but the learning process is relatively slow and analysis of the trained network is difficult (black box). Neither is it possible to extract structural knowledge (rules) from the trained neural network,

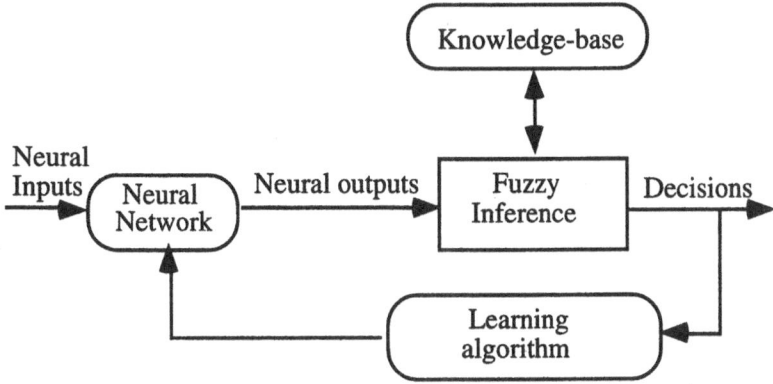

Fig. 3.2. The second model of fuzzy neural system.

nor can we integrate special information about the problem into the neural network in order to simplify the learning procedure.

Fuzzy systems are more favorable in that their behavior can be explained based on fuzzy rules and thus their performance can be adjusted by tuning the rules. But since, in general, knowledge acquisition is difficult and also the universe of discourse of each input variable needs to be divided into several intervals, applications of fuzzy systems are restricted to the fields where expert knowledge is available and the number of input variables is small.

To overcome the problem of knowledge acquisition, neural networks are extended to automatically *extract fuzzy rules from numerical data*. Cooperative approaches use neural networks to optimize certain parameters of an ordinary fuzzy system, or to preprocess data and extract fuzzy (control) rules from data.

Based upon the computational process involved in a fuzzy-neuro system, one may broadly classify the fuzzy neural structure as feedforward (static) and feedback (dynamic).

A typical fuzzy-neuro system is Berenji's ARIC (*Approximate Reasoning Based Intelligent Control*) architecture [11]. It is a neural network model of a fuzy controller and learns by updating its prediction of the physical system's behavior and fine tunes a predefined control knowledge base.

This kind of architecture allows to combine the advantages of neural networks and fuzzy controllers. The system is able to learn, and the knowledge used within the system has the form of fuzzy IF-THEN rules. By predefining these rules the system has not to learn from scratch, so it learns faster than a standard neural control system.

ARIC consists of two coupled feed-forward neural networks, the *Action-state Evaluation Network* (AEN) and the *Action Selection Network* (ASN). The ASN is a multilayer neural network representation of a fuzzy controller.

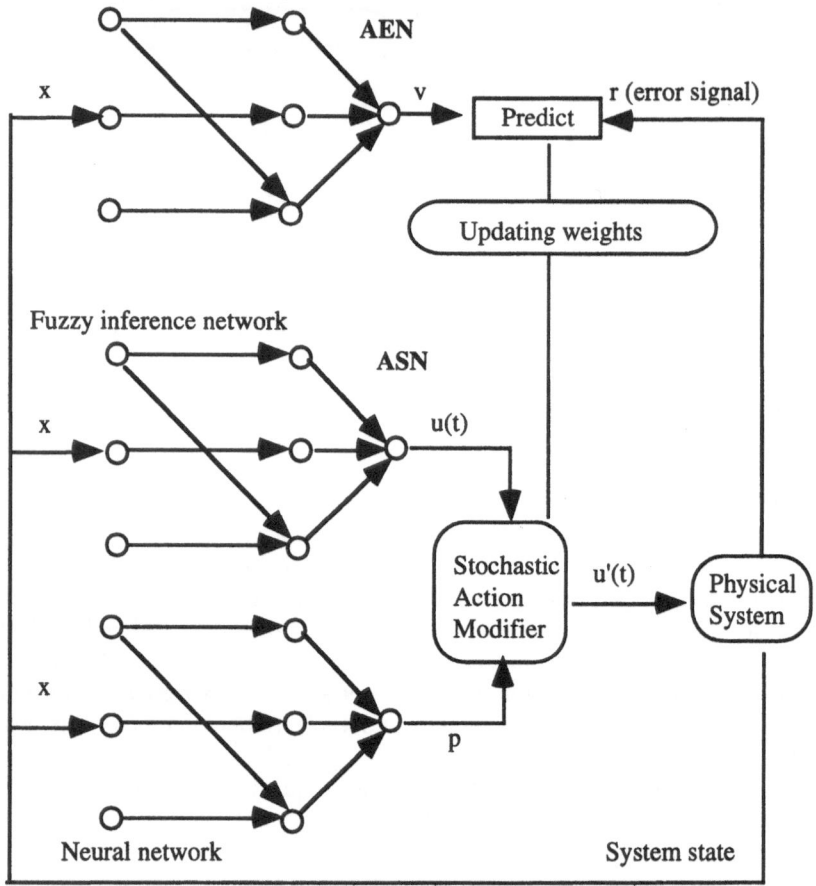

Fig. 3.3. Berenji's ARIC architecture.

In fact, it consists of two separated nets, where the first one is the fuzzy inference part and the second one is a neural network that calculates $p[t, t+1]$, a *measure of confidence associated with the fuzzy inference value* $u(t+1)$, using the weights of time t and the system state of time $t+1$. A *stochastic modifier* combines the recommended control value $u(t)$ of the fuzzy inference part and the so called "probability" value p and determines the final output value

$$u'(t) = o(u(t), p[t, t+1])$$

of the ASN. The hidden units z_i of the fuzzy inference network represent the fuzzy rules, the input units x_j the rule antecedents, and the output unit u represents the control action, that is the defuzzified combination of the conclusions of all rules (output of hidden units). In the input layer the system state variables are fuzzified. Only monotonic membership functions are used in ARIC, and the fuzzy labels used in the control rules are adjusted locally within each rule. The membership values of the antecedents of a rule are then multiplied by weights attached to the connection of the input unit to the hidden unit. The minimum of those values is its final input. In each hidden unit a special monotonic membership function representing the conclusion of the rule is stored. Because of the monotonicity of this function the crisp output value belonging to the minimum membership value can be easily calculated by the inverse function. This value is multiplied with the weight of the connection from the hidden unit to the output unit. The output value is then calculated as a weighted average of all rule conclusions.

The AEN tries to predict the system behavior. It is a feed-forward neural network with one hidden layer, that receives the system state as its input and an error signal r from the physical system as additional information. The output $v[t, t']$ of the network is viewed as a *prediction of future reinforcement*, that depends of the weights of time t and the system state of time t', where t' may be t or $t+1$. Better states are characterized by higher reinforcements. The weight changes are determined by a reinforcement procedure that uses the ouput of the ASN and the AEN. The ARIC architecture was applied to cart-pole balancing and it was shown that the system is able to solve this task [11].

3.2 Fuzzy neurons

Consider a simple neural net in Fig. 3.4. All signals and weights are real numbers. The two input neurons do not change the input signals so their output is the same as their input. The signal x_i interacts with the weight w_i to produce the product

$$p_i = w_i x_i, \ i = 1, 2.$$

The input information p_i is aggregated, by addition, to produce the input

$$\text{net} = p_1 + p_2 = w_1 x_1 + w_2 x_2,$$

to the neuron. The neuron uses its transfer function f, which could be a sigmoidal function,

$$f(x) = \frac{1}{1 + e^{-x}},$$

to compute the output

$$y = f(\text{net}) = f(w_1 x_1 + w_2 x_2).$$

This simple neural net, which employs multiplication, addition, and sigmoidal f, will be called as regular (or standard) neural net.

If we employ other operations like a t-norm, or a t-conorm, to combine the incoming data to a neuron we obtain what we call a *hybrid neural net*.

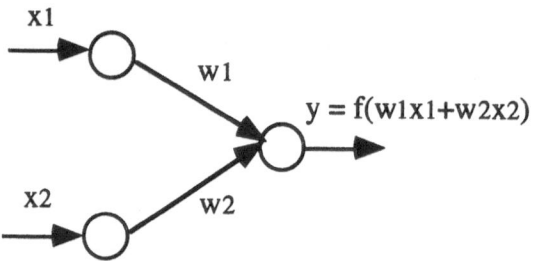

Fig. 3.4. Simple neural net.

These modifications lead to a fuzzy neural architecture based on fuzzy arithmetic operations. Let us express the inputs (which are usually membership degrees of a fuzzy concept) x_1, x_2 and the weigths w_1, w_2 over the unit interval $[0, 1]$.

A hybrid neural net may not use multiplication, addition, or a sigmoidal function (because the results of these operations are not necesserily are in the unit interval).

Definition 3.2.1 *A hybrid neural net is a neural net with crisp signals and weights and crisp transfer function. However,*

- *we can combine x_i and w_i using a t-norm, t-conorm, or some other continuous operation,*
- *we can aggregate p_1 and p_2 with a t-norm, t-conorm, or any other continuous function*
- *f can be any continuous function from input to output*

We emphasize here that all inputs, outputs and the weights of a hybrid neural net are real numbers taken from the unit interval $[0, 1]$. A processing element of a hybrid neural net is called *fuzzy neuron*. In the following we present some fuzzy neurons.

Definition 3.2.2 *(AND fuzzy neuron [81, 82])*
   The signal x_i and w_i are combined by a triangular conorm S to produce

$$p_i = S(w_i, x_i), \; i = 1, 2.$$

The input information p_i is aggregated by a triangular norm T to produce the output

$$y = AND(p_1, p_2) = T(p_1, p_2) = T(S(w_1, x_1), S(w_2, x_2))$$

of the neuron.

So, if $T = \min$ and $S = \max$ then the AND neuron realizes the min-max composition

$$y = \min\{w_1 \vee x_1, w_2 \vee x_2\}.$$

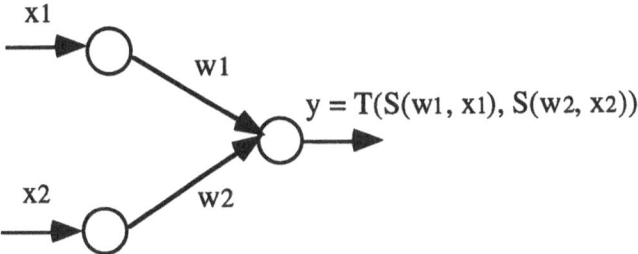

Fig. 3.5. AND fuzzy neuron.

Definition 3.2.3 *(OR fuzzy neuron [81, 82])*
 The signal x_i and w_i are combined by a triangular norm T to produce

$$p_i = T(w_i, x_i), \ i = 1, 2.$$

The input information p_i is aggregated by a triangular conorm S to produce the output

$$y = OR(p_1, p_2) = S(p_1, p_2) = S(T(w_1, x_1), T(w_2, x_2))$$

of the neuron.

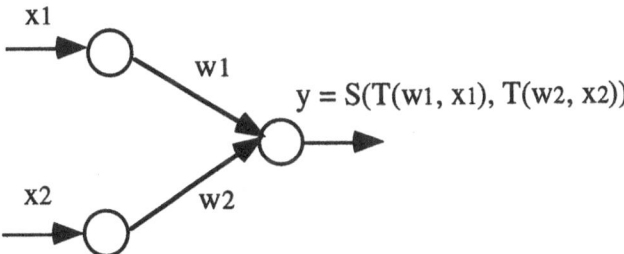

Fig. 3.6. OR fuzzy neuron.

So, if $T = \min$ and $S = \max$ then the AND neuron realizes the max-min composition

$$y = \max\{w_1 \wedge x_1, w_2 \wedge x_2\}.$$

The AND and OR fuzzy neurons realize pure logic operations on the membership values. The role of the connections is to differentiate between particular levels of impact that the individual inputs might have on the result of aggregation. We note that (i) the higher the value w_i the stronger the impact of x_i on the output y of an OR neuron, (ii) the lower the value w_i the stronger the impact of x_i on the output y of an AND neuron.

The range of the output value y for the AND neuron is computed by letting all x_i equal to zero or one. In virtue of the monotonicity property of triangular norms, we obtain

$$y \in [T(w_1, w_2), 1]$$

and for the OR neuron one derives the boundaries

$$y \in [0, S(w_1, w_2)].$$

Definition 3.2.4 *(Implication-OR fuzzy neuron [44, 46])*
The signal x_i and w_i are combined by a fuzzy implication operator I to produce

$$p_i = I(w_i, x_i) = w_i \leftarrow x_i, \ i = 1, 2.$$

The input information p_i is aggregated by a triangular conorm S to produce the output

$$y = I(p_1, p_2) = S(p_1, p_2) = S(w_1 \leftarrow x_1, w_2 \leftarrow x_2)$$

of the neuron.

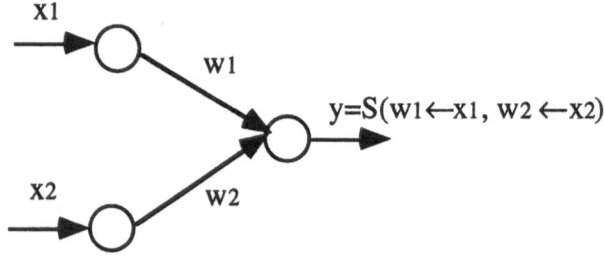

Fig. 3.7. Implication-OR fuzzy neuron.

Definition 3.2.5 *(Kwan and Cai's fuzzy neuron [126])*
The signal x_i interacts with the weight w_i to produce the product

$$p_i = w_i x_i, \ i = 1, \ldots, n$$

The input information p_i is aggregated by an agregation function h to produce the input of the neuron

$$z = h(w_1 x_1, w_2 x_2, \ldots, w_n x_n)$$

the state of the neuron is computed by

$$s = f(z - \theta)$$

where f is an activation function and θ is the activating threshold. And the m outputs of the neuron are computed by

$$y_j = g_j(s), \; j = 1, \ldots, m$$

where $g_j, \; j = 1, \ldots, m$ are the m output functions of the neuron which represent the membership functions of the input pattern $x_1, x_2, \ldots, x_n$ in all the m fuzzy sets.

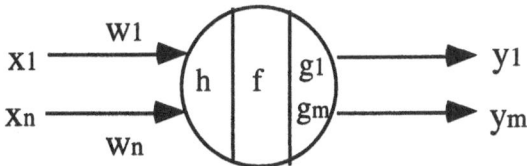

Fig. 3.8. Kwan and Cai's fuzzy neuron.

Definition 3.2.6 *(Kwan and Cai's max fuzzy neuron [126])*
The signal x_i interacts with the weight w_i to produce the product

$$p_i = w_i x_i, \; i = 1, 2.$$

The input information p_i is aggregated by the maximum conorm

$$z = \max\{p_1, p_2\} = \max\{w_1 x_1, w_2 x_2\}$$

and the j-th output of the neuron is computed by

$$y_j = g_j(f(z - \theta)) = g_j(f(\max\{w_1 x_1, w_2 x_2\} - \theta))$$

where f is an activation function.

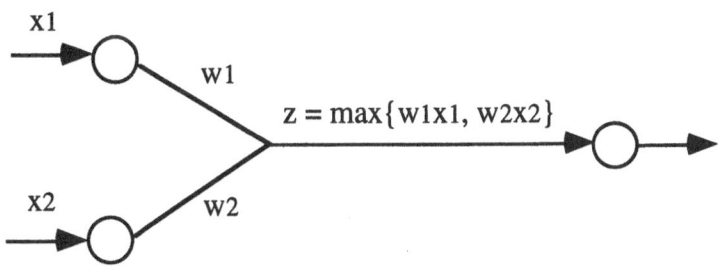

Fig. 3.9. Kwan and Cai's max fuzzy neuron.

Definition 3.2.7 *(Kwan and Cai's* min *fuzzy neurons [126])*
The signal x_i interacts with the weight w_i to produce the product

$$p_i = w_i x_i, \ i = 1, 2.$$

The input information p_i is aggregated by the minimum norm

$$y = \min\{p_1, p_2\} = \min\{w_1 x_1, w_2 x_2\}$$

and the j-th output of the neuron is computed by

$$y_j = g_j(f(z - \theta)) = g_j(f(\min\{w_1 x_1, w_2 x_2\} - \theta))$$

where f is an activation function.

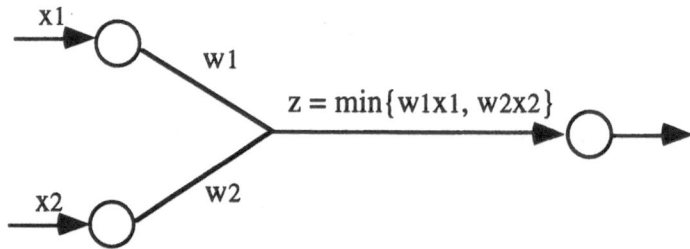

Fig. 3.10. Kwan and Cai's min fuzzy neuron.

It is well-known that regular nets are universal approximators, i.e. they can approximate any continuous function on a compact set to arbitrary accuracy. In a discrete fuzzy expert system one inputs a discrete approximation to the fuzzy sets and obtains a discrete approximation to the output fuzzy set. Usually discrete fuzzy expert systems and fuzzy controllers are continuous mappings. Thus we can conclude that given a continuous fuzzy expert system, or continuous fuzzy controller, there is a regular net that can uniformly approximate it to any degree of accuracy on compact sets. The problem with this result that it is non-constructive and only approximative. The main problem is that the theorems are existence types and do not tell you how to build the net.

Hybrid neural nets can be used to implement fuzzy IF-THEN rules in a constructive way. Following Buckley & Hayashi [33], and, Keller, Yager & Tahani [112] we will show how to construct hybrid neural nets which are computationally equivalent to fuzzy expert systems and fuzzy controllers. It should be noted that these hybrid nets are for computation and they do not have to learn anything.

Though hybrid neural nets can not use directly the standard error back-propagation algorithm for learning, they can be trained by steepest descent methods to learn the parameters of the membership functions representing the linguistic terms in the rules (supposing that the system output is a differentiable function of these parameters).

The direct fuzzification of conventional neural networks is to extend connection weigths and/or inputs and/or fuzzy desired outputs (or targets) to fuzzy numbers. This extension is summarized in Table 3.1.

Fuzzy neural net	Weights	Inputs	Targets
Type 1	crisp	fuzzy	crisp
Type 2	crisp	fuzzy	fuzzy
Type 3	fuzzy	fuzzy	fuzzy
Type 4	fuzzy	crisp	fuzzy
Type 5	crisp	crisp	fuzzy
Type 6	fuzzy	crisp	crisp
Type 7	fuzzy	fuzzy	crisp

Table 3.1. Direct fuzzification of neural networks.

Fuzzy neural networks (FNN) of *Type 1* are used in classification problem of a fuzzy input vector to a crisp class [92, 129]. The networks of *Type 2, 3* and *4* are used to implement fuzzy IF-THEN rules [101, 103]. However, the last three types in Table 3.1 are unrealistic.

- In *Type 5*, outputs are always real numbers because both inputs and weights are real numbers.
- In *Type 6* and *7*, the fuzzification of weights is not necessary because targets are real numbers.

Definition 3.2.8 *A regular fuzzy neural network is a neural network with fuzzy signals and/or fuzzy weights, sigmoidal transfer function and all the operations are defined by Zadeh's extension principle.*

Consider a simple regular fuzzy neural net in Fig. 3.11. All signals and weights are fuzzy numbers. The two input neurons do not change the input signals so their output is the same as their input. The signal X_i interacts with the weight W_i to produce the product

$$P_i = W_i X_i, \ i = 1, 2.$$

where we use the extension principle to compute P_i. The input information P_i is aggregated, by standard extended addition, to produce the input

$$\text{net} = P_1 + P_2 = W_1 X_1 + W_2 X_2,$$

to the neuron. The neuron uses its transfer function f, which is a sigmoidal function, to compute the output

$$Y = f(\text{net}) = f(W_1X_1 + W_2X_2)$$

where $f(x) = (1 + e^{-x})^{-1}$ and the membership function of the output fuzzy set Y is computed by the extension principle

$$Y(y) = \begin{cases} (W_1X_1 + W_2X_2)(f^{-1}(y)) & \text{if } 0 \le y \le 1 \\ 0 & \text{otherwise} \end{cases}$$

where $f^{-1}(y) = \ln y - \ln(1 - y)$.

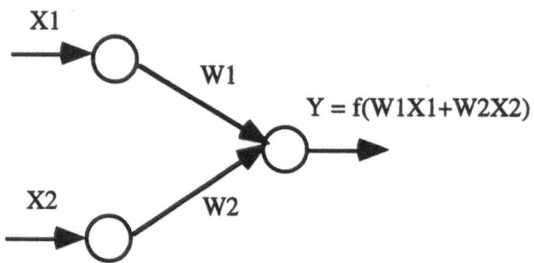

Fig. 3.11. Simple regular fuzzy neural net.

Buckley and Hayashi [31] showed that regular fuzzy neural nets are monotonic, i.e. if $X_1 \subset X_1'$ and $X_2 \subset X_2'$ then

$$Y = f(W_1X_1 + W_2X_2) \subset Y' = f(W_1X_1' + W_2X_2').$$

where f is the sigmoid transfer function, and all the operations are defined by Zadeh's extension principle.

This means that fuzzy neural nets based on the extension principle might be universal approximators only for *continuous monotonic functions*. If a fuzzy function is not monotonic there is no hope of approximating it with a fuzzy neural net which uses the extension principle.

The following example shows a continuous fuzzy function which is non-monotonic. Therefore we must abandon the extension principle if we are to obtain a universal approximator.

Example 3.2.1 *Let $f \colon \mathcal{F} \to \mathcal{F}$ be a fuzzy function defined by*

$$f(A) = (D(A, \bar{0}), 1)$$

where A is a fuzzy number, $\bar{0}$ is a fuzzy point with center zero, $D(A, \bar{0})$ denotes the Hausdorff distance between A and $\bar{0}$, and $(D(A, \bar{0}), 1)$ denotes a symmetrical triangular fuzzy number with center $D(A, \bar{0})$ and width one.

We first show that f is continuous in metric D. Let $A_n \in \mathcal{F}$ be a sequence of fuzzy numbers such that $D(A_n, A) \to 0$ if $n \to \infty$. Using the definition of metric D we have

$$D(f(A_n), f(A)) = D((D(A_n, \bar{0}), 1), (D(A, \bar{0}), 1))$$

$$= |D(A_n, \bar{0}) - D(A, \bar{0})| \leq$$

$$D(A_n, A) + D(\bar{0}, \bar{0}) = D(A_n, A)$$

which verifies the continuity of f in metric D.

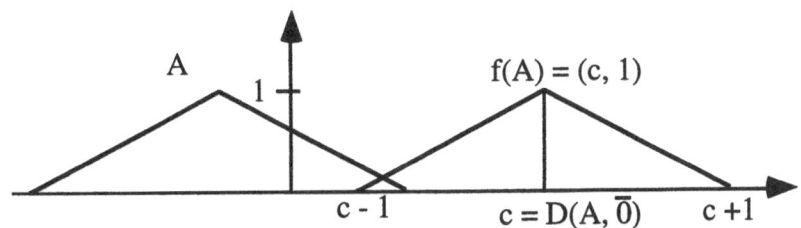

Fig. 3.12. A and $f(A)$.

Let $A, A' \in \mathcal{F}$ such that $A \subset A'$. Then $f(A) = (D(A, \bar{0}), 1)$ and $f(A') = (D(A', \bar{0}), 1)$ are both symmetrical triangular fuzzy numbers with different centers, i.e. nor $A \subset A'$ neither $A' \subset A$ can occur.

Definition 3.2.9 *A hybrid fuzzy neural network is a neural network with fuzzy signals and/or fuzzy weights. However, (i) we can combine X_i and W_i using a t-norm, t-conorm, or some other continuous operation; we can aggregate P_1 and P_2 with a t-norm, t-conorm, or any other continuous function; f can be any function from input to output*

Buckley and Hayashi [31] showed that *hybrid fuzzy neural networks* are universal approximators, i.e. they can approximate any continuous fuzzy functions on a compact domain.

Buckley, Hayashi and Czogala [25] showed that any continuous feedforward neural net can be approximated to any degree of accuracy by a discrete fuzzy expert system:

Assume that all the ν_j in the input signals and all the y_i in the output from the neural net belong to $[0, 1]$. Therefore, $o = G(\nu)$ with $\nu \in [0, 1]^n$, $o \in [0, 1]^m$ and G is continuous, represents the net. Given any input (ν) - output o pair for the net we now show how to construct the corresponding rule in the fuzzy expert system. Define fuzzy set A as $A(j) = \nu_j$, $j = 1, \ldots, n$ and zero otherwise.

Also let $C(i) = o_i$, $i = 1, \ldots, m$, and zero otherwise.

Then the rule obtained from the pair (ν, o) is

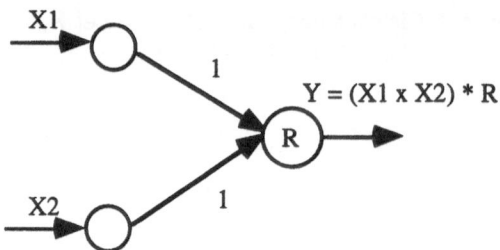

Fig. 3.13. Simple hybrid fuzzy neural net for the compositional rule of inference.

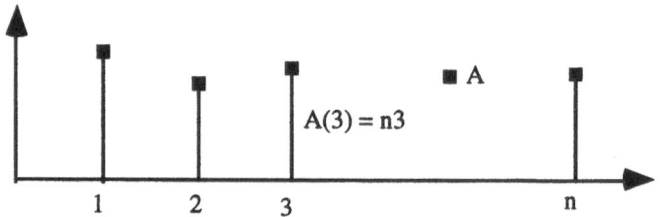

Fig. 3.14. Definition of A.

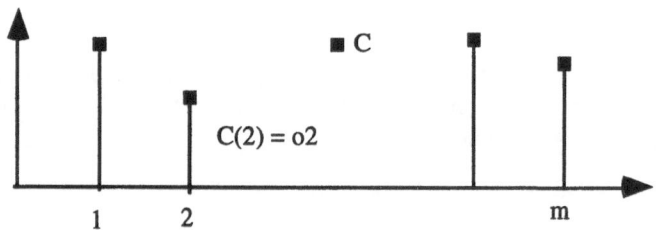

Fig. 3.15. Definition of C.

$$\Re(\nu) : \text{If } x \text{ is } A \text{ then } z \text{ is } C,$$

That is, in rule construction ν is identified with A and C.

Theorem 3.2.1 *[25] Given $\epsilon > 0$, there exists a fuzzy expert system so that*

$$\|F(u) - G(u)\| \leq \epsilon, \ \forall u \in [0,1]^n$$

where F is the input - output function of the fuzzy expert system $\Re = \{\Re(\nu)\}$.

3.3 Hybrid neural nets

Drawing heavily on Buckley and Hayashi [26] we show how to construct hybrid neural nets that are computationally identical to discrete fuzzy expert systems and the *Sugeno* and *Expert system elementary* fuzzy controller. Hybrid neural nets employ more general operations (t-norms, t-conorms, etc.) in combining signals and weights for input to a neuron.

Consider a fuzzy expert system with one block of rules

$$\Re_i : \text{If } x \text{ is } A_i \text{ then } y \text{ is } B_i, \ 1 \leq i \leq n.$$

For simplicity we have only one clause in the antecedent but our results easily extend to many clauses in the antecedent.

Given some data on x, say A', the fuzzy expert system comes up with its final conclusion y is B'. In computer applications we usually use discrete versions of the continuous fuzzy sets. Let $[\alpha_1, \alpha_2]$ contain the support of all the A_i, plus the support of all the A' we might have as input to the system. Also, let $[\beta_1, \beta_2]$ contain the support of all the B_i, plus the support of all the B' we can obtain as outputs from the system. Let $M \geq 2$ and $N \geq$ be positive integers. Let

$$x_j = \alpha_1 + (j-1)(\alpha_2 - \alpha_1)/(M-1)$$

for $1 \leq j \leq M$.

$$y_i = \beta_1 + (i-1)(\beta_2 - \beta_1)/(N-1)$$

for $1 \leq i \leq N$. The discrete version of the system is to input

$$a' = (A'(x_1), \dots, A'(x_M))$$

and obtain output $b' = (B'(y_1), \dots, B'(y_N))$.

We now need to describe the internal workings of the fuzzy expert system. There are two cases:

Case 1. Combine all the rules into one rule which is used to obtain b' from a'.

We first construct a fuzzy relation R_k to model rule

$$\Re_k : \text{If } x \text{ is } A_k, \text{ then } y \text{ is } B_k, \ 1 \leq k \leq n.$$

This is called modeling the implication and there are many ways to do this. One takes the data $A_k(x_i)$ and $B_k(y_j)$ to obtain $R_k(x_i, y_j)$ for each rule. One way to do this is

$$R_k(x_i, y_j) = \min\{A_k(x_i), B_k(y_j)\}.$$

Then we combine all the R_k into one R, which may be performed in many different ways and one procedure would be to intersect the R_k to get R. In any case, let

$$r_{ij} = R(x_i, y_j),$$

the value of R at the pair (x_i, y_j).

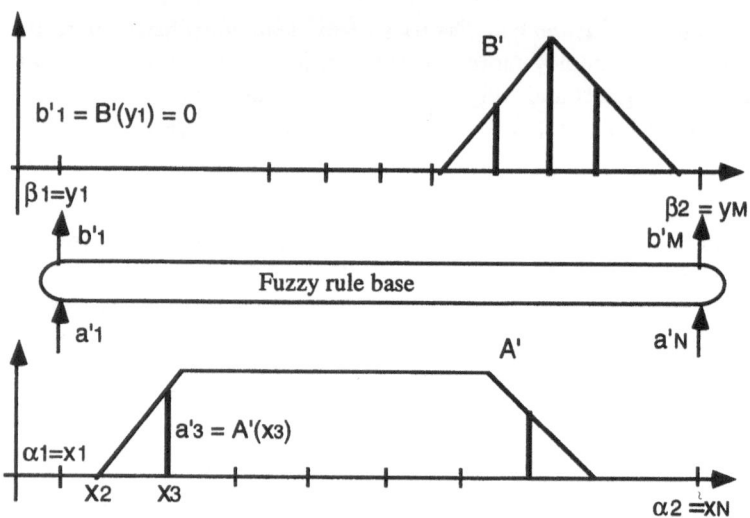

Fig. 3.16. A discrete version of fuzzy expert system.

The method of computing b' from a' is called the compositional rule of inference. Let

$$\lambda_{ij} = a'_i * r_{ij},$$

where $a'_i = A'(x_i)$ and $*$ is some method (usually a t-norm) of combining the data into λ_{ij}.

Then set $b' = (b'_1, \dots, b'_N)$ and

$$b'_j = \text{Agg}(\lambda_{1j}, \dots, \lambda_{Mj}), \ 1 \leq j \leq N,$$

for Agg a method of aggregating the information.

A hybrid neural net computationally the same as this fuzzy expert system is shown in Fig. 3.17.

We first combine the signals (a'_i) and the weights (r_{i1}) and then aggregate the data

$$a'_1 * r_{11}, \dots, a'_M * r_{M1}$$

using Agg, so the input to the neuron is b'_1. Now the transfer function is identity function $f(t) = t$, $t \in [0, 1]$ so that the output is b'_1. Similarly for all neurons, which implies the net gives b' from a'. The hybrid neural net in Fig. 3.16 provides fast parallel computation for a discrete fuzzy expert system. However, it can get too large to be useful. For example, let $[\alpha_1, \alpha_2] = [\beta_1, \beta_2] = [-10, 10]$ with discrete increments of 0.01 so that $M = N = 1000$. Then there will be: 2000 input neurons, 2000^2 connections from the input nodes to the output nodes, and 2000 output nodes.

Case 2. Fire the rules individually, given a', and combine their results into b'.

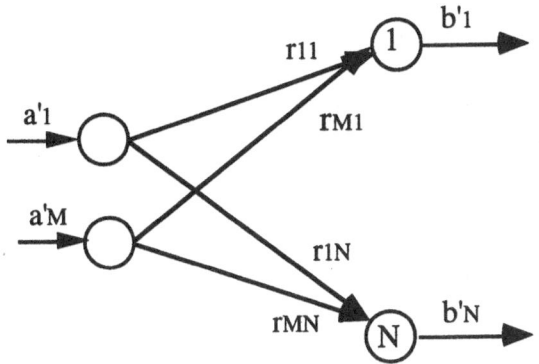

Fig. 3.17. Combine the rules first.

We compose a' with each R_k producing intermediate result

$$b'_k = (b'_{k1}, \ldots, b'_{kN}).$$

Then combine all the b'_k into b'.

- One takes the data $A_k(x_i)$ and $B_k(y_j)$ to obtain $R_k(x_i, y_j)$ for each rule. One way to do this is

$$R_k(x_i, y_j) = \min\{A_k(x_i), B_k(y_j)\}.$$

In any case, let $R_k(x_i, y_j) = r_{kij}$. Then we have $\lambda_{kij} = a'_i * r_{kij}$ and

$$b'_{kj} = \text{Agg}(\lambda_{k1j}, \ldots, \lambda_{kMj}).$$

The method of combining the b'_k would be done component wise so let

$$b'_j = Agg_1(b'_{1j}, \ldots, b'_{nj}), \ 1 \le j \le N$$

for some other aggregating operator Agg_1. A hybrid neural net computationally equal to this type of fuzzy expert system is shown in Fig. 3.18. For simplicity we have drawn the figure for $M = N = 2$.

In the hidden layer: the top two nodes operate as the first rule $\Re_1$, and the bottom two nodes model the second rule $\Re_2$. In the two output nodes: the top node, all weights are one, aggregates b'_{11} and b'_{21} using Agg_1 to produce b'_1, the bottom node, weights are one, computes b'_2.

Therefore, the hybrid neural net computes the same output b' given a' as the fuzzy expert system.

As in the previous case this hybrid net quickly gets too big to be practical at this time. Suppose there are 10 rules and

$$[\alpha_1, \alpha_2] = [\beta_1, \beta_2] = [-10, 10].$$

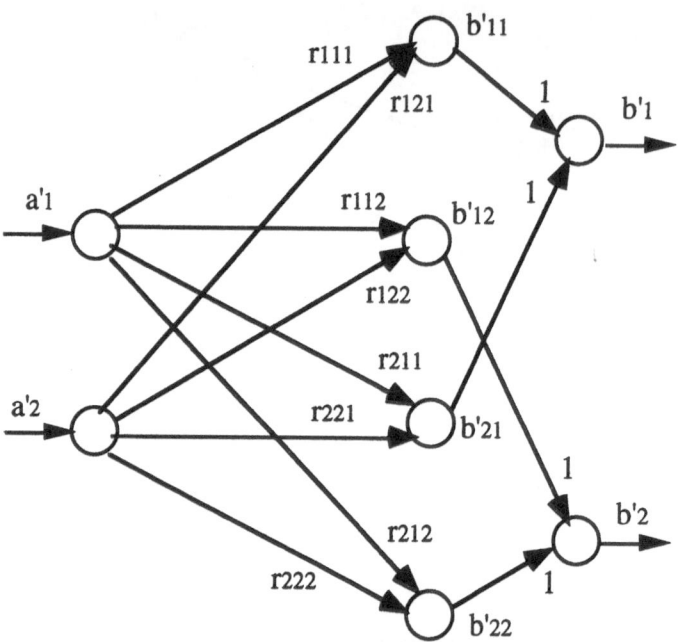

Fig. 3.18. Fire the rules first.

with discrete increments of 0.01 so that $M = N = 1000$. Then there will be: 2000 input neurons, 4 millions (10×2000^2) connections from the input nodes to the hidden layer, 2000 neurons in the hidden layer, 10×2000 connections from the hidden layer to the output neurons, and 20000 output nodes. And this hybrid net has only one clause in each rule's antecedent.

Buckley [22] identifies three basic types of elementary fuzzy controllers, *Sugeno, Expert system, and Mamdani*. We show how to build a hybrid netural net to be computationally identical to Sugeno and Expert system fuzzy controller. Actually, depending on how one computes the defuzzifier, the hybrid neural net could only be approximately the same as the Mamdani controller.

Sugeno control rules are of the type

$$\Re_i: \quad \text{If } e = A_i \text{ and } \Delta e = B_i \text{ then } z_i = \alpha_i e + \beta_i (\Delta e) + \gamma_i$$

where A_i, B_i, α_i, β_i, and γ_i are all given e is the *error*, Δe is the *change in error*.

The input to the controller is values for e and Δe and one first evaluates each rule's antecedent as follows:

$$\sigma_i = T(A_i(e), B_i(\Delta e)),$$

where T is some *t*-norm. Next, we evaluate each conclusion given e and Δe as

$$z_i = \alpha_i e + \beta_i(\Delta e) + \gamma_i,$$

The output is

$$\delta = \sum_{i=1}^{n} \sigma_i z_i \Big/ \sum_{i=1}^{n} \sigma_i.$$

A hybrid neural net computationally equivalent to the Sugeno controller is displayed in Fig. 3.19.

For simplicity, we have assumed only two rules. First consider the first hidden layer. The inputs to the top two neurons are $\alpha_1 e + \beta_1(\Delta e)$ and $\alpha_2 e + \beta_2(\Delta e)$. The transfer function in these two neurons is

$$f(x) = x + \gamma_i, \ i = 1, 2$$

so the outputs are z_i. All other neurons in Fig. 3.19 have linear activation function $f(x) = x$. The inputs to the bottom two nodes are

$$T(A_i(e), B_i(\Delta e)).$$

In the rest of the net all weights are equal to one. The output produces δ because we aggregate the two input signals using division $(\epsilon_1 + \epsilon_2)/(\sigma_1 + \sigma_2)$.

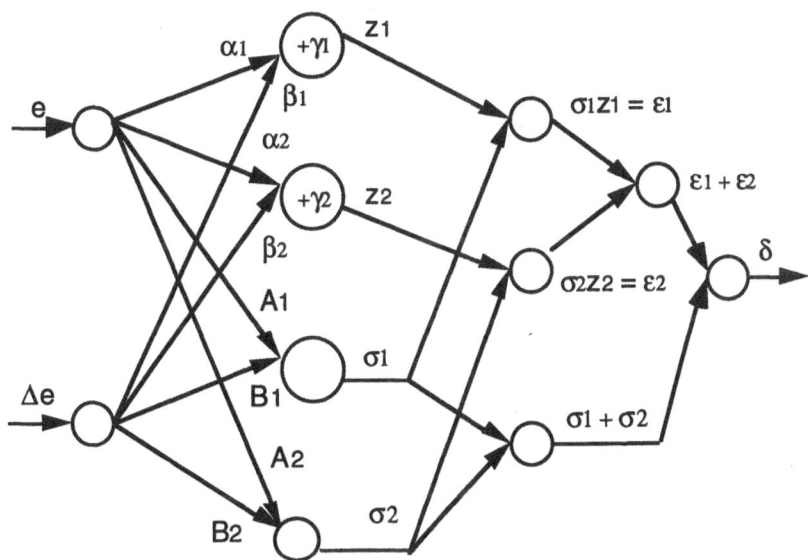

Fig. 3.19. Hybrid neural net as Sugeno controller.

The fuzzy controller based on fuzzy expert system was introduced by Buckley, Hayashi ad Czogala in [25]. The fuzzy control rules are

$$\Re_i: \text{If } e = A_i \text{ and } \Delta e = B_i \text{ then control action is } C_i$$

where C_i are triangular shaped fuzzy numbers with centers c_i, e is the *error*, Δe is the *change in error*. Given input values e and Δe each rule is evaluated producing the σ_i given by

$$\sigma_i = T(A_i(e), B_i(\Delta e)).$$

Then σ_i is assigned to the rule's consequence C_i and the controller takes all the data (σ_i, C_i) and defuzzifies to output δ. Let

$$\delta = \sum_{i=1}^{n} \sigma_i c_i \bigg/ \sum_{i=1}^{n} \sigma_i.$$

A hybrid neural net computationally identical to this controller is shown in the Fig. 3.20 (again, for simplicity we assume only two control rules).

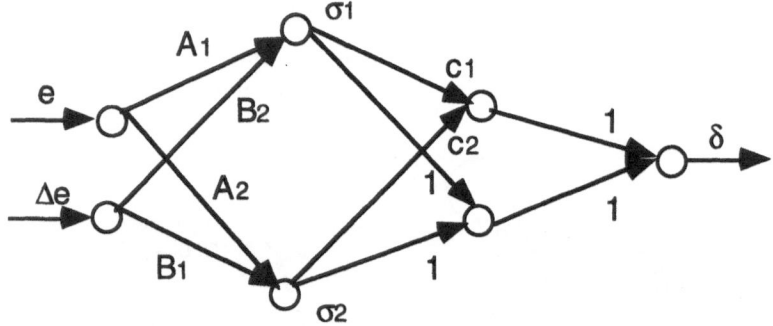

Fig. 3.20. Hybrid neural net as a fuzzy expert system controller.

The operations in this hybrid net are similar to those in Fig. 3.19.

As an example, we show how to construct a hybrid neural net (called *adaptive network* by Jang [106]) which is funcionally equivalent to Sugeno's inference mechanism.

Sugeno and Takagi use the following rules [208]

$$\Re_1 : \text{if } x \text{ is } A_1 \text{ and } y \text{ is } B_1 \text{ then } z_1 = a_1 x + b_1 y$$
$$\Re_2 : \text{if } x \text{ is } A_2 \text{ and } y \text{ is } B_2 \text{ then } z_2 = a_2 x + b_2 y$$

The firing levels of the rules are computed by

$$\alpha_1 = A_1(x_0) \times B_1(y_0)$$

$$\alpha_2 = A_2(x_0) \times B_2(y_0),$$

where the logical *and* can be modelled by any continuous t-norm, e.g

$$\alpha_1 = A_1(x_0) \wedge B_1(y_0)$$

$$\alpha_2 = A_2(x_0) \wedge B_2(y_0),$$

then the individual rule outputs are derived from the relationships

$$z_1 = a_1 x_0 + b_1 y_0, \ z_2 = a_2 x_0 + b_2 y_0$$

and the crisp control action is expressed as

$$z_0 = \frac{\alpha_1 z_1 + \alpha_2 z_2}{\alpha_1 + \alpha_2} = \beta_1 z_1 + \beta_2 z_2$$

where β_1 and β_1 are the normalized values of α_1 and α_2 with respect to the sum $(\alpha_1 + \alpha_2)$, i.e.

$$\beta_1 = \frac{\alpha_1}{\alpha_1 + \alpha_2}, \quad \beta_2 = \frac{\alpha_2}{\alpha_1 + \alpha_2}.$$

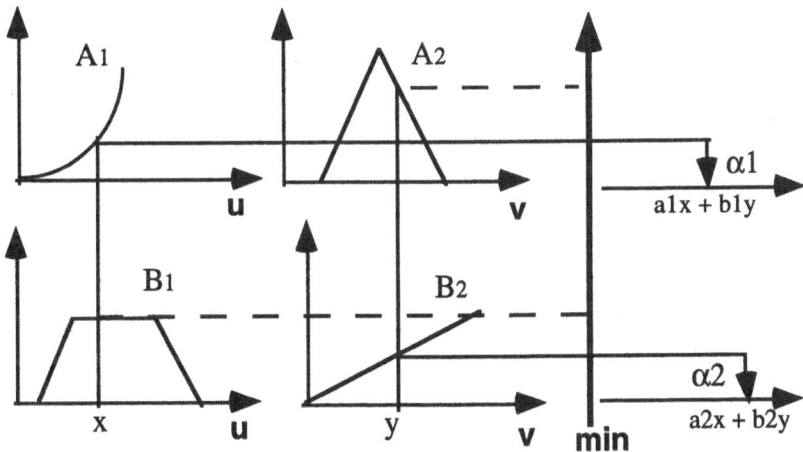

Fig. 3.21. Sugeno's inference mechanism.

A hybrid neural net computationally identical to this type of reasoning is shown in the Fig. 3.22.

For simplicity, we have assumed only two rules, and two linguistic values for each input variable.

- **Layer 1** The output of the node is the degree to which the given input satisfies the linguistic label associated to this node. Usually, we choose bell-shaped membership functions

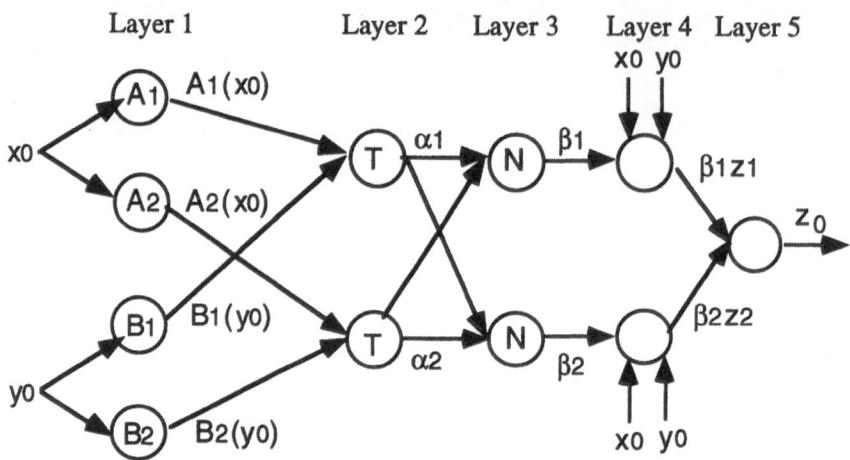

Fig. 3.22. ANFIS architecture for Sugeno's reasoning method.

$$A_i(u) = \exp\left[-\frac{1}{2}\left(\frac{u - a_{i1}}{b_{i1}}\right)^2\right],$$

$$B_i(v) = \exp\left[-\frac{1}{2}\left(\frac{v - a_{i2}}{b_{i2}}\right)^2\right],$$

to represent the linguistic terms, where

$$\{a_{i1}, a_{i2}, b_{i1}, b_{i2}\}$$

is the *parameter set*. As the values of these parameters change, the bell-shaped functions vary accordingly, thus exhibiting various forms of membership functions on linguistic labels A_i and B_i. In fact, any continuous, such as trapezoidal and triangular-shaped membership functions, are also quantified candidates for node functions in this layer. Parameters in this layer are referred to as *premise parameters*.

- **Layer 2** Each node computes the firing strength of the associated rule. The output of top neuron is

$$\alpha_1 = A_1(x_0) \times B_1(y_0) = A_1(x_0) \wedge B_1(y_0),$$

and the output of the bottom neuron is

$$\alpha_2 = A_2(x_0) \times B_2(y_0) = A_2(x_0) \wedge B_2(y_0)$$

Both node in this layer is labeled by T, because we can choose other t-norms for modeling the logical *and* operator. The nodes of this layer are called *rule nodes*.

- **Layer 3** Every node in this layer is labeled by N to indicate the normalization of the firing levels.
  The output of top neuron is the normalized (with respect to the sum of firing levels) firing level of the first rule

$$\beta_1 = \frac{\alpha_1}{\alpha_1 + \alpha_2},$$

and the output of the bottom neuron is the normalized firing level of the second rule

$$\beta_2 = \frac{\alpha_2}{\alpha_1 + \alpha_2},$$

- **Layer 4** The output of top neuron is the product of the normalized firing level and the individual rule output of the first rule

$$\beta_1 z_1 = \beta_1(a_1 x_0 + b_1 y_0),$$

The output of top neuron is the product of the normalized firing level and the individual rule output of the second rule

$$\beta_2 z_2 = \beta_2(a_2 x_0 + b_2 y_0),$$

- **Layer 5** The single node in this layer computes the overall system output as the sum of all incoming signals, i.e.

$$z_0 = \beta_1 z_1 + \beta_2 z_2.$$

If a crisp traing set $\{(x^k, y^k), \ k = 1, \dots, K\}$ is given then the parameters of the hybrid neural net (which determine the shape of the membership functions of the premises) can be learned by descent-type methods. This architecture and learning procedure is called ANFIS (adaptive-network-based fuzzy inference system) by Jang [106].

The error function for pattern k can be given by

$$E_k = (y^k - o^k)^2$$

where y^k is the desired output and o^k is the computed output by the hybrid neural net.

If the membership functions are of triangular form

$$A_i(u) = \begin{cases} 1 - \dfrac{a_{i1} - u}{a_{i2}} & \text{if } a_{i1} - a_{i2} \leq u \leq a_{i1} \\ 1 - \dfrac{u - a_{i1}}{a_{i3}} & \text{if } a_{i1} \leq u \leq a_{i1} + a_{i3} \\ 0 & \text{otherwise} \end{cases}$$

$$B_i(v) = \begin{cases} 1 - \dfrac{b_{i1} - v}{b_{i2}} & \text{if } b_{i1} - b_{i2} \le v \le b_{i1} \\[2mm] 1 - \dfrac{v - b_{i1}}{b_{i3}} & \text{if } b_{i1} \le v \le b_{i1} + b_{i3} \\[2mm] 0 & \text{otherwise} \end{cases}$$

then we can start the learning process from the initial values (see Fig. 3.23).

Generally, the initial values of the parameters are set in such a way that the membership functions along each axis satisfy ε-completeness, normality and convexity.

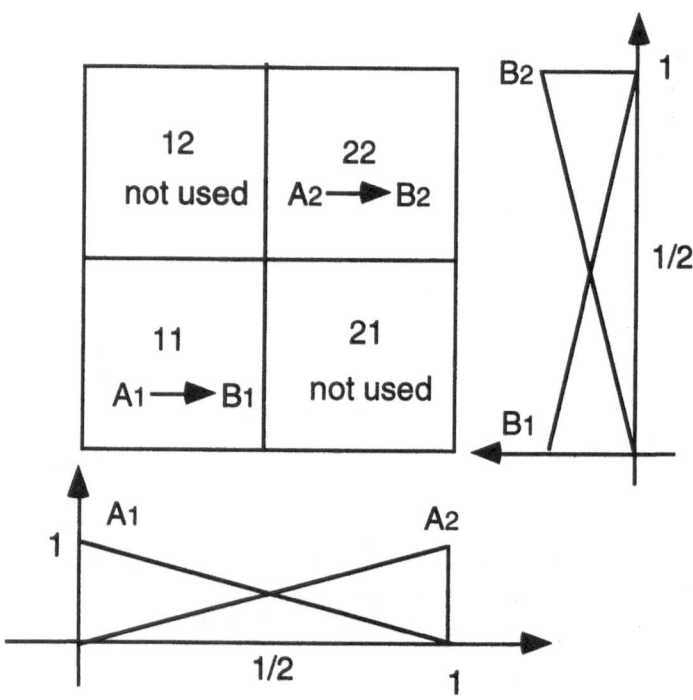

Fig. 3.23. Two-input ANFIS with four fuzzy rules.

Nauck, Klawonn and Kruse [150] initialize the network with all rules that can be constructed out of all combinations of input and output membership functions. During the learning process all hidden nodes (rule nodes) that are not used or produce counterproductive results are removed from the network.

It should be noted however, that these tuning methods have a weak point, because the convergence of tuning depends on the initial condition.

Exercise 3.3.1 *Construct a hybid neural net implementing* Tsukumato's *reasoning mechanism with two input variable, two linguistiuc values for each input variables and two fuzzy IF-THEN rules.*

Exercise 3.3.2 *Construct a hybid neural net implementing* Larsen's *reasoning mechanism with two input variable, two linguistiuc values for each input variables and two fuzzy IF-THEN rules.*

Exercise 3.3.3 *Construct a hybid neural net implementing* Mamdani's *reasoning mechanism with two input variables, two linguistiuc values for each input variable and two fuzzy IF-THEN rules.*

3.4 Computation of fuzzy logic inferences by hybrid neural net

Keller, Yager and Tahani [112] proposed the following hybrid neural network architecture for computation of fuzzy logic inferences. Each basic network structure implements a single rule in the rule base of the form

$$\text{If } x_1 \text{ is } A_1 \text{ and } \dots \text{ and } x_n \text{ is } A_n \text{ then } y \text{ is } B$$

The fuzzy sets which characterize the facts

$$x_1 \text{ is } A_1' \text{ and } \dots \text{ and } x_n \text{ is } A_n'$$

are presented to the input layer of the network.

Let $[M_1, M_2]$ contain the support of all the A_i, plus the support of all the A' we might have as input to the system. Also, let $[N_1, N_2]$ contain the support of B, plus the support of all the B' we can obtain as outputs from the system. Let $M \geq 2$ and $N \geq$ be positive integers. Let

$$\nu_j = M_1 + (j-1)(M_2 - M_1)/(M-1),$$

$$\tau_i = N_1 + (i-1)(N_2 - N_1)/(N-1)$$

for $1 \leq i \leq N$ and $1 \leq j \leq M$. The fuzzy set A_i' is denoted by

$$A_i' = \{a_{i1}', \dots, a_{iM}'\}$$

these values being the membership grades of A_i' at sampled points

$$\{\nu_1, \dots, \nu_M\}$$

over its domain of discourse.

There are two variations of the activities in the antecedent clause checking layer. In both cases, each antecedent clause of the rule determines the weights.

For the first variation, the weights w_{ij} are the fuzzy set complement of the antecedent clause, i.e., for the i-th clause

$$w_{ij} = 1 - a_{ij}$$

The weights are chosen this way because the first layer of the hybrid neural net will generate a measure of disagreement between the input possibility distribution and the antecedent clause distribution. This is done so that as the input moves away from the antecedent, the amount of disagreement will rise to one. Hence, if each node calculates the similarity between the input and the complement of the antecedent, then we will produce such a local measure of disagreement. The next layer combines this evidence.

The purpuse of the node is to determine the amount of disagreement present between the antecedent clause and the corresponding input data. If the combination at the k-th node is denoted by d_k, then

$$d_k = \max_j \{w_{kj} * a'_{kj}\} = \max_j \{(1 - a_{kj}) * a'_{kj}\}$$

where $*$ corresponds to the operations of multiplication, or minimum.

$$d_k^1 = \max_j \{(1 - a_{kj})a'_{kj}\}$$

or

$$d_k^2 = \max_j \min \{(1 - a_{kj}), a'_{kj}\}$$

The second form for the antecedent clause checking layer uses the fuzzy set A_k themselves as the weights, i.e. in this case

$$d_k^3 = \max_j |a_{kj} - a'_{kj}|$$

the sup-norm difference of the two functions A_k and A'_k.

We set the activation function to the identity, that is, the output of the node is the value obtained from the combinations of inputs and weights.

The disagreement values for each node are combined at the next level to produce an overall level of disagreement between the antecedent clauses and the input data. The disagreement values provide inhibiting signals for the firing of the rule. The weights α_i on these links correspond to the importance of the various antecedent clauses. The combination node then computes

$$1 - t = 1 - \max_i \{\alpha_i d_i\}.$$

Another option is to compute t as the weighted sum of the α_i's and d_i's.

The weights u_i on the output nodes carry the information from the consequent of rule. If the proposition "y is B" is characterized by the discrete possibility distribution

$$B = \{b_1, \dots, b_N\}$$

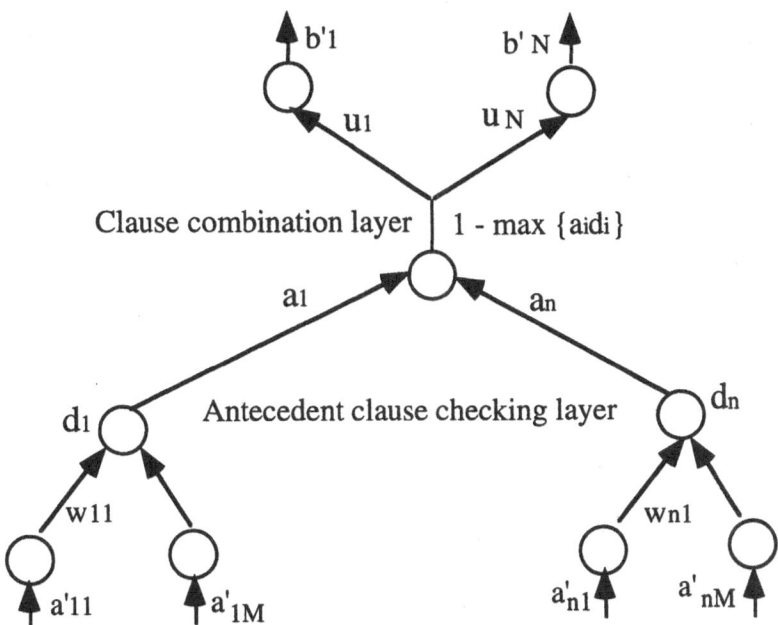

Fig. 3.24. Hybrid neural network configuration for fuzzy logic inference.

where $b_i = B(\tau_i)$ for all τ_i in the domain of B, then

$$u_i = 1 - b_i.$$

Each output node forms the value

$$b'_i = 1 - u_i(1 - t) = 1 - (1 - b_i)(1 - t) = b_i + t - b_i t$$

From this equation, it is clear that if $t = 0$, then the rule fires with conclusion
"y is B" exactly. On the other hand, if the the total disagreement is one, then
the conclusion of firing the rule is a possibility distribution composed entirely
of 1's, hence the conclusion is "y is **unknown**".

This network extends classical (crisp) logic to fuzzy logic as shown by
the following theorems. For simplicity, in each theorem we consider a single
antecedent clause rule of the form

If x is A then y is B.

Suppose A is a crsip subset of its domain of discourse. Let us denote χ_A the
characteristic function of A, i.e

$$\chi_A(u) = \begin{cases} 1 \text{ if } u \in A \\ 0 \text{ otherwise} \end{cases}$$

and let A be represented by $\{a_1, \ldots, a_M\}$, where $a_i = \chi_A(\nu_i)$.

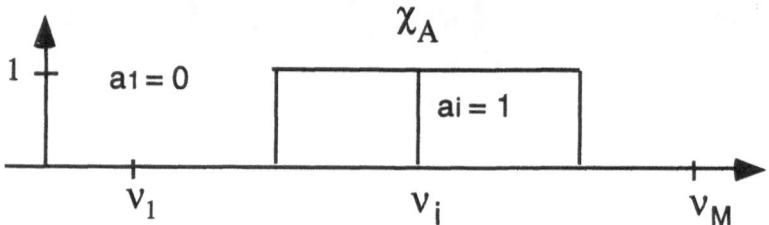

Fig. 3.25. Representation of a crisp subset A.

Theorem 3.4.1 *[112] In the single antecedent clause rule, suppose A is a crisp subset of its domain of discourse. Then the fuzzy logic inference network produces the standard modus ponens result, i.e. if the input "x is A'" is such that $A' = A$, then the network results in "y is B".*

Proof. Let A be represented by $\{a_1, \ldots, a_M\}$, where $a_i = \chi_A(\nu_i)$. Then from $A' = A$ it follows that

$$d^1 = \max_j\{(1 - a_j)a'_j\} = \max_j\{(1 - a_j)a_j\} = 0$$

since wherever $a_j = 1$ then $1 - a_j = 0$ and vice versa. Similarly,

$$d^2 = \max_j\{(1 - a_j) \wedge a'_j\} = \max_j\{(1 - a_j) \wedge a_j\} = 0.$$

Finally, for d^3 we have

$$d^3 = \max_j |a_j - a'_j| = \max_j |a_j - a_j| = 0.$$

Hence, at the combination node, $t = 0$, and so the output layer will produce

$$b'_i = b_i + t - b_i t = b_i + t(1 - b_i) = b_i.$$

Theorem 3.4.2 *[112] Consider the inference network which uses d^1 or d^2 for clause checking. Suppose that A and A' are proper crisp subsets of their domain of discourse and let $co(A) = \{x | x \notin A\}$ denote the complement of A.*

*(i) If $co(A) \cap A' \neq \emptyset$, then the network produces the result "y is **unknown**", i.e. a possibility distribution for y is equal to 1.*

(ii) If $A' \subset A$ (i.e. A' is more specific than A), then the result is "y is B".

Proof. (i) Since $co(A) \cap A' \neq \emptyset$, there exists a point ν_i in the domain such that

$$\chi_{co(A)}(\nu_i) = \chi_{A'}(\nu_i) = 1$$

In other words the weight $w_i = 1 - a_i = 1$ and $a'_i = 1$. Hence

$$d^i = \max_j\{(1 - a_j)a'_j\} = \max_j\{(1 - a_j)a_j\} = 1$$

for $i = 1, 2$. So, we have $t = 1$ at the clause combination node and

$$b'_i = b_i + t - b_i t = b_i + (1 - b_i) = 1.$$

(ii) Now suppose that $A' \subset A$. Then $A' \cap co(A) = \emptyset$, and so, $d^1 = d^2 = 0$, producing the result "y is B".

Theorem 3.4.3 [112] *Consider the inference network which uses d^3 for clause checking. Suppose that A and A' are proper crisp subsets of their domain of discourse such that $A' \neq A$, then the network result is "y is unknown".*

Proof. (i) Since $A' \neq A$, there exists a point ν_i such that

$$a_i = A(\nu_i) \neq A'(\nu_i) = a'_i.$$

Then $d^3 = \max_i\{|a_i - a'_i|\} = 1$, which ensures that the result is "y is unknown".

Theorem 3.4.4 [112] *(monotonocity theorem) Consider the single clause inference network using d^1 or d^2 for clause checking. Suppose that A, A' and A'' are three fuzzy sets such that*

$$A'' \subset A' \subset A.$$

Let the results of inference with inputs "x is A'" and "x is A''" be "y is B'" and "y is B''" respectively. Then

$$B \subset B'' \subset B'$$

that is, B'' is closer to B than B'.

Proof. For each ν_i in the domain of discourse of A,

$$0 \leq a''_j = A''(\nu_j) \leq a'_j = A'(\nu_j) \leq a_j = A(\nu_j)$$

Therefore, at the clause checking node,

$$(d^i)'' = \max_j\{(1 - a_j) * a''_j\} \leq \max_j\{(1 - a_j) * a'_j\} = (d^i)'$$

for $i = 1, 2$. Hence, $t'' \leq t'$. Finally,

$$b''_i = b_i + t'' - b_i t'' \leq b_i + t' - b_i t'$$

Clearly, from the above equations, both b''_i and b'_i are larger or equal than b_i. This completes the proof.

Intuitively, this theorem states that as the input becomes more specific, the output converges to the consequent.

Having to discretize all the fuzzy sets in a fuzzy expert system can lead to an enormous hybrid neural net as we have seen above. It is the use of an hybrid neural net that dictates the discretization because it processes real numbers. We can obtain much smaller networks if we use fuzzy neural nets.

Drawing heavily on Buckley and Hayashi [33] we represent fuzzy expert systems as *hybrid fuzzy neural networks*.

We recall that a hybrid fuzzy neural network is a neural network with fuzzy signals and/or fuzzy weights. However,

- we can combine X_i and W_i using a t-norm, t-conorm, or some other continuous operation,
- we can aggregate P_1 and P_2 with a t-norm, t-conorm, or any other continuous function
- f can be any function from input to output.

Suppose the fuzzy expert system has only one block of rules of the form

$$\Re_i: \text{If } x = A_i \text{ then } y \text{ is } B_i, \ 1 \le x \le n.$$

The input to the system is "x is A'" with final conclusion "y is B'". We again consider two cases:

- **Case 1** Combine all rules into one rule and fire.
  We first obtain a fuzzy relation R_k to model the implication in each rule, $1 \le k \le n$. Let $R_k(x, y)$ be some function of $A_k(x)$ and $B_k(y)$ for x in $[M_1, M_2]$, y in $[N_1, N_2]$. For example this function can be Mamdani's min

$$R_k(x, y) = \min\{A_k(x), B_k(y)\}.$$

Then we combine all the R_k, $1 \le k \le n$, into one fuzzy relation R on $[M_1, M_2] \times [N_1, N_2]$. For example $R(x, y)$ could be the maximum of the $R_k(x, y)$, $1 \le k \le n$

$$R(x, y) = \max R_k(x, y)$$

Given A' we compute the output B' by the compositional rule of inference as

$$B' = A' \circ R$$

For example, one could have

$$B'(y) = \sup_{M_1 \le x \le M_2} \min\{A'(x), R(x, y)\}$$

for each $y \in [N_1, N_2]$. A hybrid fuzzy neural net, the same as this fuzzy expert system, is shown in Fig. 3.26.

There is only one neuron with input weight equal to one. The transfer functions (which maps fuzzy sets into fuzzy sets) inside the neuron is the

Fig. 3.26. Combine the rules.

fuzzy relation R. So, we have input A' to the neuron with its output $B' = A' \circ R$.

We obtained the simplest possible hybrid fuzzy neural net for the fuzzy expert system. The major drawback is that there is no hardware available to implement fuzzy neural nets.

- **Case 2** Fire the rules individually and then combine their results.
  We first compose A' with each R_k to get B'_k, the conclusion of the k-th rule, and then combine all the B'_k into one final conclusion B'. Let B'_k be defined by the compositional rule of inference as

$$B'_k = A' \circ R_k$$

for all $y \in [N_1, N_2]$. Then

$$B'(y) = \mathrm{Agg}(B'_1(y), \dots, B'_n(y))$$

for some aggregation operator Agg.
A hybrid fuzzy neural net the same as this fuzzy expert system is displayed in Fig. 3.27.

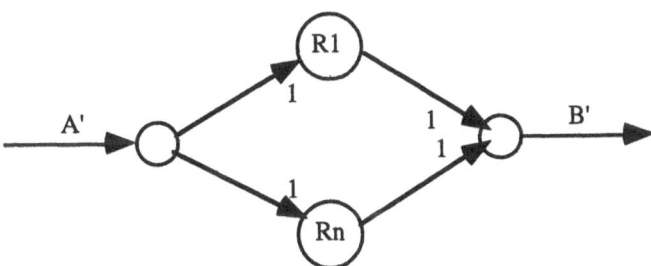

Fig. 3.27. Fire rules first.

All the weights are equal to one and the fuzzy relations R_k are the transfer functions for the neurons in the hidden layer. The input signals to the output neuron are the B'_k which are aggregated by Agg. The transfer function in the output neuron is the identity (no change) function.

3.5 Trainable neural nets for fuzzy IF-THEN rules

In this section we present some methods for implementing fuzzy IF-THEN rules by trainable neural network architectures. Consider a block of fuzzy rules

$$\Re_i : \text{If } x \text{ is } A_i, \text{ then } y \text{ is } B_i \tag{3.1}$$

where A_i and B_i are fuzzy numbers, $i = 1, \dots, n$.

Each rule in (3.1) can be interpreted as a training pattern for a multilayer neural network, where the antecedent part of the rule is the input and the consequence part of the rule is the desired output of the neural net.

The training set derived from (3.1) can be written in the form

$$\{(A_1, B_1), \dots, (A_n, B_n)\}$$

If we are given a two-input-single-output (MISO) fuzzy systems of the form

$$\Re_i : \text{If } x \text{ is } A_i \text{ and } y \text{ is } B_i, \text{ then } z \text{ is } C_i$$

where A_i, B_i and C_i are fuzzy numbers, $i = 1, \dots, n$.

Then the input/output training pairs for the neural net are the following

$$\{(A_i, B_i), C_i\}, \ 1 \le i \le n.$$

If we are given a two-input-two-output (MIMO) fuzzy systems of the form

$$\Re_i : \text{If } x \text{ is } A_i \text{ and } y \text{ is } B_i, \text{ then } r \text{ is } C_i \text{ and } s \text{ is } D_i$$

where A_i, B_i, C_i and D_i are fuzzy numbers, $i = 1, \dots, n$.

Then the input/output training pairs for the neural net are the following

$$\{(A_i, B_i), (C_i, D_i)\}, \ 1 \le i \le n.$$

There are two main approaches to implement fuzzy IF-THEN rules (3.1) by *standard error backpropagation network.*

- In the method proposed by Umano and Ezawa [192] a fuzzy set is represented by a finite number of its membership values.

  Let $[\alpha_1, \alpha_2]$ contain the support of all the A_i, plus the support of all the A' we might have as input to the system. Also, let $[\beta_1, \beta_2]$ contain the support of all the B_i, plus the support of all the B' we can obtain as outputs from the system. $i = 1, \dots, n$. Let $M \ge 2$ and $N \ge$ be positive integers. Let

  $$x_j = \alpha_1 + (j - 1)\frac{\alpha_2 - \alpha_1}{N - 1},$$

  $$y_i = \beta_1 + (i - 1)\frac{\beta_2 - \beta_1}{M - 1}$$

  for $1 \le i \le M$ and $1 \le j \le N$.

  A discrete version of the continuous training set is consists of the input/output pairs

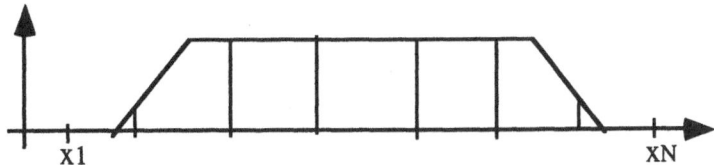

Fig. 3.28. Representation of a fuzzy number by membership values.

$$\{(A_i(x_1), \ldots, A_i(x_N)), \ (B_i(y_1), \ldots, B_i(y_M))\}$$

for $i = 1, \ldots, n$.

Using the notations $a_{ij} = A_i(x_j)$ and $b_{ij} = B_i(y_j)$ our fuzzy neural network turns into an N input and M output crisp network, which can be trained by the generalized delta rule.

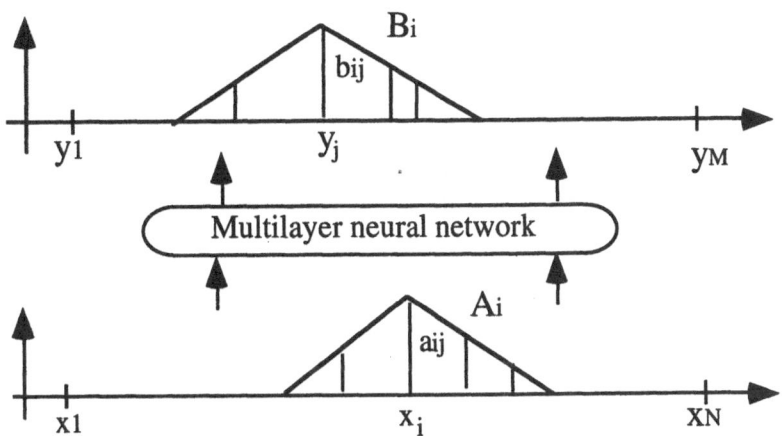

Fig. 3.29. A network trained on membership values fuzzy numbers.

Example 3.5.1 *Assume our fuzzy rule base consists of three rules*

$$\Re_1 : \quad \text{If } x \text{ is } small \text{ then } y \text{ is } negative,$$

$$\Re_2 : \quad \text{If } x \text{ is } medium \text{ then } y \text{ is } about \ zero,$$

$$\Re_3 : \quad \text{If } x \text{ is } big \text{ then } y \text{ is } positive,$$

where the membership functions of fuzzy terms are defined by

$$\mu_{small}(u) = \begin{cases} 1 - 2u & \text{if } 0 \le u \le 1/2 \\ 0 & \text{otherwise} \end{cases}$$

$$\mu_{big}(u) = \begin{cases} 2u - 1 & \text{if } 1/2 \le u \le 1 \\ 0 & \text{otherwise} \end{cases}$$

$$\mu_{medium}(u) = \begin{cases} 1 - 2|u - 1/2| & \text{if } 0 \le u \le 1 \\ 0 & \text{otherwise} \end{cases}$$

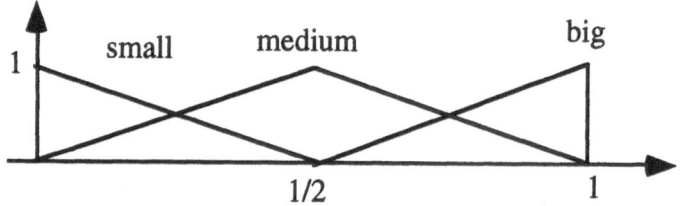

Fig. 3.30. Membership functions for *small, medium* and *big*.

$$\mu_{negative}(u) = \begin{cases} -u & \text{if } -1 \le u \le 0 \\ 0 & \text{otherwise} \end{cases}$$

$$\mu_{about\ zero}(u) = \begin{cases} 1 - 2|u| & \text{if } -1/2 \le u \le 1/2 \\ 0 & \text{otherwise} \end{cases}$$

$$\mu_{positive}(u) = \begin{cases} u & \text{if } 0 \le u \le 1 \\ 0 & \text{otherwise} \end{cases}$$

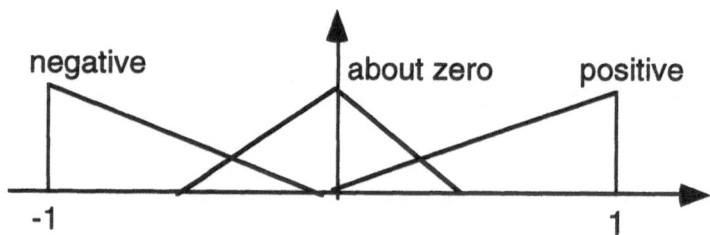

Fig. 3.31. Membership functions for negative, about zero and positive.

The training set derived from this rule base can be written in the form

$$\{(small, negative), (medium, about\ zero), (big, positive)\}$$

Let $[0, 1]$ contain the support of all the fuzzy sets we might have as input to the system. Also, let $[-1, 1]$ contain the support of all the fuzzy sets we can obtain as outputs from the system. Let $M = N = 5$ and

$$x_j = (j-1)/4$$

for $1 \leq j \leq 5$, and

$$y_i = -1 + (i-1)2/4 = -1 + (i-1)/2 = -3/2 + i/2$$

for $1 \leq i \leq M$ and $1 \leq j \leq N$. Plugging into numerical values we get $x_1 = 0$, $x_2 = 0.25$, $x_3 = 0.5$, $x_4 = 0.75$ and $x_5 = 1$; and $y_1 = -1$, $y_2 = -0.5$, $y_3 = 0$, $y_4 = 0.5$ and $y_5 = 1$.

A discrete version of the continuous training set is consists of three input/output pairs

$$\{(a_{11}, \dots, a_{15}), (b_{11}, \dots, b_{15})\}$$
$$\{(a_{21}, \dots, a_{25}), (b_{21}, \dots, b_{25})\}$$
$$\{(a_{31}, \dots, a_{35}), (b_{31}, \dots, b_{35})\}$$

where

$$a_{1j} = \mu_{small}(x_j), \quad a_{2j} = \mu_{medium}(x_j), \quad a_{3j} = \mu_{big}(x_j)$$

for $j = 1, \dots, 5$, and

$$b_{1i} = \mu_{negative}(y_i), \quad b_{2i} = \mu_{about\ zero}(y_i), \quad b_{3i} = \mu_{positive}(y_i)$$

for $i = 1, \dots, 5$. Plugging into numerical values we obtain the following training set for a standard backpropagation network

$$\{(1, 0.5, 0, 0, 0), (1, 0.5, 0, 0, 0)\}$$
$$\{(0, 0.5, 1, 0.5, 0), (0, 0, 1, 0, 0)\}$$
$$\{(0, 0, 0, 0.5, 1), (0, 0, 0, 0.5, 1)\}.$$

- **Uehara and Fujise** [191] use finite number of α-level sets to represent fuzzy numbers. Let $M \geq 2$ and let

$$\alpha_j = (j-1)/(M-1), \quad j = 1, \dots, M$$

be a partition of $[0, 1]$. Let $[A_i]^{\alpha_j}$ denote the α_j-level set of fuzzy numbers A_i

$$[A_i]^{\alpha_j} = \{u \mid A_i(u) \geq \alpha_j\} = [a_{ij}^L, a_{ij}^R]$$

for $j = 1, \dots, M$ and $[B_i]^{\alpha_j}$ denote the α_j-level set of fuzzy number B_i

$$[B_i]^{\alpha_j} = \{u \mid B_i(u) \geq \alpha_j\} = [b_{ij}^L, b_{ij}^R]$$

for $j = 1, \dots, M$. Then the discrete version of the continuous training set is consists of the input/output pairs

$$\{(a_{i1}^L, a_{i1}^R, \dots, a_{iM}^L, a_{iM}^R), (b_{i1}^L, b_{i1}^R, \dots, b_{iM}^L, b_{iM}^R)\}$$

where for $i = 1, \dots, n$.

The number of inputs and outputs depend on the number of α-level sets considered. For example, in Fig. 3.32, the fuzzy number A_i is represented by seven level sets, i.e. by a fourteen-dimensional vector of real numbers.

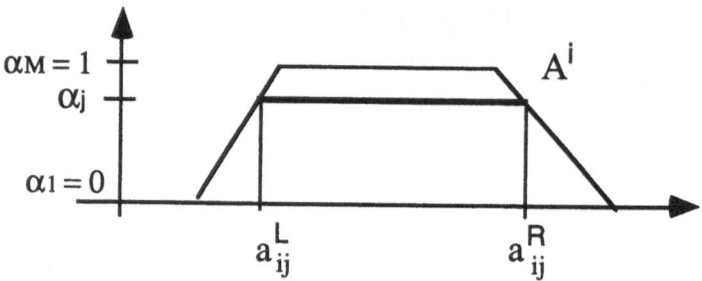

Fig. 3.32. Representation of a fuzzy number by α-level sets.

Example 3.5.2 *Assume our fuzzy rule base consists of three rules*

$$\Re_1 : \quad \text{If } x \text{ is } small \text{ then } y \text{ is } small,$$
$$\Re_2 : \quad \text{If } x \text{ is } medium \text{ then } y \text{ is } medium,$$
$$\Re_3 : \quad \text{If } x \text{ is } big \text{ then } y \text{ is } big,$$

where the membership functions of fuzzy terms are defined by

$$\mu_{small}(u) = \begin{cases} 1 - \dfrac{u - 0.2}{0.3} & \text{if } 0.2 \leq u \leq 1/2 \\ 1 & \text{if } 0 \leq u \leq 0.2 \\ 0 & \text{otherwise} \end{cases}$$

$$\mu_{big}(u) = \begin{cases} 1 - \dfrac{0.8 - u}{0.3} & \text{if } 1/2 \leq u \leq 0.8 \\ 1 & \text{if } 0.8 \leq u \leq 1 \\ 0 & \text{otherwise} \end{cases}$$

$$\mu_{medium}(u) = \begin{cases} 1 - 4|u - 1/2| & \text{if } 0.25 \leq u \leq 0.75 \\ 0 & \text{otherwise} \end{cases}$$

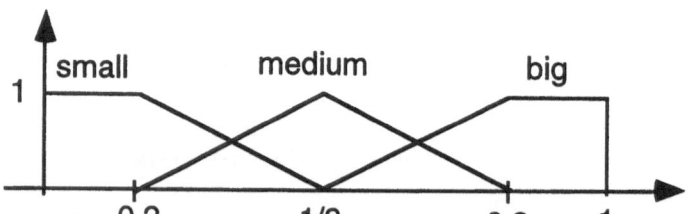

Fig. 3.33. Membership functions for *small*, *medium* and *big*.

Let $M = 6$ and let

$$\alpha_j = \frac{j-1}{4}, \; j = 1, \ldots, 6$$

be a partition of $[0, 1]$. Plugging into numerical values we get $\alpha_1 = 0$, $\alpha_2 = 0.2$, $\alpha_3 = 0.4$, $\alpha_4 = 0.6$, $\alpha_5 = 0.8$ and $\alpha_6 = 1$. Then the discrete version of the continuous training set is consists of the following three input/output pairs

$$\{(a_{11}^L, a_{11}^R, \ldots, a_{16}^L, a_{16}^R), (b_{11}^L, b_{11}^R, \ldots, b_{16}^L, b_{16}^R)\}$$

$$\{(a_{21}^L, a_{211}^R, \ldots, a_{26}^L, a_{26}^R), (b_{21}^L, b_{21}^R, \ldots, b_{26}^L, b_{26}^R)\}$$

$$\{(a_{31}^L, a_{31}^R, \ldots, a_{36}^L, a_{36}^R), (b_{31}^L, b_{31}^R, \ldots, b_{36}^L, b_{36}^R)\}$$

where

$$[a_{1j}^L, a_{1j}^R] = [b_{1j}^L, b_{1j}^R] = [small]^{\alpha_j}$$

$$[a_{2j}^L, a_{2j}^R] = [b_{2j}^L, b_{2j}^R] = [medium]^{\alpha_j}$$

and

$$[a_{3j}^L, a_{3j}^R] = [b_{3j}^L, b_{3j}^R] = [big]^{\alpha_j}.$$

It is easy to see that $a_{1j}^L = b_{1j}^L = 0$ and $a_{3j}^R = b_{3j}^R = 1$ for $1 \leq j \leq 6$. Plugging into numerical values we obtain the following training set

$$\{(0, 0.5, 0, 0.44, 0, 0.38, 0, 0.32, 0, 0.26, 0, 0.2),$$

$$(0, 0.5, 0, 0.44, 0, 0.38, 0, 0.32, 0, 0.26, 0, 0.2)\}$$

$$\{(0.5, 1, 0.56, 1, 0.62, 1, 0.68, 1, 0.74, 1, 0.8, 1),$$

$$(0.5, 1, 0.56, 1, 0.62, 1, 0.68, 1, 0.74, 1, 0.8, 1)\}$$

$$\{(0.25, 0.75, 0.3, 0.7, 0.35, 0.65, 0.4, 0.6, 0.45, 0.55, 0.5, 0.5),$$

$$(0.25, 0.75, 0.3, 0.7, 0.35, 0.65, 0.4, 0.6, 0.45, 0.55, 0.5, 0.5)\}.$$

Exercise 3.5.1 *Assume our fuzzy rule base consists of three rules*

$\Re_1$:   *If x_1 is small and x_2 is small then y is small,*

$\Re_2$:   *If x_1 is medium and x_2 is medium then y is medium,*

$\Re_3$:   *If x_1 is big and x_2 is big then y is big,*

where the membership functions of fuzzy terms are defined by

$$\mu_{small}(u) = \begin{cases} 1 - \dfrac{u - 0.2}{0.3} & \text{if } 0.2 \leq u \leq 1/2 \\ 1 & \text{if } 0 \leq u \leq 0.2 \\ 0 & \text{otherwise} \end{cases}$$

$$\mu_{big}(u) = \begin{cases} 1 - \dfrac{0.8 - u}{0.3} & \text{if } 1/2 \le u \le 0.8 \\ 1 & \text{if } 0.8 \le u \le 1 \\ 0 & \text{otherwise} \end{cases}$$

$$\mu_{medium}(u) = \begin{cases} 1 - 4|u - 1/2| & \text{if } 0.25 \le u \le 0.75 \\ 0 & \text{otherwise} \end{cases}$$

Assume that $[0, 1]$ *contains the support of all the fuzzy sets we might have as input and output for the system. Derive training sets for standard backpropagation network from 10 selected membership values of fuzzy terms.*

Exercise 3.5.2 *Assume our fuzzy rule base consists of three rules*

$\Re_1$:   *If x_1 is small and x_2 is big then y is small,*

$\Re_2$:   *If x_1 is medium and x_2 is small then y is medium,*

$\Re_3$:   *If x_1 is big and x_2 is medium then y is big,*

where the membership functions of fuzzy terms are defined by

$$\mu_{small}(u) = \begin{cases} 1 - \dfrac{u - 0.2}{0.3} & \text{if } 0.2 \le u \le 1/2 \\ 1 & \text{if } 0 \le u \le 0.2 \\ 0 & \text{otherwise} \end{cases}$$

$$\mu_{big}(u) = \begin{cases} 1 - \dfrac{0.8 - u}{0.3} & \text{if } 1/2 \le u \le 0.8 \\ 1 & \text{if } 0.8 \le u \le 1 \\ 0 & \text{otherwise} \end{cases}$$

$$\mu_{medium}(u) = \begin{cases} 1 - 4|u - 1/2| & \text{if } 0.25 \le u \le 0.75 \\ 0 & \text{otherwise} \end{cases}$$

Assume that $[0, 1]$ *contains the support of all the fuzzy sets we might have as input and output for the system. Derive training sets for standard backpropagation network from 10 selected values of α-level sets of fuzzy terms.*

3.6 Implementation of fuzzy rules by regular FNN of Type 2

Ishibuchi, Kwon and Tanaka [96] proposed an approach to implement of fuzzy IF-THEN rules by training neural networks on fuzzy training patterns.

Assume we are given the following fuzzy rules

$$\Re_p : \text{If } x_1 \text{ is } A_{p1} \text{ and } \ldots \text{ and } x_n \text{ is } A_{pn} \text{ then } y \text{ is } B_p$$

where A_{ij} and B_i are fuzzy numbers and $p = 1, \ldots, m$.

The following training patterns can be derived from these fuzzy IF-THEN rules

$$\{(X_1, B_1), \ldots, (X_m, B_m)\} \tag{3.2}$$

where $X_p = (A_{p1}, \ldots, A_{pn})$ denotes the antecedent part and the fuzzy target output B_p is the consequent part of the rule.

Our learning task is to train a neural network from fuzzy training pattern set (3.2) by a regular fuzzy neural network of *Type 2* from Table 3.1.

Ishibuchi, Fujioka and Tanaka [90] propose the following extension of the standard backpropagation learning algorithm:

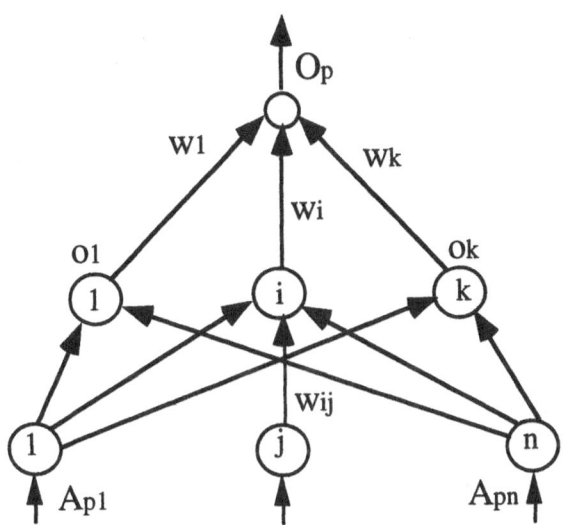

Fig. 3.34. Regular fuzzy neural network architecture of *Type 2*.

Suppose that A_p, the p-th training pattern, is presented to the network. The output of the i-th hidden unit, o_i, is computed as

$$o_{pi} = f\left(\sum_{j=1}^{n} w_{ij} A_{pj} \right).$$

For the output unit

$$O_p = f\left(\sum_{i=1}^{k} w_i o_{pi} \right)$$

where $f(t) = 1/(1 + \exp(-t))$ is a unipolar transfer function. It should be noted that the input-output relation of each unit is defined by the extension principle.

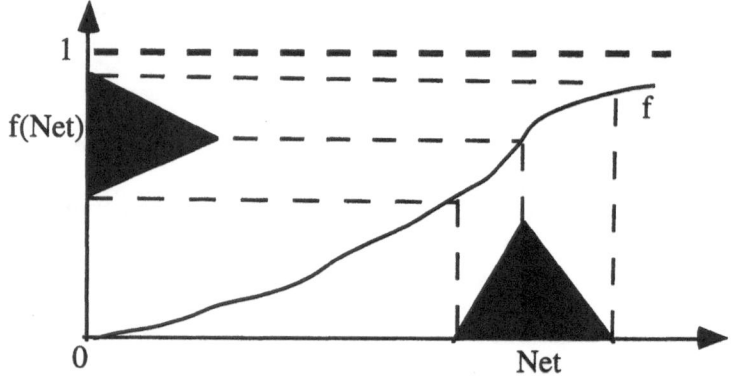

Fig. 3.35. Fuzzy input-output relation of each neuron.

Let us denote the α-level sets of the computed output O_p by

$$[O_p]^\alpha = [O_p^L(\alpha), O_p^R(\alpha)], \; \alpha \in [0, 1]$$

where $O_p^L(\alpha)$ denotes the left-hand side and $O_p^R(\alpha)$ denotes the right-hand side of the α-level sets of the computed output.

Since f is strictly monoton increasing we have

$$[O_p]^\alpha = \left[f\left(\sum_{i=1}^k w_i o_{pi} \right) \right]^\alpha = \left[f\left(\sum_{i=1}^k [w_i o_{pi}]^L(\alpha) \right), f\left(\sum_{i=1}^k [w_i o_{pi}]^R(\alpha) \right) \right],$$

where

$$[o_{pi}]^\alpha = \left[f\left(\sum_{j=1}^n w_{ij} A_{pj} \right) \right]^\alpha$$

$$= \left[f\left(\sum_{j=1}^n [w_{ij} A_{pj}]^L(\alpha) \right), f\left(\sum_{j=1}^n [w_{ij} A_{pj}]^R(\alpha) \right) \right].$$

The α-level sets of the target output B_p are denoted by

$$[B_p]^\alpha = [B_p^L(\alpha), B_p^R(\alpha)], \; \alpha \in [0, 1]$$

where $B_p^L(\alpha)$ denotes the left-hand side and $B_p^R(\alpha)$ denotes the right-hand side of the α-level sets of the desired output.

A cost function to be minimized is defined for each α-level set as follows

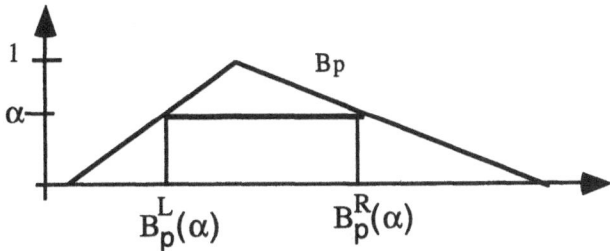

Fig. 3.36. An α-level set of the target output pattern B_p.

$$e_p(\alpha) := e_p^L(\alpha) + e_p^R(\alpha)$$

where

$$e_p^L(\alpha) = \frac{1}{2}(B_p^L(\alpha) - O_p^L(\alpha))^2$$

$$e_p^R(\alpha) = \frac{1}{2}(B_p^R(\alpha) - O_p^R(\alpha))^2$$

i.e. $e_p^L(\alpha)$ denotes the error between the left-hand sides of the α-level sets of the desired and the computed outputs, and $e_p^R(\alpha)$ denotes the error between the left right-hand sides of the α-level sets of the desired and the computed outputs.

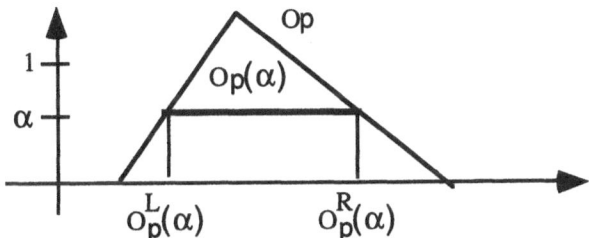

Fig. 3.37. An α-level set of the computed output pattern O_p.

Then the error function for the p-th training pattern is

$$e_p = \sum_\alpha \alpha e_p(\alpha) \tag{3.3}$$

Theoretically this cost function satisfies the following equation if we use infinite number of α-level sets in (3.3).

$$e_p \to 0 \text{ if and only if } O_p \to B_p$$

From the cost function $e_p(\alpha)$ the following learning rules can be derived

$$w_i := w_i - \eta\alpha\frac{\partial e_p(\alpha)}{\partial w_i},$$

for $i = 1, \ldots, k$ and

$$w_{ij} := w_{ij} - \eta\alpha\frac{\partial e_p(\alpha)}{\partial w_{ij}}$$

for $i = 1, \ldots, k$ and $j = 1, \ldots, n$.

The Reader can find the exact calculation of the partial derivatives

$$\frac{\partial e_p(\alpha)}{\partial w_i} \text{ and } \frac{\partial e_p(\alpha)}{\partial w_{ij}}$$

in [94], pp. 95-96.

3.7 Implementation of fuzzy rules by regular FNN of Type 3

Following [103] we show how to implement fuzzy IF-THEN rules by regular fuzzy neural nets of *Type 3* (fuzzy input/output signals and fuzzy weights) from Table 3.1.

Assume we are given the following fuzzy rules

$$\Re_p : \text{If } x_1 \text{ is } A_{p1} \text{ and } \ldots \text{ and } x_n \text{ is } A_{pn} \text{ then } y \text{ is } B_p$$

where A_{ij} and B_i are fuzzy numbers and $p = 1, \ldots, m$.

The following training patterns can be derived from these fuzzy IF-THEN rules

$$\{(X_1, B_1), \ldots, (X_m, B_m)\}$$

where $X_p = (A_{p1}, \ldots, A_{pn})$ denotes the antecedent part and the fuzzy target output B_p is the consequent part of the rule.

The output of the i-th hidden unit, o_i, is computed as

$$o_{pi} = f\left(\sum_{j=1}^n W_{ij} A_{pj}\right).$$

For the output unit

$$O_p = f\left(\sum_{i=1}^k W_i o_{pi}\right),$$

where A_{pj} is a fuzzy input, W_i and W_{ij} are fuzzy weights of triangular form and $f(t) = 1/(1 + \exp(-t))$ is a unipolar transfer function.

The fuzzy outputs for each unit is numerically calculated for the α-level sets of fuzzy inputs and weights. Let us denote the α-level set of the computed output O_p by

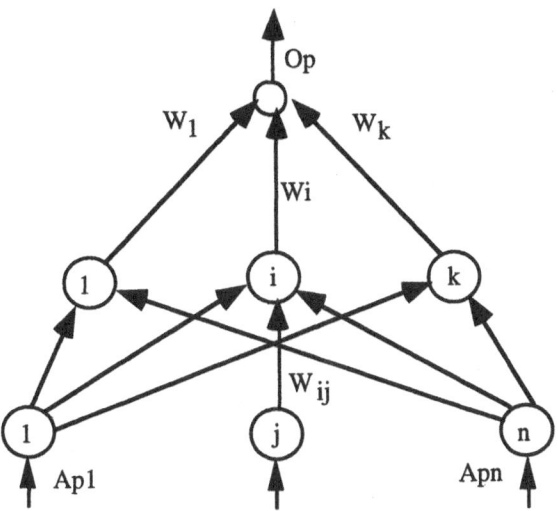

Fig. 3.38. Regular fuzzy neural network architecture of *Type 3*.

$$[O_p]^\alpha = [O_p^L(\alpha), O_p^R(\alpha)],$$

the α-level set of the target output B_p are denoted by

$$[B_p]^\alpha = [B_p^L(\alpha), B_p^R(\alpha)],$$

the α-level sets of the weights of the output unit are denoted by

$$[W_i]^\alpha = [W_i^L(\alpha), W_i^R(\alpha)],$$

the α-level sets of the weights of the hidden unit are denoted by

$$[W_{ij}]^\alpha = [W_{ij}^L(\alpha), W_{ij}^R(\alpha)],$$

for $\alpha \in [0,1]$, $i = 1, \ldots, k$ and $j = 1, \ldots, n$. Since f is strictly monoton increasing we have

$$[o_{pi}]^\alpha = [f(\sum_{j=1}^n W_{ij} A_{pj})]^\alpha = [f(\sum_{j=1}^n [W_{ij} A_{pj}]^L(\alpha)), f(\sum_{j=1}^n [W_{ij} A_{pj}]^R(\alpha))]$$

$$[O_p]^\alpha = [f(\sum_{i=1}^k W_i o_{pi})]^\alpha = [f(\sum_{i=1}^k [W_i o_{pi}]^L(\alpha)), f(\sum_{i=1}^k [W_i o_{pi}]^R(\alpha))]$$

A cost function to be minimized is defined for each α-level set as follows

$$e_p(\alpha) := e_p^L(\alpha) + e_p^R(\alpha)$$

where

$$e_p^L(\alpha) = \frac{1}{2}(B_p^L(\alpha) - O_p^L(\alpha))^2, \quad e_p^R(\alpha) = \frac{1}{2}(B_p^R(\alpha) - O_p^R(\alpha))^2$$

i.e. $e_p^L(\alpha)$ denotes the error between the left-hand sides of the α-level sets of the desired and the computed outputs, and $e_p^R(\alpha)$ denotes the error between the left right-hand sides of the α-level sets of the desired and the computed outputs.

Then the error function for the p-th training pattern is

$$e_p = \sum_\alpha \alpha e_p(\alpha) \tag{3.4}$$

Let us derive a learning algorithm of the fuzzy neural network from the error function $e_p(\alpha)$. Since fuzzy weights of hidden neurons are supposed to be of symmetrical triangular form, they can be represented by three parameters $W_{ij} = (w_{ij}^1, w_{ij}^2, w_{ij}^3)$. where w_{ij}^1 denotes the lower limit, w_{ij}^2 denotes the center and w_{ij}^3 denotes the upper limit of W_{ij}.

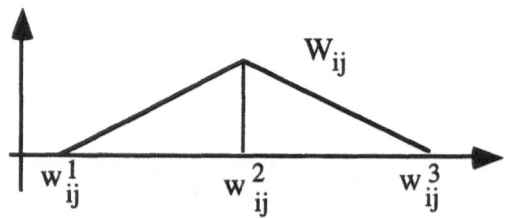

Fig. 3.39. Representation of W_{ij}.

Similarly, the weights of the output neuron can be represented by three parameter $W_i = (w_i^1, w_i^2, w_i^3)$, where w_i^1 denotes the lower limit, w_i^2 denotes the center and w_i^3 denotes the upper limit of W_i.

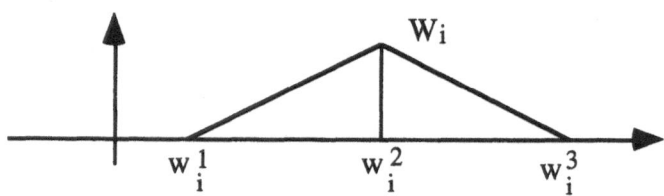

Fig. 3.40. Representation of W_i.

From simmetricity of W_{ij} and W_i it follows that

$$w_{ij}^2 = \frac{w_{ij}^1 + w_{ij}^3}{2}, \quad w_i^2 = \frac{w_i^1 + w_i^3}{2}$$

for $1 \leq j \leq n$ and $1 \leq i \leq k$.

From the cost function $e_p(\alpha)$ the following weights adjustments can be derived

$$\Delta w_i^1(t) = -\eta \frac{\partial e_p(\alpha)}{\partial w_i^1} + \beta \Delta w_i^1(t-1)$$

$$\Delta w_i^3(t) = -\eta \frac{\partial e_p(\alpha)}{\partial w_i^3} + \beta \Delta w_i^3(t-1)$$

where η is a learning constant, β is a momentum constant and t indexes the number of adjustments, for $i = 1, \ldots, k$, and

$$w_{ij}^1(t) = -\eta \frac{\partial e_p(\alpha)}{\partial w_{ij}^1} + \beta \Delta w_{ij}^1(t-1)$$

$$w_{ij}^3(t) = -\eta \frac{\partial e_p(\alpha)}{\partial w_{ij}^3} + \beta \Delta w_{ij}^3(t-1)$$

where η is a learning constant, β is a momentum constant and t indexes the number of adjustments, for $i = 1, \ldots, k$ and $j = 1, \ldots, n$.

The explicit calculation of above derivatives can be found in ([103], pp. 291-292).

The fuzzy weight $W_{ij} = (w_{ij}^1, w_{ij}^2, w_{ij}^3)$ is updated by the following rules

$$w_{ij}^1(t+1) = w_{ij}^1(t) + \Delta w_{ij}^1(t)$$

$$w_{ij}^3(t+1) = w_{ij}^3(t) + \Delta w_{ij}^3(t)$$

$$w_{ij}^2(t+1) = \frac{w_{ij}^1(t+1) + w_{ij}^3(t+1)}{2},$$

for $i = 1, \ldots, k$ and $j = 1, \ldots, n$. The fuzzy weight $W_i = (w_i^1, w_i^2, w_i^3)$ is updated in a similar manner, i.e.

$$w_i^1(t+1) = w_i^1(t) + \Delta w_i^1(t)$$

$$w_i^3(t+1) = w_i^3(t) + \Delta w_i^3(t)$$

$$w_i^2(t+1) = \frac{w_i^1(t+1) + w_i^3(t+1)}{2},$$

for $i = 1, \ldots, k$.

After the adjustment of W_i it can occur that its lower limit may become larger than its upper limit. In this case, we use the following simple heuristics

$$w_i^1(t+1) := \min\{w_i^1(t+1), w_i^3(t+1)\}$$

$$w_i^3(t+1) := \max\{w_i^1(t+1), w_i^3(t+1)\}.$$

We employ the same heuristics for W_{ij}.

$$w_{ij}^1(t+1) := \min\{w_{ij}^1(t+1), w_{ij}^3(t+1)\}$$
$$w_{ij}^3(t+1) := \max\{w_{ij}^1(t+1), w_{ij}^3(t+1)\}.$$

Let us assume that m input-output pairs

$$\{(X_1, B_1), \dots, (X_m, B_m)\}$$

where $X_p = (A_{p1}, \dots, A_{pn})$, are given as training data.

We also assume that M values of α-level sets are used for the learning of the fuzzy neural network.

Summary 3.7.1 *In this case, the learning algorithm can be summarized as follows:*

- **Step 1** *Fuzzy weights are initialized at small random values, the running error E is set to 0 and $E_{\max} > 0$ is chosen*
- **Step 2** *Repeat* **Step 3** *for $\alpha = \alpha_1, \alpha_2, \dots, \alpha_M$.*
- **Step 3** *Repeat the following procedures for $p = 1, 2, \dots, m$. Propagate X_p through the network and calculate the α-level set of the fuzzy output vector O_p. Adjust the fuzzy weights using the error function $e_p(\alpha)$.*
- **Step 4** *Cumulative cycle error is computed by adding the present error to E.*
- **Step 5** *The training cycle is completed. For $E < E_{\max}$ terminate the training session. If $E > E_{\max}$ then E is set to 0 and we initiate a new training cycle by going back to* **Step 2.**

3.8 Tuning fuzzy control parameters by neural nets

Fuzzy inference is applied to various problems. For the implementation of a fuzzy controller it is necessary to determine membership functions representing the linguistic terms of the linguistic inference rules.

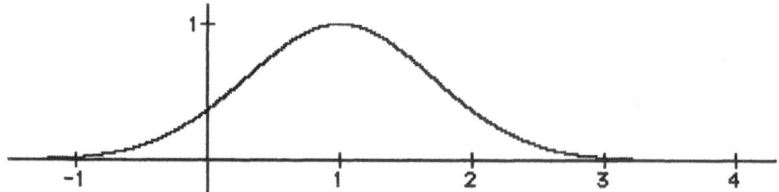

Fig. 3.41. Gaussian membership function for "x is approximately one".

For example, consider the linguistic term *approximately one*. Obviously, the corresponding fuzzy set should be a unimodal function reaching its maximum at the value one. Neither the shape, which could be triangular or Gaussian, nor the range, i.e. the support of the membership function is uniquely determined by *approximately one*.

Generally, a control expert has some idea about the range of the membership function, but he would not be able to argue about small changes of his specified range.

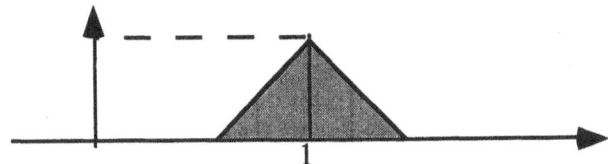

Fig. 3.42. Triangular membership function for "x is approximately one".

The effectivity of the fuzzy models representing nonlinear input-output relationships depends on the fuzzy partition of the input space.
Therefore, the tuning of membership functions becomes an import issue in fuzzy control. Since this tuning task can be viewed as an optimization problem neural networks and *genetic algorithms* [104] offer a possibility to solve this problem.

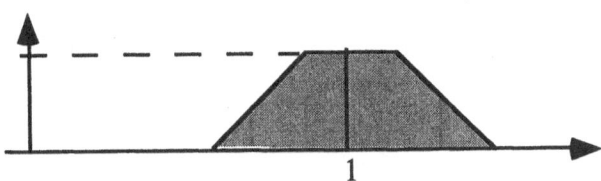

Fig. 3.43. Trapezoidal membership function for "x is approximately one".

A straightforward approach is to assume a certain shape for the membership functions which depends on different parameters that can be learned by a neural network. This idea was carried out in [162] where the membership functions are assumed to be symmetrical triangular functions depending on two parameters, one of them determining where the function reaches its maximum, the order giving the width of the support. Gaussian membership functions were used in [89].

Both approaches require a set training data in the form of correct input-output tuples and a specification of the rules including a preliminary definition of the corresponding membership functions.

We describe a simple method for learning of membership functions of the antecedent and consequent parts of fuzzy IF-THEN rules.

.Suppose the unknown nonlinear mapping to be realized by fuzzy systems can be represented as

$$y^k = f(x^k) = f(x_1^k, \ldots, x_n^k) \tag{3.5}$$

for $k = 1, \ldots, K$, i.e. we have the following training set

$$\{(x^1, y^1), \ldots, (x^K, y^K)\}$$

For modeling the unknown mapping in (3.5), we employ simplified fuzzy IF-THEN rules of the following type

$$\Re_i: \text{if } x_1 \text{ is } A_{i1} \text{ and } \ldots \text{ and } x_n \text{ is } A_{in} \text{ then } y = z_i, \tag{3.6}$$

$i = 1, \ldots, m$, where A_{ij} are fuzzy numbers of triangular form and z_i are real numbers. In this context, the word *simplified* means that the individual rule outputs are given by crisp numbers, and therefore, we can use their weighted sum (where the weights are the firing strengths of the corresponding rules) to obtain the overall system output.

Let o^k be the output from the fuzzy system corresponding to the input x^k. Suppose the firing level of the i-th rule, denoted by α_i, is defined by Larsen's product operator

$$\alpha_i = \prod_{j=1}^{n} A_{ij}(x_j^k)$$

(one can define other t-norm for modeling the logical connective *and*), and the output of the system is computed by the discrete center-of-gravity defuzzification method as

$$o^k = \frac{\sum_{i=1}^{m} \alpha_i z_i}{\sum_{i=1}^{m} \alpha_i}.$$

We define the measure of error for the k-th training pattern as usually

$$E_k = \frac{1}{2}(o^k - y^k)^2$$

where o^k is the computed output from the fuzzy system $\Re$ corresponding to the input pattern x^k and y^k is the desired output, $k = 1, \ldots, K$.

The steepest descent method is used to learn z_i in the consequent part of the fuzzy rule $\Re_i$. That is,

$$z_i(t+1) = z_i(t) - \eta \frac{\partial E_k}{\partial z_i} = z_i(t) - \eta(o^k - y^k)\frac{\alpha_i}{\alpha_1 + \cdots + \alpha_m},$$

for $i = 1, \ldots, m$, where η is the learning constant and t indexes the number of the adjustments of z_i.

Suppose that every linguistic variable in (3.6) can have seven linguistic terms

$$\{NB, NM, NS, ZE, PS, PM, PB\}$$

and their membership function are of triangular form characterized by three parameters (center, left width, right width). Of course, the membership functions representing the linguistic terms $\{NB, NM, NS, ZE, PS, PM, PB\}$ can vary from input variable to input variable, e.g. the linguistic term "Negative Big" can have maximum n different representations.

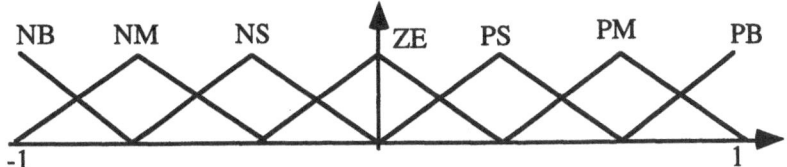

Fig. 3.44. Initial linguistic terms for the input variables.

The parameters of triangular fuzzy numbers in the premises are also learned by the steepest descent method.

We illustrate the above tuning process by a simple example. Consider two fuzzy rules of the form (3.6) with one input and one output variable

$$\Re_1 : \text{if } x \text{ is } A_1 \text{ then } y = z_1$$
$$\Re_2 : \text{if } x \text{ is } A_2 \text{ then } y = z_2$$

where the fuzzy terms A_1 "small" and A_2 "big" have sigmoid membership functions defined by

$$A_1(x) = \frac{1}{1 + \exp(b_1(x - a_1))},$$

$$A_2(x) = \frac{1}{1 + \exp(b_2(x - a_2))},$$

where a_1, a_2, b_1 and b_2 are the parameter set for the premises.

Let x be the input to the fuzzy system. The firing levels of the rules are computed by

$$\alpha_1 = A_1(x) = \frac{1}{1 + \exp(b_1(x - a_1))},$$

$$\alpha_2 = A_2(x) = \frac{1}{1 + \exp(b_2(x - a_2))},$$

and the output of the system is computed by the discrete center-of-gravity defuzzification method as

$$o = \frac{\alpha_1 z_1 + \alpha_2 z_2}{\alpha_1 + \alpha_2} = \frac{A_1(x)z_1 + A_2(x)z_2}{A_1(x) + A_2(x)}.$$

Suppose further that we are given a training set

$$\{(x^1, y^1), \ldots, (x^K, y^K)\}$$

obtained from the unknown nonlinear function f.

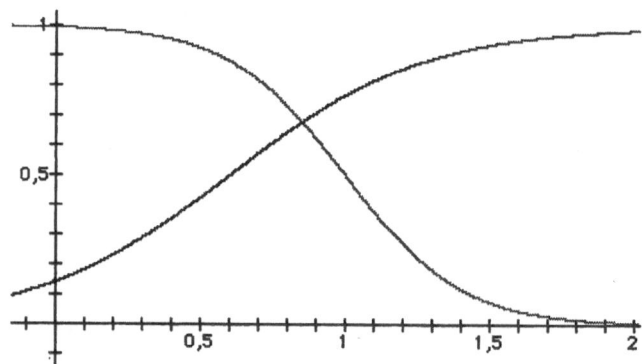

Fig. 3.45. Initial sigmoid membership functions.

Our task is construct the two fuzzy rules with appropriate membership functions and consequent parts to generate the given input-output pairs.

That is, we have to learn the following parameters

- a_1, b_1, a_2 and b_2, the parameters of the fuzzy numbers representing the linguistic terms "small" and "big",
- z_1 and z_2, the values of consequent parts.

We define the measure of error for the k-th training pattern as usually

$$E_k = E_k(a_1, b_1, a_2, b_2, z_1, z_2) = \frac{1}{2}(o^k(a_1, b_1, a_2, b_2, z_1, z_2) - y^k)^2$$

where o^k is the computed output from the fuzzy system corresponding to the input pattern x^k and y^k is the desired output, $k = 1, \ldots, K$.

The steepest descent method is used to learn z_i in the consequent part of the i-th fuzzy rule. That is,

$$z_1(t+1) = z_1(t) - \eta \frac{\partial E_k}{\partial z_1} = z_1(t) - \eta \frac{\partial}{\partial z_1} E_k(a_1, b_1, a_2, b_2, z_1, z_2) =$$

$$z_1(t) - \eta(o^k - y^k)\frac{\alpha_1}{\alpha_1 + \alpha_2} = z_1(t) - \eta(o^k - y^k)\frac{A_1(x^k)}{A_1(x^k) + A_2(x^k)}$$

$$z_2(t+1) = z_2(t) - \eta \frac{\partial E_k}{\partial z_2} = z_2(t) - \eta \frac{\partial}{\partial z_2} E_k(a_1, b_1, a_2, b_2, z_1, z_2) =$$

$$z_2(t) - \eta(o^k - y^k) \frac{\alpha_2}{\alpha_1 + \alpha_2} = z_2(t) - \eta(o^k - y^k) \frac{A_2(x^k)}{A_1(x^k) + A_2(x^k)}$$

where $\eta > 0$ is the learning constant and t indexes the number of the adjustments of z_i.

In a similar manner we can find the shape parameters (center and slope) of the membership functions A_1 and A_2.

$$a_1(t+1) = a_1(t) - \eta \frac{\partial E_k}{\partial a_1},$$

$$b_1(t+1) = b_1(t) - \eta \frac{\partial E_k}{\partial b_1}$$

$$a_2(t+1) = a_2(t) - \eta \frac{\partial E_k}{\partial a_2},$$

$$b_2(t+1) = b_2(t) - \eta \frac{\partial E_k}{\partial b_2}$$

where $\eta > 0$ is the learning constant and t indexes the number of the adjustments of the parameters. We show now how to compute analytically the partial derivative of the error function E_k with respect to a_1, the center of the fuzzy number A_1.

$$\frac{\partial E_k}{\partial a_1} = \frac{\partial}{\partial a_1} E_k(a_1, b_1, a_2, b_2, z_1, z_2)$$

$$= \frac{1}{2} \frac{\partial}{\partial a_1} (o^k(a_1, b_1, a_2, b_2, z_1, z_2) - y^k)^2$$

$$= \left(o^k - y^k\right) \frac{\partial o^k}{\partial a_1},$$

where

$$\frac{\partial o^k}{\partial a_1} = \frac{\partial}{\partial a_1} \left[\frac{A_1(x^k)z_1 + A_2(x^k)z_2}{A_1(x^k) + A_2(x^k)} \right]$$

$$= \frac{\partial}{\partial a_1} \left[\left(\frac{z_1}{1 + \exp(b_1(x^k - a_1))} + \frac{z_2}{1 + \exp(b_2(x^k - a_2))} \right) \Big/ \right.$$

$$\left. \left(\frac{1}{1 + \exp(b_1(x^k - a_1))} + \frac{1}{1 + \exp(b_2(x^k - a_2))} \right) \right]$$

$$= \frac{\partial}{\partial a_1} \left[\frac{z_1[1 + \exp(b_2(x^k - a_2))] + z_2[1 + \exp(b_1(x^k - a_1))]}{2 + \exp(b_1(x^k - a_1)) + \exp(b_2(x^k - a_2))} \right]$$

$$= \frac{-b_1 z_2 \epsilon_2 (2 + \epsilon_1 + \epsilon_2) + b_1 \epsilon_1 (z_1(1 + \epsilon_2) + z_2(1 + \epsilon_1))}{(2 + \epsilon_1 + \epsilon_2)^2}$$

where we have used the notations

$$\epsilon_1 = \exp(b_1(x^k - a_1))$$

and

$$\epsilon_2 = \exp(b_2(x^k - a_2)).$$

The learning rules are simplified if we use the following fuzzy partition

$$A_1(x) = \frac{1}{1 + \exp(-b(x - a))}, \quad A_2(x) = \frac{1}{1 + \exp(b(x - a))}$$

where a and b are the *shared* parameters of A_1 and A_2. In this case the equation

$$A_1(x) + A_2(x) = 1$$

holds for all x from the domain of A_1 and A_2.

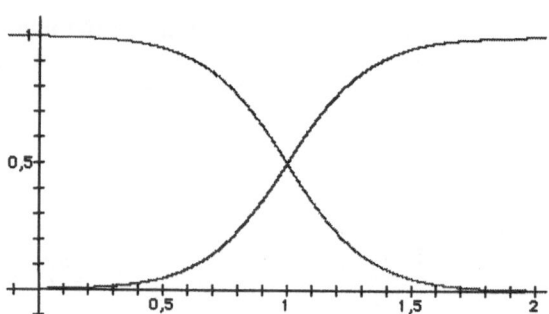

Fig. 3.46. Symmetrical membership functions.

The weight adjustments are defined as follows

$$z_1(t + 1) = z_1(t) - \eta \frac{\partial E_k}{\partial z_1} = z_1(t) - \eta(o^k - y^k)A_1(x^k)$$

$$z_2(t + 1) = z_2(t) - \eta \frac{\partial E_k}{\partial z_2} = z_2(t) - \eta(o^k - y^k)A_2(x^k)$$

$$a(t + 1) = a(t) - \eta \frac{\partial E_k(a, b)}{\partial a}$$

$$b(t+1) = b(t) - \eta \frac{\partial E_k(a,b)}{\partial b}$$

where

$$\frac{\partial E_k(a,b)}{\partial a} = (o^k - y^k)\frac{\partial o^k}{\partial a}$$

$$= (o^k - y^k)\frac{\partial}{\partial a}[z_1 A_1(x^k) + z_2 A_2(x^k)]$$

$$= (o^k - y^k)\frac{\partial}{\partial a}[z_1 A_1(x^k) + z_2(1 - A_1(x^k))]$$

$$= (o^k - y^k)(z_1 - z_2)\frac{\partial A_1(x^k)}{\partial a}$$

$$= (o^k - y^k)(z_1 - z_2)b A_1(x^k)(1 - A_1(x^k))$$

$$= (o^k - y^k)(z_1 - z_2)b A_1(x^k)A_2(x^k),$$

and

$$\frac{\partial E_k(a,b)}{\partial b} = (o^k - y^k)(z_1 - z_2)\frac{\partial A_1(x^k)}{\partial b}$$

$$= -(o^k - y^k)(z_1 - z_2)(x^k - a)A_1(x^k)A_2(x^k).$$

Jang [106] showed that fuzzy inference systems with simplified fuzzy IF-THEN rules are universal approximators, i.e. they can approximate any continuous function on a compact set to arbitrary accuracy. It means that the more fuzzy terms (and consequently more rules) are used in the rule base, the closer is the output of the fuzzy system to the desired values of the function to be approximated.

A method which can cope with *arbitrary membership functions* for the input variables is proposed in [75, 188, 189]. The training data have to be divided into r disjoint clusters $R_1, \dots, R_r$. Each cluster R_i corresponds to a control rule R_i. Elements of the clusters are tuples of input-output values of the form (x, y) where x can be a vector $x = (x_1, \dots, x_n)$ of n input variables.

This means that the rules are not specified in terms of linguistic variables, but in the form of crisp input-output tuples.

A multilayer perceptron with n input units, some hidden layers, and r output units can be used to learn these clusters. The input data for this learning task are the input vectors of all clusters, i.e. the set

$$\{x \mid \exists i\, \exists y : (x, y) \in R_i\}.$$

The target output $t_{u_i}(x)$ for input x at output unit u_i is defined as

$$t_{u_i}(x) = \begin{cases} 1 \text{ if there exists } y \text{ such that } (x, y) \in R_i \\ 0 \text{ otherwise} \end{cases}$$

After the network has learned its weights, arbitrary values for x can be taken as inputs. Then the output at output unit u_i can be interpreted as the degree to which x matches the antecedent of rule R_i, i.e. the function

$$x \rightarrow o_{u_i}$$

is the membership function for the fuzzy set representing the linguistic term on the left-hand side of rule R_i.

In case of a *Mamdani* type fuzzy controller the same technique can be applied to the output variable, resulting in a neural network which determines the fuzzy sets for the right-hand sides of the rules.

For *Sugeno* type fuzzy controller, where each rule yields a crisp output value together with a number, specifying the matching degree for the antecedent of the rule, another technique can be applied. For each rule R_i a neural network is trained with the input-output tuples of the set R_i. Thus these r neural networks determine the crisp output values for the rules $R_i, \ldots, R_r$.

These neural networks can also be used to eliminate unnecessary input variables in the input vector x for the rules $R_1, \ldots, R_r$ by neglecting one input variable in one of the rules and comparing the control result with the one, when the variable is not neglected. If the performance of the controller is not influenced by neglecting input variable x_j in rule R_i, x_j is unnecessary for R_i and can be left out.

ANFIS (Adaptive Neural Fuzzy Inference Systems) [106] is a great example of an architecture for tuning fuzzy system parameters from input/output pairs of data. The fuzzy inference process is implemented as a generalized neural network, which is then tuned by gradient descent techniques. It is capable of tuning antecedent parameters as well as consequent parameters of fuzzy rules which use a softened trapezoidal membership function. It has been applied to a variety of problems, including chaotic time series prediction and the IRIS cluster learning problem.

These tuning methods has a weak point, because the convergence of tuning depends on the initial condition. Ishigami, Fukuda, Shibata and Arai [104] present a hybrid auto-tuning method of fuzzy inference using genetic algorithms and the generalized delta learning rule, which guarantees the optimal structure of the fuzzy model.

3.9 Fuzzy rule extraction from numerical data

Fuzzy systems and neural networks are widely used for function approximation. When comparing these two technologies, fuzzy systems are more favorable in that their behavior can be explained based on fuzzy rules and thus their performance can be adjusted by tuning the rules. But since, in general, knowledge acquisition is difficult and also the universe of discourse of each input variable needs to be divided into several intervals, applications of fuzzy

systems are restricted to the fields where expert knowledge is available and
the number of input variables is small. To overcome the problem of knowledge acquisition, several methods for extracting fuzzy rules from numerical
data have been developed.

In the previous section we described, how neural networks could be used
to optimize certain parameters of a fuzzy rule base.

*We assumed that the fuzzy IF-THEN rules where already specified in lin-
guistic form or as a crisp clustering of a set of correct input-output tuples.*

If we are given a set of crisp input-output tuples we can try to extract
fuzzy (control) rules from this set. This can either be done by *fuzzy clustering
methods* [16] or by using neural networks.

The input vectors of the input-output tuples can be taken as inputs for a
Kohonen self-organizing map, which can be interpreted in terms of linguistic
variables [166]. The main idea for this interpretation is to refrain from the
winner take-all principle after the weights for the self-organizing map are
learned. Thus each output unit u_i is from being the 'winner' given input
vector x, a matching degree $\mu_i(x)$ can be specified, yielding the degree to
which x satisfies the antecedent of the corresponding rule.

Finally, in order to obtain a Sugeno type controller, to each rule (output
unit) a crisp control output value has to be associated. Following the idea of
the Sugeno type controller, we could choose the value

$$\sum_{(x,y)\in S} \sum_i \mu_i(x)y \bigg/ \sum_{(x,y)\in S} \mu_i(x)$$

where S is the set of known input-output tuples for the controller and i
indexes the rules.

Another way to obtain directly a fuzzy clustering is to apply the modified
Kohonen network proposed in [15].

Kosko uses another approach to generate fuzzy-if-then rules from existing data [121]. Kosko shows that fuzzy sets can be viewed as points in a
multidimensional unit hypercube. This makes it possible to use *fuzzy associative memories* (FAM) to represent fuzzy rules. Special adaptive clustering
algorithms allow to learn these representations (AFAM).

In [181] fuzzy rules with variable fuzzy regions (hyperboxes) are extracted
for classification problems. This approach has a potential applicability to
problems having a high-dimensional input space. But because the overlap of
hyperboxes of different classes must be resolved by dynamically expanding,
splitting and contracting hyperboxes, the approach is difficult to apply to the
problems in which several classes overlap.

Abe and Lan [2] suggest a method for extracting fuzzy rules for pattern
classification. The fuzzy rules with variable fuzzy regions were defined by
activation hyperboxes which show the existence region of data for a class
and inhibition hyperboxes which inhibit the existence of the data for that

class. These rules were extracted *directly* from numerical data by recursively resolving overlaps between two classes.

Abe and Lan [3] present a method for extracting fuzzy rules *directly* from numerical data for function approximation. Suppose that the unknown function has a one-dimensional output y and an m-dimensional input vector x. First we divide, $[M_1, M_2]$, the universe of discourse of y into n intervals as follows:

$$[y_0, y_1], (y_1, y_2], \ldots, (y_{n-1}, y_n]$$

where $y_0 = M_1$ and $y_n = M_2$. We call the i-th interval the output interval i. Using the input data whose outputs are in the output interval i, we recursively define the input region that generates output in the output interval i.

Namely, first we determine activation hyperboxes, which define the input region corresponding to the output interval i, by calculating the minimum and maximum values of input data for each output interval.

If the activation hyperbox for the output interval i overlaps with the activation hyperbox for the output interval j, the overlapped region is defined as an inhibition hyperbox.

If the input data for output intervals i or/and j exist in the inhibition hyperbox, within this inhibition hyperbox, we define one or two additional activation hyperboxes; moreover, if two activation hyperboxes are defined and they overlap, we further define an additional inhibition hyperbox: this process is repeated until the overlap is resolved. Fig. 3.45 illustrates this process schematically.

Based on an activation hyperbox or based on an activation hyperbox and its corresponding inhibition hyperbox (if generated), a fuzzy rule is defined. Fig. 3.46 shows a fuzzy system architecture, including a fuzzy inference net which calculates degrees of membership for output intervals and a defuzzifier.

For an input vector x, degrees of membership for output intervals 1 to n are calculated in the inference net and then the output y is calculated by defuzzifier using the degrees of membership as inputs.

The fuzzy inference net consists of four layers at most. The inference net is sparsely connected. Namely, different output intervals have different units for the second to fourth layers and there is no connection among units of different output intervals.

- The second layer units consist of fuzzy rules which calculate the degrees of membership for an input vector x.
- The third layer units take the maximum values of inputs from the second layer, which are the degrees of membership generated by resolving overlaps between two output intervals. The number of third layer units for the output interval i is determined by the number of output intervals whose input spaces overlap with that of the output interval i. Therefore, if there is no overlap between the input space of the output interval i and that of any other output intervals, the output interval i and that of any other

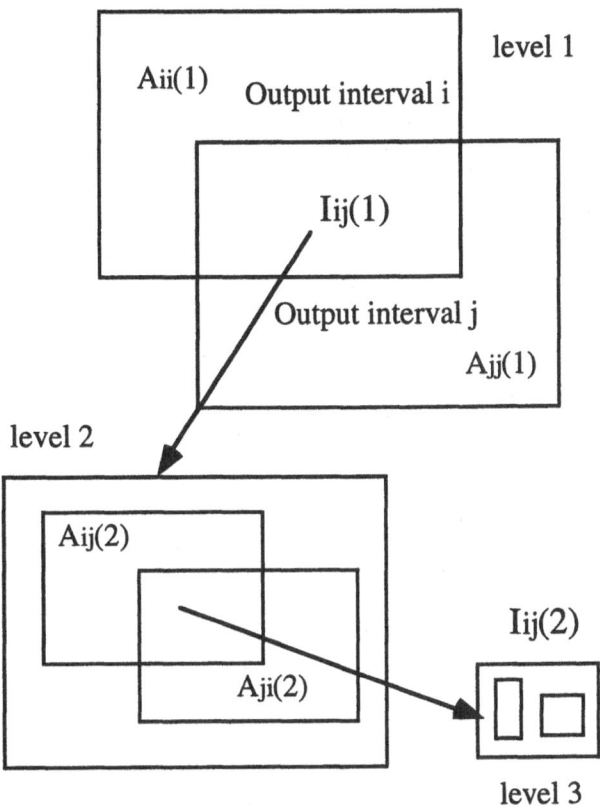

Fig. 3.47. Recursive definition of activation and inhibition hyperboxes.

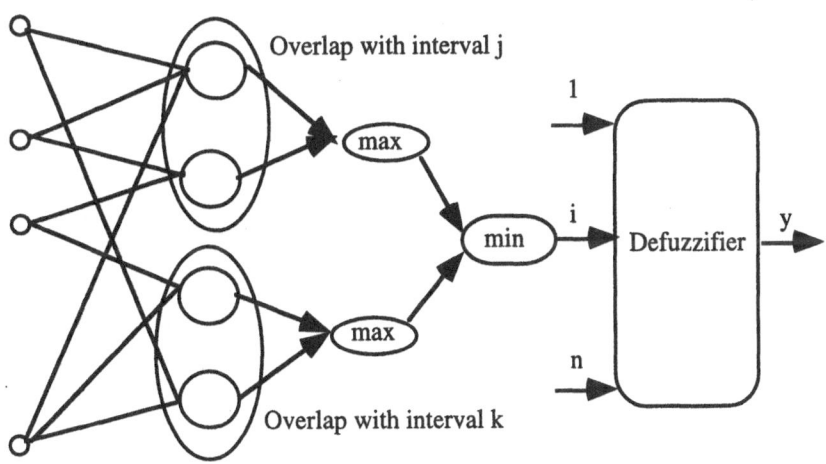

Fig. 3.48. Architecture of Abe and Lan's fuzzy inference system.

output intervals, the network for the output interval i is reduced to two layers.

- The fourth layer unit for the output interval i takes the minimum value among the maximum values generated by the preceding layer, each of them is associated with an overlap between two output intervals. Therefore, if the output interval i overlaps with only one output interval, the network for the output interval i is reduced to three layers. Calculation of a minimum in the fourth layer resolves overlaps among more than two output intervals. Thus in the process of generating hyperboxes, we need to resolve only an overlap between two output intervals at a time.

3.10 Neuro-fuzzy classifiers

Conventional approaches of pattern classification involve clustering training samples and associating clusters to given categories. The complexity and limitations of previous mechanisms are largely due to the lacking of an effective way of defining the boundaries among clusters. This problem becomes more intractable when the number of features used for classification increases. On the contrary, fuzzy classification assumes the boundary between two neighboring classes as a continuous, overlapping area within which an object has partial membership in each class. This viewpoint not only reflects the reality of many applications in which categories have fuzzy boundaries, but also provides a simple representation of the potentially complex partition of the feature space. In brief, we use fuzzy IF-THEN rules to describe a classifier. Assume that K patterns $x_p = (x_{p1}, \ldots, x_{pn})$, $p = 1, \ldots, K$ are given from two classes, where x_p is an n-dimensional crisp vector. Typical fuzzy classification rules for $n = 2$ are like

> If x_{p1} is *small* and x_{p2} is *very large*
> then $x_p = (x_{p1}, x_{p2})$ belongs to Class C_1
> If x_{p1} is *large* and x_{p2} is *very small*
> then $x_p = (x_{p1}, x_{p2})$ belongs to Class C_2

where x_{p1} and x_{p2} are the features of pattern (or object) p, *small* and *very large* are linguistic terms characterized by appropriate membership functions. The *firing level* of a rule

$$\Re_i : \text{If } x_{p1} \text{ is } A_i \text{ and } x_{p2} \text{ is } B_i \text{ then } x_p = (x_{p1}, x_{p2}) \text{ belongs to Class } C_i$$

with respect to a given object x_p is interpreted as the degree of belongness of x_p to C_i. This firing level, denoted by α_i, is usually determined as

$$\alpha_i = T(A_i(x_{p1}), A_2(x_{p2})),$$

where T is a triangular norm modeling the logical connective *and*.

As such, a fuzzy rule gives a meaningful expression of the qualitative aspects of *human recognition*. Based on the result of pattern matching between rule antecedents and input signals, a number of fuzzy rules are triggered in parallel with various values of firing strength. Individually invoked actions are considered together with a combination logic. Furthermore, we want the system to have *learning ability* of updating and fine-tuning itself based on newly coming information.

The task of *fuzzy classification* is to generate an appropriate fuzzy partition of the feature space . In this context the word *appropriate* means that the number of misclassified patterns is very small or zero. Then the rule base should be optimized by deleting rules which are not used.

Consider a two-class classification problem shown in Fig. 3.49. Suppose that the fuzzy partition for each input feature consists of three linguistic terms {*small, medium, big*} which are represented by triangular membership functions.

Both initial fuzzy partitions in Fig. 3.49 satisfy 0.5-completeness for each input variable, and a pattern x_p is classified into Class j if there exists at least one rule for Class j in the rule base whose firing strength (defined by the minimum t-norm) with respect to x_p is bigger or equal to 0.5. So a rule is created by finding for a given input pattern x_p the combination of fuzzy sets, where each yields the highest degree of membership for the respective input feature. If this combination is not identical to the antecedents of an already existing rule then a new rule is created.

However, it can occur that if the fuzzy partition is not set up correctly, or if the number of linguistic terms for the input features is not large enough, then some patterns will be missclassified.

The following 9 rules can be generated from the initial fuzzy partitions shown in Fig. 3.49:

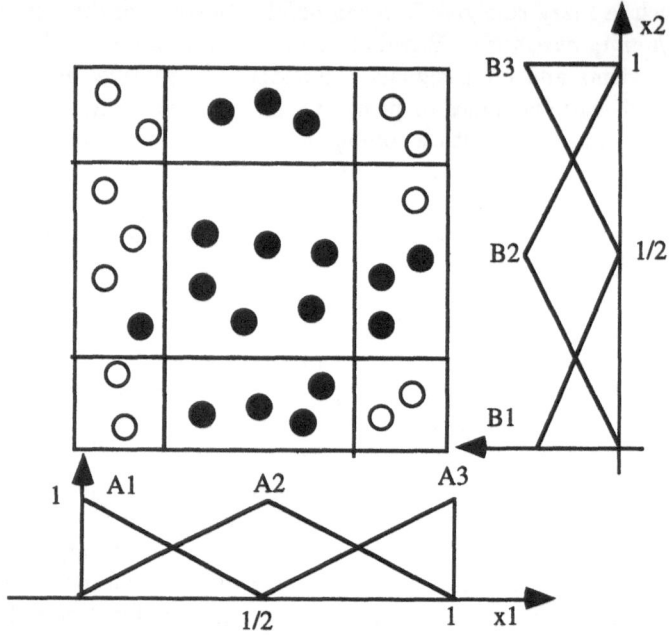

Fig. 3.49. Initial fuzzy partition with 9 fuzzy subspaces and 2 misclassified patterns. Closed and open circles represent the given pattens from Class 1 and Class 2, respectively.

$\Re_1$:   If x_1 is *small* and x_2 is *big*
          then $x_p = (x_1, x_2)$ belongs to Class C_1

$\Re_2$:   If x_1 is *small* and x_2 is *medium*
          then $x_p = (x_1, x_2)$ belongs to Class C_1

$\Re_3$:   If x_1 is *small* and x_2 is *small*
          then $x_p = (x_1, x_2)$ belongs to Class C_1

$\Re_4$:   If x_1 is *big* and x_2 is *small*
          then $x_p = (x_1, x_2)$ belongs to Class C_1

$\Re_5$:   If x_1 is *big* and x_2 is *big*
          then $x_p = (x_1, x_2)$ belongs to Class C_1

$\Re_6$:   If x_1 is *medium* and x_2 is *small*
          then $x_p = (x_1, x_2)$ belongs to Class C_2

$\Re_7$:   If x_1 is *medium* and x_2 is *medium*
          then $x_p = (x_1, x_2)$ belongs to Class C_2

$\Re_8$:   If x_1 is *medium* and x_2 is *big*
          then $x_p = (x_1, x_2)$ belongs to Class C_2

$\Re_9$:   If x_1 is *big* and x_2 is *medium*
          then $x_p = (x_1, x_2)$ belongs to Class C_2

where we have used the linguistic terms *small* for A_1 and B_1, *medium* for A_2 and B_2, and *big* for A_3 and B_3.

However, the same rate of error can be reached by noticing that if "x_1 is medium" then the pattern (x_1, x_2) belongs to Class 2, independently from the value of x_2, i.e. the following 7 rules provides the same classification result

$\Re_1$:   If x_1 is *small* and x_2 is *big* then x_p belongs to Class C_1

$\Re_2$:   If x_1 is *small* and x_2 is *medium* then x_p belongs to Class C_1

$\Re_3$:   If x_1 is *small* and x_2 is *small* then x_p belongs to Class C_1

$\Re_4$:   If x_1 is *big* and x_2 is *small* then x_p belongs to Class C_1

$\Re_5$:   If x_1 is *big* and x_2 is *big* then x_p belongs to Class C_1

$\Re_6$:   If x_1 is *medium* then x_p belongs to Class C_2

$\Re_7$:   If x_1 is *big* and x_2 is *medium* then x_p belongs to Class C_2

Fig. 3.50 is an example of fuzzy partitions (3 linguistic terms for the first input feature and 5 for the second) which classify correctly the patterns.

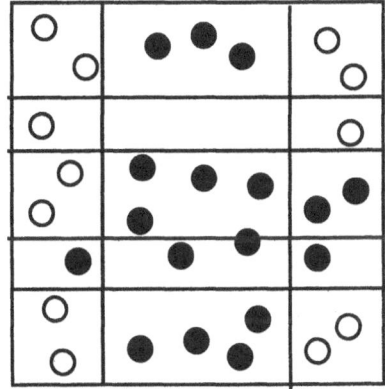

Fig. 3.50. Appropriate fuzzy partition with 15 fuzzy subspaces.

As an other example, Let us consider a two-class classification problem [102]. In Fig. 3.51 closed and open rectangulars represent the given from Class 1 and Class 2, respectively.

If one tries to classify all the given patterns by fuzzy rules based on a simple fuzzy grid, a fine fuzzy partition and $(6 \times 6 = 36)$ rules are required.

However, it is easy to see that the patterns from Fig. 3.51 may be correctly classified by the following five fuzzy IF-THEN rules

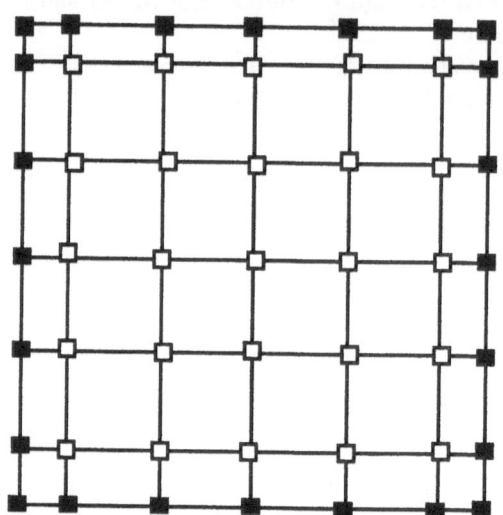

Fig. 3.51. A two-dimensional classification problem.

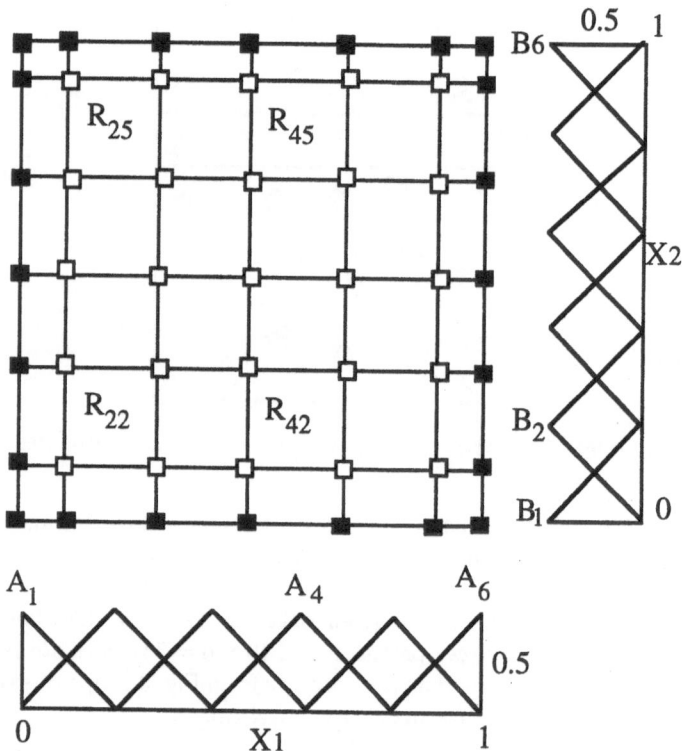

Fig. 3.52. Fuzzy partition with 36 fuzzy subspaces.

$\Re_1$:   If x_1 is *very small* then Class 1,

$\Re_2$:   If x_1 is *very large* then Class 1,

$\Re_3$:   If x_2 is *very small* then Class 1,

$\Re_4$:   If x_2 is *very large* then Class 1,

$\Re_5$:   If x_1 is *not very small* and x_1 is *not very large* and
        x_2 is *not very small* and x_2 is *not very large* then Class 2

Sun and Jang [186] propose an adaptive-network-based fuzzy classifier to solve fuzzy classification problems.

Fig. 3.53 demonstrates this classifier architecture with two input variables x_1 and x_2. The training data are categorized by two classes C_1 and C_2. Each input is represented by two linguistic terms, thus we have four rules.

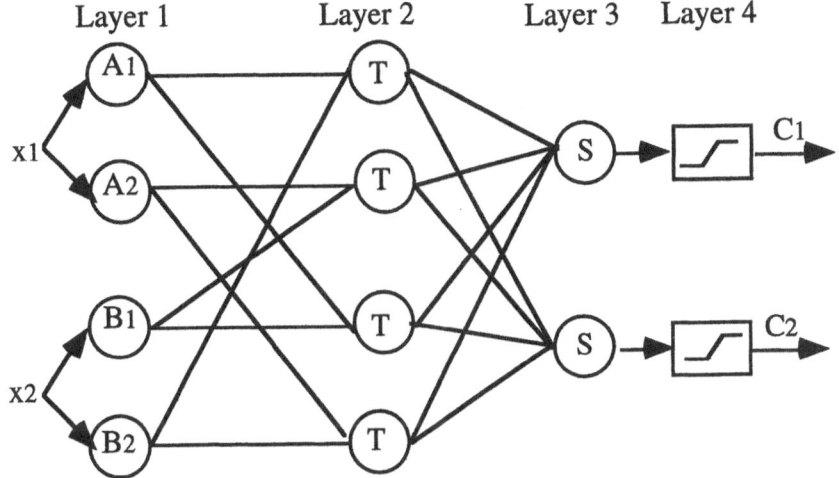

Fig. 3.53. An adaptive-network-based fuzzy classifier.

- **Layer 1** The output of the node is the degree to which the given input satisfies the linguistic label associated to this node. Usually, we choose bell-shaped membership functions

$$A_i(u) = \exp\left[-\frac{1}{2}\left(\frac{u - a_{i1}}{b_{i1}}\right)^2\right],$$

$$B_i(v) = \exp\left[-\frac{1}{2}\left(\frac{v - a_{i2}}{b_{i2}}\right)^2\right],$$

to represent the linguistic terms, where

$$\{a_{i1}, a_{i2}, b_{i1}, b_{i2}\}$$

is the *parameter set*. As the values of these parameters change, the bell-shaped functions vary accordingly, thus exhibiting various forms of membership functions on linguistic labels A_i and B_i. In fact, any continuous, such as trapezoidal and triangular-shaped membership functions, are also quantified candidates for node functions in this layer. The initial values of the parameters are set in such a way that the membership functions along each axis satisfy ϵ-completeness, normality and convexity. The parameters are then tuned with a descent-type method.

- **Layer 2**   Each node generates a signal corresponding to the conjuctive combination of individual degrees of match. The output signal is the firing strength of a fuzzy rule with respect to an object to be categorized.

  In most pattern classification and query-retrival systems, the conjuction operator plays an important role and its interpretation context-dependent. Since does not exist a single operator that is suitable for all applications, we can use parametrized t-norms to cope with this dynamic property of classifier design. For example, we can use Hamacher's t-norm with parameter $\gamma \geq 0$

$$H_\gamma(a, b) = \frac{ab}{\gamma + (1 - \gamma)(a + b - ab)},$$

or Yager's t-norm with parameter $p > 0$

$$Y_p(a, b) = 1 - \min\{0, [(1 - a)^p + (1 - b)^p]^{1/p}\}.$$

All nodes in this layer is labeled by T, because we can choose any t-norm for modeling the logical *and* operator. The nodes of this layer are called *rule nodes*.

Features can be combined in a compensatory way. For instance, we can use the generalized p-mean proposed by Dyckhoff and Pedrycz:

$$\left(\frac{x^p + y^p}{2}\right)^{1/p}, \ p \geq 1.$$

We take the linear combination of the firing strengths of the rules at *Layer 3* and apply a sigmoidal function at *Layer 4* to calculate the degree of belonging to a certain class.

If we are given the training set

$$\{(x^k, y^k), \ k = 1, \dots, K\}$$

where x^k refers to the k-th input pattern and

$$y^k = \begin{cases} (1, 0)^T & \text{if } x^k \text{ belongs to Class 1} \\ (0, 1)^T & \text{if } x^k \text{ belongs to Class 2} \end{cases}$$

then the parameters of the hybrid neural net (which determine the shape of the membership functions of the premises) can be learned by descent-type methods. This architecture and learning procedure is called ANFIS (adaptive-network-based fuzzy inference system) by Jang [106].

The error function for pattern k can be defined by

$$E_k = \frac{1}{2}[(o_1^k - y_1^k)^2 + (o_2^k - y_2^k)^2]$$

where y^k is the desired output and o^k is the computed output by the hybrid neural net.

3.11 FULLINS

Sugeno and Park [184] proposed a framework of learning based on indirect linguistic instruction, and the performance of a system to be learned is improved by the evaluation of rules.

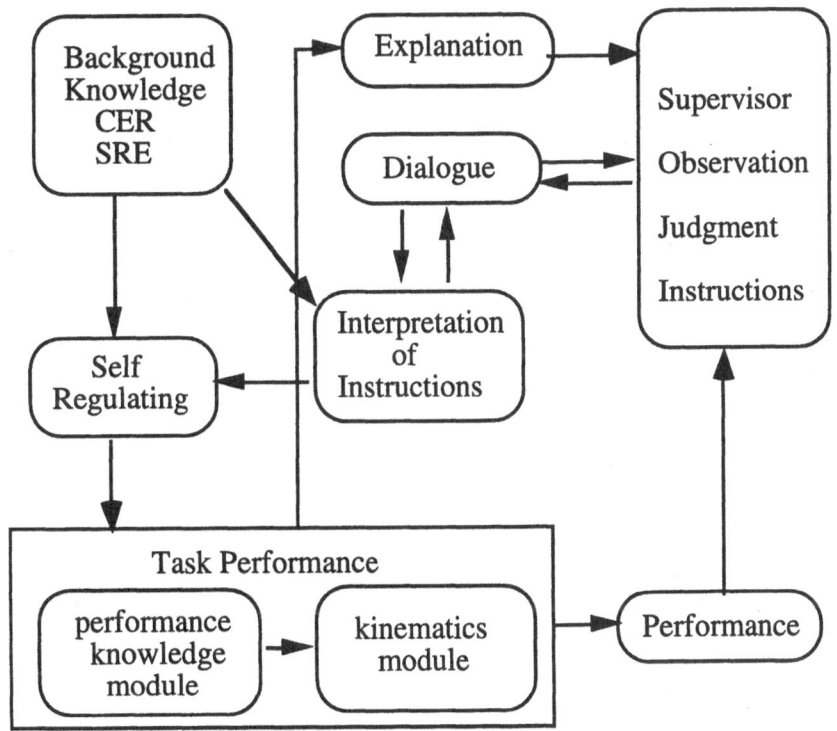

Fig. 3.54. The framework of FULLINS.

FULLINS (Fuzzy learning based on linguistic instruction) is a mechanism for learning through interpreting meaning of language and its concepts. It has the following components:

- **Task Performance Functional Component**
  Performs tasks achieving a given goal. Its basic knowledge performing the tasks is modified through the Self-Regulating Component when a supervisor's linguistic instruction is given.
- **Dialogue Functional Component**
  An interface to interpret linguistic instructions through dialogue with a supervisor.
- **Explanation Functional Component**
  Explains to a supervisor the elements of a process in performing the tasks by the Task Performance Component.
- **Background Knowledge Functional Component**
  Interpret instructions in the *Interpretation Component* and modifies the basic performance knowledge in the *Self-Regulating Component*.
- **Interpretation Functional Component**
  Interprets instructions by the meaning elements using the Background Knowledge and the Dialogue Components. An instruction is assumed to have some *meaning elements*, and a meaning element is associated with values called trends.
- **Self-Regulating Functional Component**
  The basic performance knowledge is regulated using the evaluation rules constructed by the searched meaning elements and the background knowledge.

Following Sugeno and Park [184] we describe now the method of linguistic instructions. A direct instruction is in the form of entire methods or individual IF-THEN rules. An individual IF-THEN rule is a rule (basic performance knowledge) to perform a given goal. An entire method is a set of rules that work together to satisfy a goal. It is difficult for supervisor to give direct instructions, because the instructions must be based on precise structure and components of rules to reach a given goal. An indirect instruction is not a part of the basic performance knowledge prepared for the system. An indirect instruction is not given in any specific format, and the contents of the instruction have macroscopic properties. In FULLINS indirect instructions are interpreted by meaning elements and their trends. For example, in a driving school does not give minute instructions about the steering angle of the wheel, the degree of stepping on the accelerator, etc. when he teaches a student higher level driving techniques such as turning around a curve, safe driving, etc. After explanating and demonstrating some driving techniques, the instructor gives the student a macroscopic indirect linguistic instruction based on his judgement and evaluation of the student's performance.

E.g. "Turning around a curve", the instructor judges the performance of a student's driving techniques and then gives an instruction to him like this

If you approach near the turning point, turn slowly, turn smoothly, etc.

If an instruction is given, the student interprets it with his internal knowledge through dialogue with the instructor: Turn around the curve slowly, step on the brake slightly or step on the accelerator weakly.

Indirect instructions L_i have three components:

$$L_i = [AP][LH_i][AW]$$

where LH stands for *Linguistic hedges*, AW for *Atomic words*, AP for *Auxiliary Phrases*.

An indirect instruction in a driving school is:

$$L_i = [\text{Turn around a curve}(AP)][\text{more}((LH)][\text{slowly}(AW)]$$

where LH and AP are prepared in the system. AW is interpreted in a combination of meaning elements and their trends.

The meaning of an instruction is restricted by the attached linguistic hedges.

$$L_i = [\text{If you approach near the turning point, turn slowly, turn smoothly}]$$

Then the following dialogue can take place between the instructor and the student

- SYS: Do you want me **"press the accelerator weaker than before"**?
- SV: RIGHT
- SYS: Do you want me **"steering wheels less than before"**?
- SV: RIGHT

The supervisor's instruction is interpreted through two questions, because there exists causal relation between brake and accelerator in the above assumption:

$L_i = $ [**press the brake slightly** (m_1)] and [**press on accelerator slowly** (m_2)] and [**turn steering small** (m_3)].

Instructions are entered by the supervisor's input-key in FULLINS and by the instructor's voice in the driving school. Meaning elements are words or phrases to interpret indirect linguistic instructions. In case of driving school, the meaning elements are

- *[degree of strength pressing on brake]*
- *[degree of strength pressing on accelerator]*
- *[degree of steering]*.

The three trends of [**degree of strength pressing on accelerator**] are:

- [**Press on the accelerator strongly**] (Positive trend (+)),
- [**Press on accelerator weakly**] (Negative trend (-)),
- [**Accelerator has nothing to do with the instruction**] (No trend (0)).

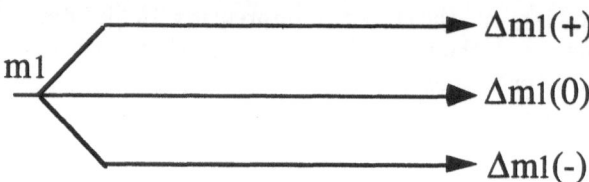

Fig. 3.55. Three trends of the meaning element m_1.

Trends of meaning elements can be defined for an element m_1 as follows

- $\Delta m_1(+)$: m_1 contributes to a meaning element of an instruction with trend (+)
- $\Delta m_1(-)$: m_1 contributes to a meaning element of an instruction with trend (-)
- $\Delta m_1(0)$: m_1 does not act on a meaning element of an instruction

A set of meaning elements consists of a set of dependent and independent meaning elements. If the meaning element and its trend, [**press on brake slightly**] is selected, then [**press on accelerator weakly**] is also selected without having any dialogue with the instructor. The causal net is

$$\Delta m_1(+) \rightarrow \Delta m_2(-)$$

For example, if we have

$$\Delta m_1(+) \rightarrow \Delta m_3(+) \rightarrow \Delta m_4(-), \ \Delta m_5(+) \rightarrow \Delta m_6(+)$$

then the corresponding causal net is shown in Fig. 3.56.

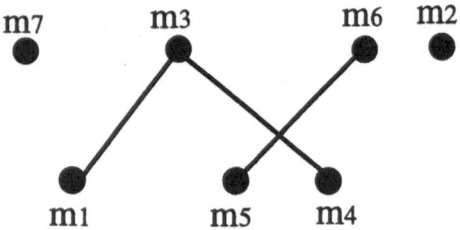

Fig. 3.56. Causal net.

Meaning elements are searched through dialogue between the system and the supervisor, and then Dialogue Meaning Elements Set and the Linguistic Instruction Knowledge Base are used. Linguistic Instruction Knowledge Base consists of two memory modules: Atomic Words Memory and Linguistic Hedges Memory module. Atomic Words Memory is a module in which some atomic words are memorized. Some linguistic hedges are memorized with each weight in Linguistic Hedges Memory:

[(non, 0), (slightly, 0.2), (rather, 0.4), (more, 0.6), (pretty, 0.8), (very, 1.0)].

LH entered together with AW is matched with each linguistic hedge prepared in Linguistic Hedges Memory, then the weight allocated on the hedge is selected. The meaning of the instruction is restricted by LH: the consequent parts of the evaluation rule constructed by the searched meaning elements are modified by the weight of LH. The interpretation of linguistic instruction

$$L_i = (LH_i)(\Delta m_1(+))$$

is the following

- $L_i = [\text{Drive}(AP_1)][\text{more}(LH_1)][\text{fast}(AW_1)][\text{than before}(AP_2)]$
- $AW_i = \Delta m_1(+)$: Press on acceleration strongly

The amount of information about a learning object is increased, as the number of supervisor's instructions increases. If a supervisor's instruction is given, Atomic Words Memory is checked if AW_i of the instruction does or does not exist in it. If the same one exists in Atomic Words Memory then AW_i is interpreted by the matched meaning element without searching meaning element through dialogue.

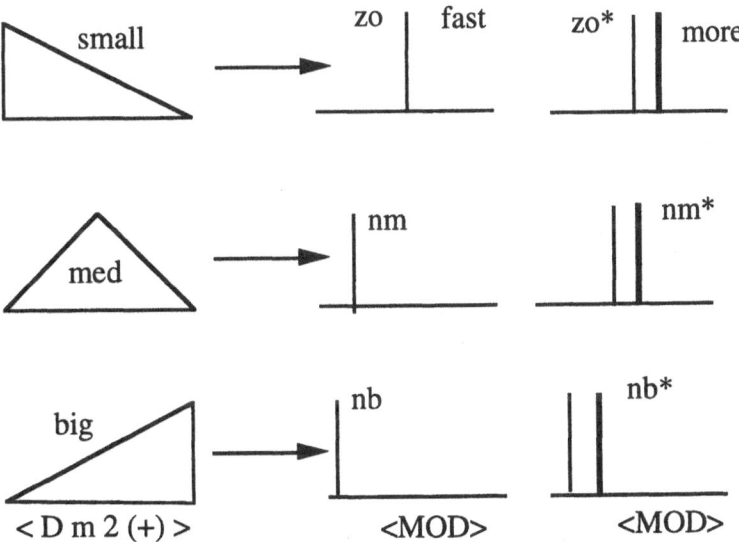

Fig. 3.57. Modifying the consequent parts of the rule.

The evaluation rule is constructed by a combination of the meaning element and its trend using Constructing evaluation Rules. The meaning of linguistic instruction is restricted by modifying the consequent part of the evaluation rule by

$$\Delta H = W_{LH_i} \cdot \Delta R$$

where ΔR is the maximum value for shifting the consequence parameter by LH_i. Fig. 3.57 shows an evaluation rule, where MOD is the modifying value of the parameter for the consequent part in the basic performance rule. The modifying value by linguistic hedge [**more**] is

$$\Delta H = 0.6 \cdot \Delta R.$$

The membership functions of the consequent part [**fast**] are "zo", "nm" and "nb" and the ultimate membership functions of the consequent part are "zo^*", "nm^*" and "nb^*".

Sugeno and Park illustrate the capabilities of FULLINS by controlling an unmanned helicopter. The control rules are constructed by using the acquired knowledge from a pilot's experience and knowledge. The objective flight modes are *Objective Line Following Flight System* and *Figure Eight Flight System*.

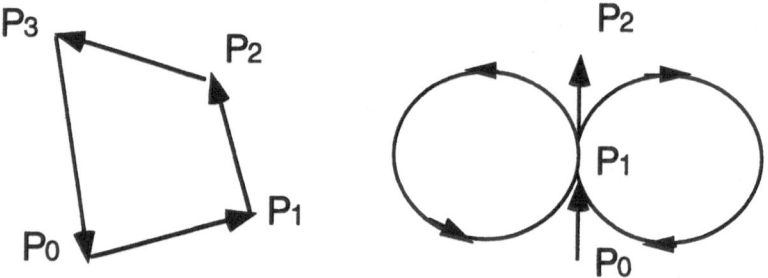

Fig. 3.58. Objective Line Following and Eight Flight Systems

The performance of the figure eight flight is improved by learning from the supervisor's instructions. The measure of performance of the is given by the following goals: following the objective line $\overline{P_1 P_2}$, adjusting of the diameter of the right turning circle, and making both diameters small or large simultaneously.

3.12 Applications of fuzzy neural systems

The first applications of fuzzy neural networks to consumer products appeared on the (Japanese and Korean) market in 1991. Some examples include *air conditioners, electric carpets, electric fans, electric thermo-pots, desk-type electric heaters, forced-flue kerosene fan heaters, kerosene fan heaters, microwave ovens, refrigerators, rice cookers, vacuum cleaner, washing machines, clothes driers, photocopying machines, and word processors.*

Neural networks are used to design membership functions of fuzzy systems that are employed as decision-making systems for controlling equipment. Although fuzzy logic can encode expert knowledge directly using rules with linguistic labels, it usually takes a lot of time to design and tune the membership functions which quantitatively define these linquistic labels. Neural network learning techniques can automate this process and substantially reduce development time and cost while improving performance. The idea of using neural networks to design membership functions was proposed by Takagi and Hayashi [188]. This was followed by applications of the gradient descent method to the tuning of parameters that define the shape and position of membership functions. This system tuning is equivalent to learning in a feed-forward network. This method has been widely used to design triangle membership functions, Gaussian membership functions, sigmoidal membership functions, and bell-shaped membership functions. Simple function shapes, such as triangles, are used for most actual products. The center and widths of the membership functions are tuned by a gradient method, reducing the error between the actual fuzzy system output and the desired output. Figure 3.59 is an example of this type of neural network usage.

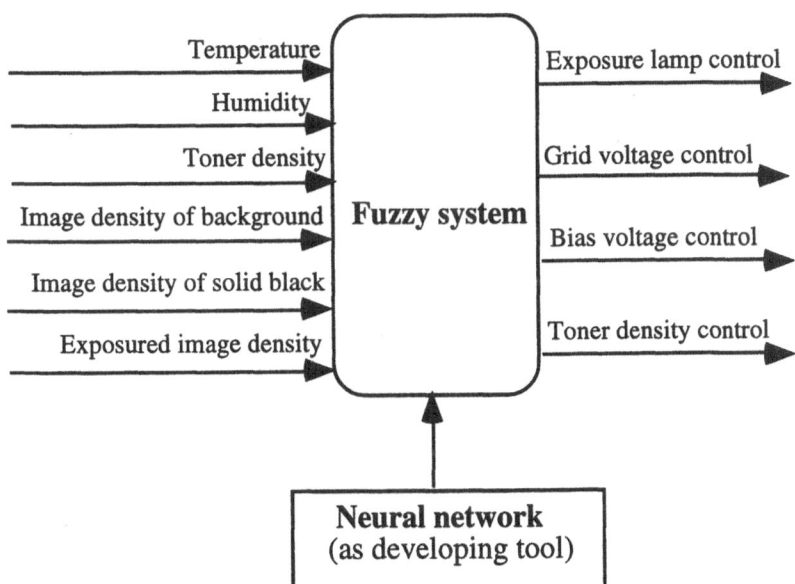

Fig. 3.59. Photocopier machine (*Matsushita Electric*).

Nikko Securities uses a neural network to improve the rating of convertible bonds [163]. The system learns from the reactions of an expert rating instruction, which can change according to the actual economic situation. It analyzes the results, and then uses them to give advice. Their system consists of a seven-layer neural network. The neural network's internal connections and synaptic weights are initialized using the knowledge in the fuzzy logic system; it then learns by backpropagation learning and changes its symbolic representation accordingly. This representation is then returned to a fuzzy logic representation, and the system has acquired knowledge. The system can then give advice based on the knowledge it has acquired. Such a system is called a *neural fuzzy system*.

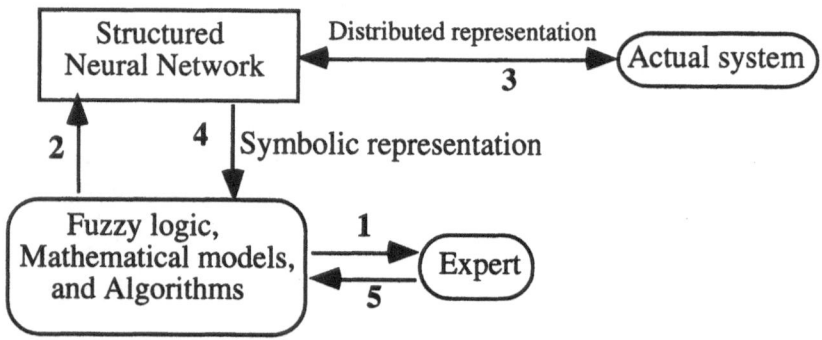

Fig. 3.60. Neural fuzzy system.

1   Translate expert's knowledge into a symbolic representation.

2   Initialize the neural net by symbolic representation.

3   Decrease errors between actual system and neural net by learning.

4   Translate the distributed representation based on the structure of the neural network.

5   Acquire knowledge from the modified symbolic representation.

This seven-layer system had a ratio of correct answers of 96%. A similar learning system with a conventional three-layered neural network had a ratio of correct answers of 84%, and was difficult to understand the internal representation. The system learned 40 times faster than the three-layer system. This comparison is evidence of the effectiveness of neural fuzzy systems.

Another way to combine fuzzy systems and neural networks is to connect them up serially. In the *Sanyo electric fan* [176], the fan must rotate toward the user - which requires calculating the direction of the remote controller. Three infrared sensors in the fan's body detect the strengths of the signal from a remote controller. First, the distance to the remote is calculated by a fuzzy system. Then, this distance value and the ratios of sensor outputs are used by a neural network to compute the required direction. The latter calculation is done by a neural net because neither mathematical models nor fuzzy reasoning proved good at carrying out this function. The final product has an error of $\pm 4^o$ as opposed to the $\pm 10^o$ error of statistical regression methods [161].

Sanyo uses neural networks for adjusting auto-exposure in their photocopying machine. Moreover, the toner control of this machine is controlled by fuzzy reasoning. *Ricoh Co.* has applied two neural networks to control electrostatic latent image conditions at necessary potential, as well as a neural network to figure optimum developing bias voltage from image density, temperature, humidity, and copy volume [147].

To make the control smoother, more sensitive, and more accurate one has to incorporate more sensor data. This becomes more complicated as the input space increases in dimension. In this approach, a neural net handles the larger set of sensor inputs and corrects the output of a fuzzy system (which was designed earlier for the old set of inputs). A complete redesigning of the fuzzy system is thus avoided. This leads to substantial savings in development time (and cost), since redesigning the membership functions, which becomes more difficult as the number of inputs increases, is obviated.

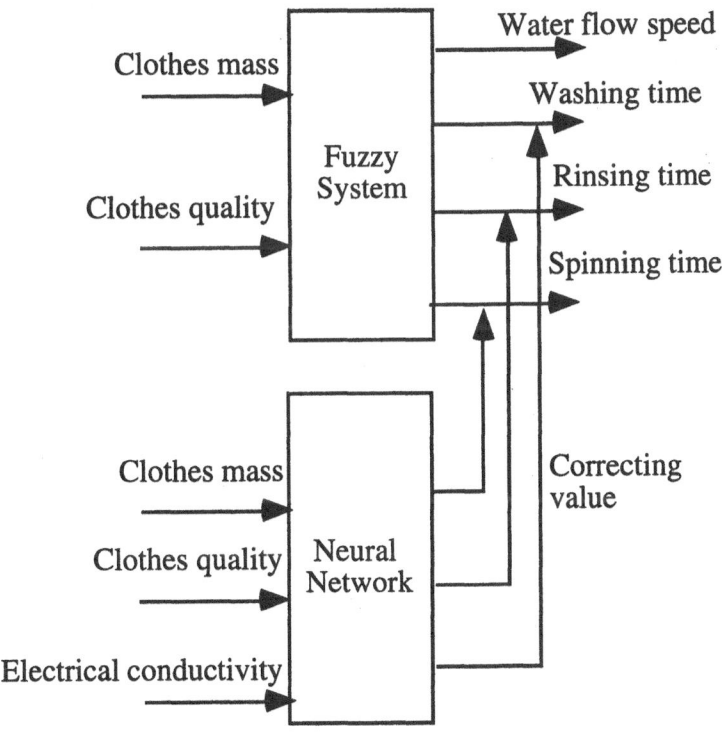

Fig. 3.61. Schematic of a Hitachi washing machine.

Figure 3.61 shows the schematic of a *Hithachi washing mashine* [84]. The fuzzy system shown in the upper part was part of the first model. Later, an improved model incorporated extra information by using a correcting neural net, as shown. The additional input (fed only to the net) is electrical conductivity, which is used to measure the opacity/transparency of the water.

Toshiba produces washing mashines which have a similar control system [158]. *Sanyo* uses a similar approach in its washing machine, although some of the inputs/outputs are different.

ANFIS (Adaptive Neural Fuzzy Inference Systems) [106] is a great example of an architecture for tuning fuzzy system parameters from input/output pairs of data. The fuzzy inference process is implemented as a generalized neural network, which is then tuned by gradient descent techniques. It is capable of tuning antecedent parameters as well as consequent parameters of fuzzy rules which use a softened trapezoidal membership function. It has been applied to a variety of problems, including chaotic timeseries prediction and the IRIS cluster learning problem.

This relationship, which is not analytically known, has been defined in terms of a fuzzy rule set extracted from the data points. The rule extraction was done using ANFIS, a fuzzy neural system that combines the advantages of fuzzy systems and neural networks. As a fuzzy system, it does not require a large data set and it provides transparency, smoothness, and representation of prior knowledge. As a neural system, it provides parametric adaptability.

Bibliography

1. S. Abe and M.-S. Lan, A classifier using fuzzy rules extracted directly from numerical data, in: *Proceedings of IEEE Internat. Conf. on Fuzzy Systems*, San Francisco,1993 1191-1198.
2. S. Abe and M.-S. Lan, Fuzzy rules extraction directly from numerical data for function approximation, *IEEE Trans. Syst., Man, and Cybernetics*, 25(1995) 119-129.
3. S. Abe and M.-S. Lan, A method for fuzzy rule extraction directly from numerical data and its application to pattern classification, *IEEE Transactions on Fuzzy Systems*, 3(1995) 18-28.
4. Constantin von Altrock, *Fuzzy Logic and Neuro Fuzzy Applications Explained*, (Prentice-Hall, Englewood Cliffs, 1995).
5. Constantin von Altrock, *Fuzzy Logic and NeuroFuzzy Applications in Business and Finance*, (Prentice-Hall, Englewood Cliffs, 1996).
6. F. Aminzadeh and M. Jamshidi eds., *Fuzzy Sets, Neural Networks, and Distributed Artificial Intelligence* (Prentice-Hall, Englewood Cliffs, 1994).
7. P.E. An, S. Aslam-Mir, M. Brown, and C.J. Harris, A reinforcement learning approach to on-line optimal control, in: *Proc. of IEEE International Conference on Neural Networks*, Orlando, Fl, 1994 2465-2471.
8. K. Asakawa and H. Takagi, Neural Networks in Japan *Communications of ACM*, 37(1994) 106-112.
9. K.. Asai, M. Sugeno and T. Terano, *Applied Fuzzy Systems* (Academic Press, New York, 1994).
10. A. Bastian, Handling the nonlinearity of a fuzzy logic controller at the transition between rules, *Fuzzy Sets and Systems*, 71(1995) 369-387.
11. H.R. Berenji, A reinforcement learning-based architecture for fuzzy logic control, *Int. Journal Approximate Reasoning*, 6(1992) 267-292.
12. H.R. Berenji and P. Khedkar, Learning and tuning fuzzy logic controllers through reinforcements, *IEEE Transactions on Neural Networks*, 3(1992) 724-740.
13. H.R. Berenji, R.N. Lea, Y. Jani, P. Khedkar, A.Malkani and J. Hoblit, Space shuttle attitude control by reinforcement learning and fuzzy logic, in: *Proc. IEEE Internat. Conf. on Fuzzy Systems*, San Francisco,1993 1396-1401.
14. H.R. Berenji, Fuzzy systems that can learn, in: J.M. Zurada, R.J. Marks and C.J. Robinson eds., *Computational Intelligence: Imitating Life* (IEEE Press, New York, 1994) 23-30.
15. J.C. Bezdek, E.C. Tsao and N.K. Pal, Fuzzy Kohonen clustering networks, in: *Proc. IEEE Int. Conference on Fuzzy Systems 1992*, San Diego, 1992 1035-1043.
16. J.C. Bezdek and S.K. Pal eds., *Fuzzy Models for Pattern Recognition* (IEEE Press, New York, 1992).

17. S.A. Billings, H.B. Jamaluddin, and S. Chen. Properties of neural networks with application to modelling nonlinear systems, *Int. J. Control*, 55(1992)193–224.

18. A. Blanco, M. Delgado and I. Requena, Improved fuzzy neural networks for solving relational equations, *Fuzzy Sets and Systems*, 72(1995) 311-322.

19. G Bordogna and G. Pasi, A user-adaptive neural network supporting a rule-based relevance feedback, *Fuzzy Sets and Systems*, (82)(1996) 201-211.

20. M. Brown and C.J. Harris, A nonlinear adaptive controller: A comparison between fuzzy logic control and neurocontrol. *IMA J. Math. Control and Info.*, 8(1991) 239–265.

21. M. Brown and C. Harris, *Neurofuzzy Adaptive Modeling and Control* (Prentice-Hall, Englewood Cliffs, 1994).

22. J.J. Buckley, Theory of the fuzzy controller: An introduction, *Fuzzy Sets and Systems*, 51(1992) 249-258.

23. J.J. Buckley and Y. Hayashi, Fuzzy neural nets and applications, *Fuzzy Systems and AI*, 1(1992) 11-41.

24. J.J. Buckley, Approximations between nets, controllers, expert systems and processes, in: *Proceedings of 2nd Internat. Conf. on Fuzzy Logic and Neural Networks*, Iizuka, Japan, 1992 89-90.

25. J.J. Buckley, Y. Hayashi and E. Czogala, On the equivalence of neural nets and fuzzy expert systems, *Fuzzy Sets and Systems*, 53(1993) 129-134.

26. J.J.Buckley, Sugeno type controllers are universal controllers, *Fuzzy Sets and Systems*, 53(1993) 299-304.

27. J.J. Buckley and Y. Hayashi, Numerical relationships between neural networks, continuous functions, and fuzzy systems, *Fuzzy Sets and Systems*, 60(1993) 1-8.

28. J.J. Buckley and Y. Hayashi, Hybrid neural nets can be fuzzy controllers and fuzzy expert systems, *Fuzzy Sets and Systems*, 60(1993) 135-142.

29. J.J. Buckley and E. Czogala, Fuzzy models, fuzzy controllers and neural nets, *Arch. Theoret. Appl. Comput. Sci.*, 5(1993) 149-165.

30. J.J. Buckley and Y. Hayashi, Can fuzzy neural nets approximate continuous fuzzy functions? *Fuzzy Sets and Systems*, 61(1993) 43-51.

31. J.J .Buckley and Y. Hayashi, Fuzzy neural networks, in: L.A. Zadeh and R.R. Yager eds., *Fuzzy Sets, Neural Networks and Soft Computing* (Van Nostrand Reinhold, New York, 1994) 233-249.

32. J.J .Buckley and Y. Hayashi, Fuzzy neural networks: A survey, *Fuzzy Sets and Systems*, 66(1994) 1-13.

33. J.J .Buckley and Y. Hayashi, Neural nets for fuzzy systems, *Fuzzy Sets and Systems*, 71(1995) 265-276.

34. J.J .Buckley, E. Eslami and Y. Hayashi, Solving fuzzy equations using neural nets, *Fuzzy Sets and Systems*, 86(1997) 271-278.

35. J.J .Buckley and E. Eslami, Neural net solutions to fuzzy problems: The quadratic equation, *Fuzzy Sets and Systems*, 86(1997) 289-298

36. G.A.Capenter et al, Fuzzy ARTMAP: A neural network architecture for incremental supervised learning of analog multidimensional maps, *IEEE Transactions on Neural Networks*, 3(1992) 698-713.

37. S. Chen, S.A. Billings, and P.M. Grant, Recursive hybrid algorithm for nonlinear system identification using radial basis function networks, *Int. J. Control*, 55(1992) 1051–1070.

38. F.C. Chen and M.H. Lin, On the learning and convergence of radial basis networks, in: *Proc. IEEE Int. Conf. Neural Networks*, San Francisco, 1993 983–988.

39. Young-Jeng Chen and Ching-Cheng Teng, Rule combination in a fuzzy neural network, *Fuzzy Sets and Systems*, (82)(1996) 161-166.

40. Sung-Bae Cho, Pattern recognition with neural networks combined by genetic algorithm, *Fuzzy Sets and Systems*, 103(1999) 339-347.
41. E. Cox, Adaptive fuzzy systems, *IEEE Spectrum*, , February 1993, 27–31.
42. E. Cox, The Fuzzy system Handbook. *A Practitioner's Guide to Building, Using, and Maintaining Fuzzy Systems* (Academic Press, New York, 1994).
43. D. Dumitrescu, Fuzzy training procedures I, *Fuzzy Sets and Systems*, 56(1993) 155-169.
44. P. Eklund, H. Virtanen and T. Riisssanen, On the fuzzy logic nature of neural nets, in: *Proccedings of Neuro-Nimes*, 1991 293–300.
45. P. Eklund and F. Klawonn, A Formal Framework for Fuzzy Logic Based Diagnosis, in: R.Lowen and M.Roubens eds., *Proccedings of the Fourth IFSA Congress*, vol. *Mathematics*, Brussels, 1991, 58-61.
46. P. Eklund, M. Fogström and J. Forsström, A generic neuro-fuzzy tool for developing medical decision support, in: P. Eklund ed., *Proceedings of MEPP'92, International Seminar on Fuzzy Control through Neural Interpretations of Fuzzy Sets* (Åbo Akademis tryckeri, Åbo, 1992) 1–27.
47. P. Eklund, F. Klawonn, and D. Nauck, Distributing errors in neural fuzzy control. in: *Proceedings of the 2nd Internat Conference on Fuzzy Logic and Neural Networks*, Iizuka, Japan, 1992 1139–1142.
48. P. Eklund and F. Klawonn, Neural fuzzy logic programming, *IEEE transactions on Neural Networks* 3(1992) 815-818.
49. P. Eklund, Neural Logic: A Basis for Second Generation Fuzzy Controllers, in: U.Höhle and E.P.Klement eds., *Proceedings of 14th Linz Seminar on Fuzzy Set Theory*, Johannes Kepler Universität, 1992 19-23.
50. P. Eklund and R. Fullér, A neuro-fuzzy approach to medical diagnostics, in:*Proceedings of EUFIT'93 Conference*, September 7-10, 1993, Aachen, Germany (Verlag der Augustinus Buchhandlung, Aachen, 1993) 810-813.
51. P. Eklund, J. Forsström, A. Holm, M.. Nyström, and G. Selén, Rule generation as an alternative to knowledge acquisition: A systems architecture for medical informatics, *Fuzzy Sets and Systems*, 66(1994) 195-205.
52. P. Eklund, Network size versus preprocessing, in: R.R. Yager and L.A. Zadeh eds., *Fuzzy Sets, Neural Networks and Soft Computing* (Van Nostrand, New York, 1994) 250-264.
53. P. Eklund, A generic system for developing medical decision support, *Fuzzy Systems A.I. Rep. Letters*, 3(1994) 71-78.
54. P. Eklund and J. Forsström, Computational intelligence for laboratory information systems, *Scand. J. Clin. Lab. Invest.*, 55 Suppl. 222 (1995) 75-82.
55. A.O. Esogbue, A fuzzy adaptive controller using reinforcement learning neural networks, in: *Proc. IEEE Internat. Conf. on Fuzzy Systems*, San Francisco,1993 178–183.
56. J. Forsström, P. Eklund, H. Virtanen, J. Waxlax and J. Lähdevirta, DiagaiD: A Connectionists Approach to Determine the Information Value of Clinical Data, *Artificial Intelligence in Medicine*, 3 (1991) 193-201.
57. T. Fukuda and T. Shibata, Fuzzy-neuro-GA based intelligent robotics, in: J.M. Zurada, R.J. Marks and C.J. Robinson eds., *Computational Intelligence: Imitating Life* (IEEE Press, New York, 1994) 352-363.
58. M. Furukawa and T. Yamakawa, The design algorithms of membership functions for a fuzzy neuron, *Fuzzy Sets and Systems*, 71(1995) 329-343.
59. S. Gallant, *Neural Network Learning and Expert Systems*, MIT Press, Cambridge, Mass., USA, 1993
60. A. Geyer-Schulz, *Fuzzy rule based Expert Systems and Genetic Learning* (Physica-Verlag, Berlin, 1995).

61. S. Giove, M. Nordio and A. Zorat, An Adaptive Fuzzy Control for Automatic Dialysis, in: E.P. Klement and W. Slany eds., *Fuzzy Logic in Artificial Intelligence*, (Springer-Verlag, Berlin 1993) 146-156.

62. P.Y. Glorennec, Learning algorithms for neuro-fuzzy networks, in: A. Kandel and G. Langholz eds., *Fuzzy Control Systems* (CRC Press, New York, 1994) 4-18.

63. A. Gonzalez, R. Perez and J.L. Verdegay, Learning the structure of a fuzzy rule: A genetic approach, *Fuzzy Systems A.I.Rep. Letters*, 3(1994) 57-70.

64. S. Goonatilake and S. Khebbal eds., *Intelligent Hybrid Systems*, John Wiley and Sons, New York 1995.

65. M.M. Gupta and J. Qi, On fuzzy neuron models, in: *Proceedings of International Joint Conference on Neural Networks*, Seattle, 1991 431-436.

66. M.M. Gupta and J. Qi, On fuzzy neuron models, in: L.A. Zadeh and J. Kacprzyk eds., *Fuzzy Logic for the Management of Uncertainty* (J. Wiley, New York, 1992) 479-491.

67. M.M. Gupta, Fuzzy logic and neural networks, *Proc. 2nd Internat. Conf. on Fuzzy logic and Neural Networks*, Iizuka, Japan, 1992 157-160.

68. M.M. Gupta and M.B. Gorzalczany, Fuzzy neuro-computation technique and its application to modeling and control, in: *Proc. IEEE Internat. Conf on Fuzzy Systems*, San Diego, 1992 1271-1274.

69. M.M. Gupta and D.H. Rao, On the principles of fuzzy neural networks, *Fuzzy Sets and Systems*, 59(1993) 271-279.

70. S.K. Halgamuge and M. Glesner, Neural networks in designing fuzzy systems for real world applications, *Fuzzy Sets and Systems*, 65(1994) 1-12.

71. C.J. Harris, C.G. Moore, and M. Brown, *Intelligent control, aspects of fuzzy logic and neural networks* (World Scientific Press, 1993).

72. C.J. Harris ed., *Advances in Intelligent Control* (Taylor and Francis, London, 1994).

73. Y. Hayashi, J.J. Buckley and E. Czogala, Systems engineering applications of fuzzy neural networks, *Journal of Systems Engineering*, 2(1992) 232-236.

74. Y. Hayashi, J.J. Buckley and E. Czogala, Fuzzy neural controller, in: *Proc. IEEE Internat. Conf on Fuzzy Systems*, San Diego, 1992 197-202.

75. Y. Hayashi, H. Nomura, H. Yamasaki and N. Wakami, Construction of fuzzy inference rules by NFD and NDFL, *International Journal of Approximate Reasoning*, 6(1992) 241-266.

76. Y. Hayashi, Neural expert system using fuzzy teaching input, in: *Proc. IEEE Internat. Conf on Fuzzy Systems*, San Diego, 1992 485-491.

77. Y. Hayashi, J.J. Buckley and E. Czogala, Fuzzy neural network with fuzzy signals and weight, *International Journal of Intelligent Systems*, 8(1992) 527-537.

78. Y. Hayashi, J.J. Buckley and E. Czogala, Direct fuzzification of neural network and fuzzified delta rule, *Proc. 2nd Internat. Conf. on Fuzzy logic and Neural Networks*, Iizuka, Japan, 1992 73-76.

79. Y. Hayashi and J.J. Buckley, Direct fuzzification of neural networks, in: *Proceedings of 1st Asian Fuzzy Systems Symposium*, Singapore, 1993 560-567.

80. Y. Hayashi and J.J. Buckley, Approximations between fuzzy expert systems and neural networks, *International Journal of Approximate Reasoning*, 10(1994) 63-73.

81. K. Hirota and W. Pedrycz, Knowledge-based networks in classification problems, *Fuzzy Sets and Systems*, 51(1992) 1-27.

82. K. Hirota and W. Pedrycz, OR/AND neuron in modeling fuzzy set connectives, *IEEE Transactions on Fuzzy Systems*, 2(994) 151-161.

83. K. Hirota and W. Pedrycz, Fuzzy modelling environment for designing fuzzy controllers, *Fuzzy Sets and Systems*, 70(1995) 287-301.

84. Hitachi, Neuro and fuzzy logic automatic washing machine and fuzzy logic drier, *Hitachi News Rel., No. 91-024* (Feb. 26, 1991). Hitachi, 1991 (in Japanese).

85. S. Horikowa, T. Furuhashi and Y. Uchikawa, On fuzzy modeling using fuzzy neural networks with the backpropagation algorithm, *IEEE Transactions on Neural Networks*, 3(1992).

86. S. Horikowa, T. Furuhashi and Y. Uchikawa, On identification of structures in premises of a fuzzy model using a fuzzy neural network, in: *Proc. IEEE International Conference on Fuzzy Systems*, San Francisco, 1993 661-666.

87. L. Huang, Bai-Ling Zhang and Qian Huang, Robust interval regression analysis using neural networks, *Fuzzy Sets and Systems*, 97(1998) 337-347.

88. K.J. Hunt, D. Sbarbaro-Hofer, R. Zbikowski and P.J. Gawthrop, Neural networks for control systems - a survey, *Automatica*, 28(1992) 1083–1112.

89. H. Ichihashi, Iterative fuzzy modelling and a hierarchical network, in: R.Lowen and M.Roubens eds., *Proceedings of the Fourth IFSA Congress, Vol. Engineering*, Brussels, 1991 49-52.

90. H. Ishibuchi, R. Fujioka and H. Tanaka, An architecture of neural networks for input vectors of fuzzy numbers, in: *Proc. IEEE Internat. Conf on Fuzzy Systems*, San Diego, 1992 1293-1300.

91. H. Ishibuchi, K. Nozaki and H. Tanaka, Distributed representation of fuzzy rules and its application to pattern classification, *Fuzzy Sets and Systems*, 52(1992) 21-32.

92. H. Ishibuchi and H. Tanaka, Approximate pattern classification using neural networks, in: R.Lowen and M.Roubens eds., *Fuzzy Logic: State of the Art* (Kluwer, Dordrecht, 1993) 225-236.

93. H. Ishibuchi, K. Nozaki and H. Tanaka, Efficient fuzzy partition of pattern space for classification problems, *Fuzzy Sets and Systems*, 59(1993) 295-304.

94. H. Ishibuchi, R. Fujioka and H. Tanaka, Neural networks that learn from fuzzy IF-THEN rules, *IEEE Transactions on Fuzzy Systems*, 1(993) 85-97.

95. H. Ishibuchi, H. Okada and H. Tanaka, Fuzzy neural networks with fuzzy weights and fuzzy biases, in: *Proc. IEEE Internat. Conference on Neural Networks*, San Francisco, 1993 447-452.

96. H. Ishibuchi, K. Kwon and H. Tanaka, Implementation of fuzzy IF-THEN rules by fuzzy neural networks with fuzzy weights, in: *Proceedings of EUFIT'93 Conference*, September 7-10, 1993 Aachen, Germany, Verlag der Augustinus Buchhandlung, Aachen, 1993 209-215.

97. H. Ishibuchi, K. Kwon and H. Tanaka, Learning of fuzzy neural networks from fuzzy inputs and fuzzy targets, in: *Proc. 5th IFSA World Congress*, Seoul, Korea, 1993 147-150.

98. H. Ishibuchi, K. Nozaki and H. Tanaka, Empirical study on learning in fuzzy systems, in: *Proc. 2nd IEEE Internat. Conference on Fuzzy Systems*, San Francisco, 1993 606-611.

99. H. Ishibuchi, K. Nozaki, N. Yamamato and H. Tanaka, Genetic operations for rule selection in fuzzy classification systems, in: *Proc. 5th IFSA World Congress*, Seoul, Korea, 1993 15-18.

100. H. Ishibuchi, K. Nozaki, N. Yamamato, Selecting fuzzy rules by genetic algorithm for classification problems, in: *Proc. 2nd IEEE Internat. Conference on Fuzzy Systems*, San Francisco, 1993 1119-1124.

101. H. Ishibuchi, H. Okada and H. Tanaka, Interpolation of fuzzy IF-THEN rules by neural networks, *International Journal of Approximate Reasoning*, 10(1994) .3-27.

102. H. Ishibuchi, K. Nozaki, N. Yamamato and H. Tanaka, Construction of fuzzy classification systems with rectangular fuzzy rules using genetic algorithms, *Fuzzy Sets and Systems*, 65(1994) 237-253.

103. H. Ishibuchi, K. Kwon and H. Tanaka, A learning algorithm of fuzzy neural networks with triangular fuzzy weights, *Fuzzy Sets and Systems*, 71(1995) 277-293.

104. H. Ishigami, T. Fukuda, T. Shibita and F. Arai, Structure optimization of fuzzy neural network by genetic algorithm, *Fuzzy Sets and Systems*, 71(1995) 257-264.

105. Masumi Ishikawa and Teppei Moriyama, Prediction of time series by a structural learning of neural networks, *Fuzzy Sets and Systems*, (82)(1996) 167-176.

106. J.-S. Roger Jang, ANFIS: Adaptive-network-based fuzzy inference system, *IEEE Trans. Syst., Man, and Cybernetics*, 23(1993) 665-685.

107. Jyh-Shing Roger Jang, Chuen-Tsai Sun and Eiji Mizutani, *Neuro-Fuzzy and Soft Computing: A Computational Approach to Learning and Machine Intelligence*, Prentice Hall, 1996.

108. Nikola K. Kasabov, Learning fuzzy rules and approximate reasoning in fuzzy neural networks and hybrid systems, *Fuzzy Sets and Systems*, (82)(1996) 135-149.

109. N.K. Kasabov, R.I. Kilgour and S.J. Sinclair, From hybrid adjustable neuro-fuzzy systems to adaptive connectionist-based systems for phoneme and word recognition, *Fuzzy Sets and Systems*, 103(1999) 349-367.

110. N.K. Kasabov and R.Kozma, *Neuro-Fuzzy Techniques for Intelligent Information Systems*, Studies Fuzziness and Soft Computing, Vol. 30, Physica-Verlag, Heidelberg, 1999.

111. J.M. Keller and D. Hunt, Incorporating fuzzy membership functions into the perceptron algorithm, *IEEE Transactions on Pattern. Anal. Mach. Intell.*, 7(1985) 693-699.

112. J.M. Keller, R.R. Yager and H.Tahani, Neural network implementation of fuzzy logic, *Fuzzy Sets and Systems*, 45(1992) 1-12.

113. J.M. Keller and H.Tahani, Backpropagation neural networks for fuzzy logic, *Information Sciences*, 6(1992) 205-221.

114. J.M. Keller and H.Tahani, Implementation of conjunctive and disjunctive fuzzy logic rules with neural networks, *International Journal of Approximate Reasoning*, 6(1992) 221-240.

115. J.M. Keller, R. Krishnapuram, Z.H. Chen and O. Nasraoui, Fuzzy additive hybrid operators for network-based decision making, *International Journal of Intelligent Systems* 9(1994) 1001-1023.

116. P.S. Khedkar, Learning as adaptive interpolation in neural fuzzy systems, in: J.M. Zurada, R.J. Marks and C.J. Robinson eds., *Computational Intelligence: Imitating Life* (IEEE Press, New York, 1994) 31-42.

117. Y.S. Kim and S. Mitra, An adaptive integrated fuzzy clustering model for pattern recognition, *Fuzzy Sets and Systems*, 65(1994) 297-310.

118. F. Klawonn and V. Novák, The relation between inference and interpolation in the framework of fuzzy systems, *Fuzzy Sets and Systems*, (81)(1996) 331-354.

119. F. Klawonn and R. Kruse, Constructing a fuzzy controller from data, *Fuzzy Sets and Systems*, (85)(1997) 177-193.

120. S.G. Kong and B. Kosko, Adaptive fuzzy systems for backing up a truck-and-trailer, *IEEE Transactions on Neural Networks*, 3(1992) 211-223.

121. B. Kosko, *Neural Networks and Fuzzy Systems* (Prentice-Hall, Englewood Cliffs, 1992).

122. R. Krishnapuram and J. Lee, Fuzzy-set-based hierarchical networks for information fusion in computer vision, *Neural Networks*, 5(1992) 335-350.

123. R. Kruse, J. Gebhardt and R. Palm eds., *Fuzzy Systems in Computer Science* (Vieweg, Braunschweig, 1994).

124. D.C. Kuncicky, A fuzzy interpretation of neural networks, in: *Proceedings of 3rd IFSA Congress*, 1989 113–116.

125. R.J. Kuo and P.H. Cohen, Manufacturing process control through integration of neural networks and fuzzy model, *Fuzzy Sets and Systems*, 98(1998) 15-31.

126. H.K. Kwan and Y.Cai, A fuzzy neural network and its application to pattern recognition, *IEEE Transactions on Fuzzy Systems*, 3(1994) 185-193.

127. S.C. Lee and E.T. Lee, Fuzzy sets and neural networks, *Journal of Cybernetics* 4(1974) 83-103.

128. S.C. Lee and E.T. Lee, Fuzzy neural networks, *Math. Biosci.* 23(1975) 151-177.

129. H.-M. Lee and W.-T. Wang, A neural network architecture for classification of fuzzy inputs, *Fuzzy Sets and Systems*, 63(1994) 159-173.

130. M. Lee, S.Y. Lee and C.H. Park, Neuro-fuzzy identifiers and controllers, *J. of Intelligent Fuzzy Systems*, 6(1994) 1-14.

131. K.-M. Lee, D.-H. Kwang and H.L. Wang, A fuzzy neural network model for fuzzy inference and rule tuning,*International Journal of Uncertainty, Fuzziness and Knowledge-Based Systems*, 3(1994) 265-277.

132. Xiaozhong Li and Da Ruan, Novel neural algorithms based on fuzzy δ rules for solving fuzzy relation equations: Part I, *Fuzzy Sets and Systems*, (90)(1997) 11-23.

133. Wei Li, Chenyu Ma and F.M. Wahl, A neuro-fuzzy system architecture for behavior-based control of a mobile robot in unknown environments, *Fuzzy Sets and Systems*, 87(1997) 133-140.

134. Xiaozhong Li, Da Ruan, Novel neural algorithms based on fuzzy δ rules for solving fuzzy relation equations: Part II, *Fuzzy Sets and Systems*, 103(1999) 473-486.

135. C.T. Lin and C.S.G. Lee, Neural-network-based fuzzy logic control and decision system, *IEEE Transactions on Computers*, 40(1991) 1320-1336.

136. Y. Lin and G.A. Cunningham III, A new approach to fuzzy-neural system modeling, *IEEE Transactions on Fuzzy systems*, 3(1995) 190-198.

137. C.T. Lin and Y.C. Lu, A neural fuzzy system with linguistic teaching signals, *IEEE Transactions on Fuzzy Systems*, 3(1995) 169-189.

138. Chin-Teng Lin and C.S. George Lee, *Neural Fuzzy Systems: A Neuro-Fuzzy Synergism to Intelligent Systems*, (Prentice Hall, Englewood Cliffs, New York, 1996).

139. R.J. Machado and A.F. Rocha, A hybrid architecture for fuzzy connectionist expert systems, in: A. Kandel and G. Langholz eds., *Hybrid Architectures for Intelligent Systems* (CRC Press, Boca Raton, FL, 1992).

140. R.A. Marques Pereira, L. Mich and L. Gaio, Curve reconstruction with dynamical fuzzy grading and weakly continuous constraints, in: *Proceedings of the 2nd Workshop on Current Issues in Fuzzy Technologies*, Trento, June 1992, (Dipartimento di Informatica e Studi Aziendali, Universitá di Trento 1993) 77-85.

141. L. Medsker, *Hybrid Neural Network and Expert Systems* (Kluwer Academic Publishers, Boston, 1994).

142. K. Michels, Numerical stability analysis for a fuzzy or neural network controller, *Fuzzy Sets and Systems*, 89(1997) 335-350

143. S.Mitra and S.K.Pal, Neuro-fuzzy expert systems: overview with a case study, in: S.Tzafestas and A.N. Venetsanopoulos eds., *Fuzzy Reasoning in Information, Decision and Control Systems* (Kluwer, Dordrecht, 1994) 121-143.

144. S.Mitra and S.K.Pal, Self-organizing neural network as a fuzzy classifier, *IEEE Trans. Syst., Man, and Cybernetics*, 24(1994) 385-399.

145. S.Mitra and S.K.Pal, Fuzzy multi-layer perceptron, inferencing and rule generation, *IEEE Transactions on Neural Networks*, 6(1995) 51-63.

146. S.Mitra, Fuzzy MLP based expert system for medical diagnosis, *Fuzzy sets and Systems*, 65(1994) 285-296.

147. T. Morita, M. Kanaya and T. Inagaki, Photo-copier image density control using neural network and fuzzy theory. in: *Proceedings of the Second International Workshop on Industrial Fuzzy Control and Intelligent Systems*, 1992 10-16.

148. D. Nauck, F. Klawonn and R. Kruse, Fuzzy sets, fuzzy controllers and neural networks, *Wissenschaftliche Zeitschrift der Humboldt-Universität zu Berlin*, reihe Medizin, 41(1992) 99-120.

149. D. Nauck and R. Kruse, A fuzzy neural network learning fuzzy control rules and membership functions by fuzzy error backpropagation, in: *Proceedings of IEEE Int. Conference on Neural Networks*, San Francisco, 1993 1022-1027.

150. D. Nauck, F. Klawonn and R. Kruse, Combining neural networks and fuzzy controllers, in: E.P. Klement and W. Slany eds., *Fuzzy Logic in Artificial Intelligence*, (Springer-Verlag, Berlin, 1993) 35-46.

151. D. Nauck and R. Kruse, NEFCON-I: An X-Window based simulator for neural fuzzy controllers, in: *Proceedings of IEEE Int. Conference on Neural Networks*, Orlando, 1994 1638-1643.

152. D. Nauck, Fuzzy neuro systems: An overview, in: R. Kruse, J. Gebhardt and R. Palm eds., *Fuzzy systems in Computer Science* (Vieweg, Wiesbaden, 1994) 91-107.

153. D. Nauck, Building neural fuzzy controllers with NEFCON-I. in: R. Kruse, J. Gebhardt and R. Palm eds., *Fuzzy systems in Computer Science* (Vieweg, wiesbaden, 1994) 141-151.

154. D. Nauck, F. Klawonn and R. Kruse, *Neurale Netze und Fuzzy-Systeme* (Vieweg, wiesbaden, 1994).

155. D. Nauck and R. Kruse, NEFCLASS - A neuro-fuzzy approach for the classification of data, in: K.M. George et al eds., *Applied Computing, Proceedings of the 1995 ACM Symposium on Applied Computing*, Nashville, February 26-28, 1995, ACM Press, 1995.

156. D. Nauck and R. Kruse, Designing Neuro-Fuzzy Systems Through Backpropagation, in: W.Pedrycz ed., *Fuzzy Modelling: Paradigms and Practice*, Kluwer, 1996 203-228.

157. D. Nauck, F. Klawonn and R. Kruse, *Foundations on Neuro-Fuzzy Systems*, Wiley, Chichester, 1997.

158. R. Narita, H. Tatsumi and H. Kanou, Application of neural networks to household applications. *Toshba Rev. 46*, 12 (December 1991) 935-938. (in Japanese)

159. J. Nie and D. Linkens, *Fuzzy Neural Control - Principles, Algorithms and Applications* (Prentice-Hall, Englewood Cliffs, 1995).

160. J. Nie J. Nie, T.H. Lee and D.A. Linkens, A note on the integration of fuzzy systems with neural networks under a TLTT framework, *Fuzzy Sets and Systems*, 87(1997) 277-289.

161. Nikkei Electronics, New trend in consumer electronics: Combining neural networks and fuzzy logic, *Nikkei Elec.*, 528(1991) 165-169 (In Japanese).

162. H. Nomura, I. Hayashi and N. Wakami, A learning method of fuzzy inference rules by descent method, in: *Proceedings of the IEEE International Conference on Fuzzy Systems*, San Diego, 1992 203-210.

163. H. Okada, N. Watanabe, A. Kawamura and K. Asakawa, Initializing multilayer neural networks with fuzzy logic. in: *Proceedings of the International Joint Conference on Neural Networks*, Baltimore, 1992 239-244.

164. Ralf Östermark, A fuzzy neural network algorithm for multigroup classification, *Fuzzy Sets and Systems*, 105(1999) 113-122.

165. S.K.Pal and S.Mitra, Fuzzy versions of Kohonen's net and MLP-based classification: Performance evaluation for certain nonconvex decision regions, *Information Sciences*, 76(1994) 297-337.
166. W. Pedrycz and W.C. Card, Linguistic interpretation of self-organizing maps, in: *Proceedings of the IEEE International Conference on Fuzzy Systems*, San Diego, 1992 371-378.
167. W. Pedrycz, *Fuzzy Control and Fuzzy Systems* (Wiley, New York, 1993).
168. W. Pedrycz, *Fuzzy Sets Engineering* (CRC Press, Boca Raton, 1995).
169. C. Posey, A.Kandel and G. Langholz, Fuzzy hybrid systems, in: A. Kandel and G. Langholz eds., *Hybrid architectures for Intelligent Systems* (CRC Press, Boca Raton, Florida, 1992) 174-196.
170. G.V.S. Rajau and J Zhou, Adaptive hierarchical fuzzy controller, *IEEE Trans. Syst., Man, and Cybernetics*, 23(1993) 973-980.
171. A.L. Ralescu ed., *Fuzzy Logic in Artificial Intelligence, Proc. IJCAI'93 Workshop*, Chambéry, France, Lecture Note in artificial Intelligence, Vol. 847 (Springer, Berlin, 1994).
172. J. Rasmussen, Diagnostic reasoning in action, *IEEE Trans. Syst., Man, and Cybernetics*, 23(1993) 981-992.
173. I. Requena and M. Delgado, R-FN: A model of fuzzy neuron, in: *Proc. 2nd Int. Conf. on Fuzzy Logic & Neural Networks*, Iizuka, Japan, 1992 793-796.
174. R.A.Ribeiro, H.-J. Zimmermann, R.R.Yager and J. Kacprzyk eds., *Soft Computing in Financial Engineering*, Studies in Fuzziness and Soft Computing, Vol. 28, Springer-Verlag, Berlin/Heidelberg, 1999.
175. T. Riissanen, An Experiment with Clustering, *Proceedings MEPP92, International Seminar on Fuzzy Control through Neural Interpretations of Fuzzy Sets*, Mariehamn, Åland, June 15-19, 1992, Åbo Akademi tryckeri, Åbo, 1992, 57-65.
176. Sanyo, Electric fan series in 1991, *Sanyo News Rel.*, (March 14, 1991). Sanyo, 1991 (In Japanese).
177. E. Sanchez, Fuzzy logic knowledge systems and artificial neural networks in medicine and biology, in: R.R. Yager and L.A. Zadeh eds., *An Introduction to Fuzzy Logic Applications in Intelligent Systems* (Kluwer, Boston, 1992) 235-251.
178. J.D. Schaffer, Combinations of genetic algorithms with neural networks or fuzzy systems, in: J.M. Zurada, R.J. Marks and C.J. Robinson eds., *Computational Intelligence: Imitating Life* (IEEE Press, New York, 1994) 371-382.
179. R. Serra and G. Zanarini, *Complex Systems and Cognitive Processes* (Springer Verlag, Berlin, 1990).
180. J.J. Shann and H.C. Fu, A fuzzy neural network for rule acquiring on fuzzy control system, *Fuzzy Sets and Systems*, 71(1995) 345-357.
181. P. Simpson, Fuzzy min-max neural networks: 1.Classification, *IEEE Transactions on Neural Networks*, 3(1992) 776-786.
182. P. Simpson, Fuzzy min-max neural networks: 2.Clustering, *IEEE Transactions on Fuzzy systems*, 1(1993) 32-45.
183. Mu-Chun Su, Ching-Tang Hsieh and Chieh-Ching Chin, A neuro-fuzzy approach to speech recognition without time alignment, *Fuzzy Sets and Systems*, 98(1998) 33-41
184. M. Sugeno and G.-K. Park, An approach to linguistic instruction based learning, *International Journal of Uncertainty, Fuzziness and Knowledge-Based Systems*, 1(1993) 19-56.
185. S.M. Sulzberger, N.N.Tschichold-Gürman and S.J. Vestli, FUN: Optimization of fuzzy rule based systems using neural networks, in: *Proc. IEEE Int. Conference on Neural Networks*, San Francisco, 1993 312-316.
186. C.-T. Sun and J.-S. Jang, A neuro-fuzzy classifier and its applications, in: *Proc. IEEE Int. Conference on Neural Networks*, San Francisco, 1993 94-98.

187. H. Takagi, Fusion technology of fuzzy theory and neural networks - survey and future directions, in: *Proc. First Int. Conf. on Fuzzy Logic & Neural Networks*, 1990 13–26.

188. H. Takagi and I. Hayashi, NN-driven fuzzy reasoning. *International Journal of Approximate Reasoning*, 3(1991) 191-212.

189. H. Takagi, N. Suzuki, T. Koda and Y. Kojima, neural networks designed on approximate reasoning architecture and their applications, *IEEE Transactions on Neural Networks*, 3(1992) 752-760.

190. I.B.Turksen, Fuzzy expert systems for IE/OR/MS, *Fuzzy Sets and Systems*, 51(1992) 1-27.

191. K. Uehara and M. Fujise, Learning of fuzzy inference criteria with artificial neural network, in: *Proc. 1st Int. Conf. on Fuzzy Logic & Neural Networks*, Iizuka, Japan, 1990 193-198.

192. M. Umano and Y, Ezawa, Execution of approximate reasoning by neural network, *Proceedings of FAN Symposium*, 1991 267-273 (in Japanese).

193. H. Virtanen, Combining and incrementing fuzzy evidence - Heuristic and formal approaches to fuzzy logic programming, in: R.Lowen and M.Roubens eds., *Proccedings of the fourth IFSA Congress, vol. Mathematics*, Brussels, 1991 200-203.

194. L.-X. Wang and J.M. Mendel, Generating fuzzy rules by learning from examples, *IEEE Trans. Syst., Man, and Cybernetics*, 22(1992) 1414-1427.

195. Xiaomei Wang and James M. Keller, Human-based spatial relationship generalization through neural/fuzzy approaches, *Fuzzy Sets and Systems*, 101(1999) 5-20.

196. H. Watanabe et al., Application of fuzzy discriminant analysis for diagnosis of valvular heart disease, *IEEE Transactions on Fuzzy Systems*, 2(1994) 267- 276.

197. P.J. Werbos, Neurocontrol and fuzzy logic: connections and designs, *International Journal of Approximate Reasoning*, 6(1992) 185-219.

198. R.R. Yager, Using fuzzy logic to build neural networks, in: R.Lowen and M.Roubens eds., *Proceedings of the Fourth IFSA Congress, Vol. Artifical intelligence*, Brussels, 1991 210-213.

199. R.R. Yager, Implementing fuzzy logic controllers using a neural network framework, *Fuzzy Sets and Systems*, 48(1992) 53-64.

200. R.R. Yager and L.A. Zadeh eds., *Fuzzy Sets, Neural Networks, and Soft Computing* (Van Nostrand Reinhold, New York, 1994).

201. T. Yamakawa, A neo fuzzy neuron and its applications to system identification and prediction of chaotic behaviour, in: J.M. Zurada, R.J. Marks and C.J. Robinson eds., *Computational Intelligence: Imitating Life* (IEEE Press, New York, 1994) 383-395.

202. J. Yan, M. Ryan and J. Power, *Using Fuzzy Logic - Towards Intelligent Systems* (Prentice-Hall, Englewood Cliffs, 1994).

203. Y. Yam and K.S. Leung eds., *Future Directions of Fuzzy Theory and Systems* (World Scientific, Singapore, 1994).

204. L.A. Zadeh and J. Kacprzyk eds., *Computing with Words in Information/Intelligent Systems 1*, Studies Fuzziness and Soft Computing, Vol. 33, Physica-Verlag, Heidelberg, 1999.

205. L.A. Zadeh and J. Kacprzyk eds., *Computing with Words in Information/Intelligent Systems 2*, Studies Fuzziness and Soft Computing, Vol. 34, Physica-Verlag, Heidelberg, 1999.

4. Appendix

4.1 Case study: A portfolio problem

Suppose that the value of our portfolio depends on the currency fluctuations on the global finance market. Our knowledge is given in the form of fuzzy if-then rules, where all of the linguistic values for the exchange rates and the portfolio values are represented by sigmoidal fuzzy numbers. It is relatively easy to create fuzzy if-then rules for portfolio evaluation, however it is time-consuming and difficult to fine-tune them.

Following Carlsson and Fullér [35, 40] in this Section we compute the crisp portfolio values by Tsukomoto's inference mechanism and introducing some reasonable interdependences among the linguistic terms we show a simple method for tuning the membership functions in the rules. For modeling the partially known causal link between the exchange rates and the portfolio value we employ fuzzy if-then rules of the following type

$$\Re_i: \text{ if } x_1 \text{ is } A_{i1} \text{ and } \dots \text{ and } x_n \text{ is } A_{in} \text{ then } PV \text{ is } C_i$$

where PV is the linguistic variable for the portfolio value, $x_1, \dots, x_n$ are the linguistic variables for exchange rates having effects on the portfolio value. Each x_i has two linguistic terms *low* and *high*, denoted by L_i and H_i, which satisfy the equality $L_i(t) + H_i(t) = 1$ for each t. The portfolio value can have four terms: *very big* (VB), *big* (B), *small* (S) and *very small* (VS). It is clear that the value of the portfolio can not be negative (in the worst case we lose everything). The membership functions for the portfolio are supposed to satisfy the properties $B(t) + S(t) = 1$, $VS(t) = S(t+c)$ and $VB(t) = B(t-c)$ for some constant c and for each t.

We believe that two linguistic terms $\{low, high\}$ are sufficient for exchange rates, because the term "exchange rate is medium" can be derived from the terms "exchange rate is low" and "exchange rate is high". In a similar manner we consider our portfolio value as *small* if its value is smaller or exceeds a little bit the value of our investment, and *big* if its value is definitively bigger than our investment. The term "portfolio value is medium" is rapidly changing and can be derived from other terms.

Under these assumptions it seems to be reasonable to derive the daily portfolio values from the actual exchange rates and from the rule-base $\Re =$

$\{\Re_1, \dots, \Re_m\}$ by using Tsukomoto's reasoning mechanism, which requires monoton membership functions for all linguistic terms. Consider a simple case with the following three fuzzy if-then rules in our knowledge-base:

$\Re_1 :$ **if** x_1 **is** L_1 **and** x_2 **is** L_2 **and** x_3 **is** L_3 **then** PV **is** VB

$\Re_2 :$ **if** x_1 **is** H_1 **and** x_2 **is** H_2 **and** x_3 **is** L_3 **then** PV **is** B

$\Re_3 :$ **if** x_1 **is** H_1 **and** x_2 **is** H_2 **and** x_3 **is** H_3 **then** PV **is** S

where x_1, x_2 and x_3 denote the exchange rates between USD and DEM, USD and SEK, and USD and FIM, respectively. The rules are interpreted as:

$\Re_1$: If the US dollar is weak against the German mark Swedish crown and the Finnish mark then our portfolio value is very big.

$\Re_2$: If the US dollar is strong against the German mark and the Swedish crown and the US dollar is weak against the Finnish mark then our portfolio value is big.

$\Re_3$: If the US dollar is strong against the German mark the Swedish crown and the Finnish mark then our portfolio value is small.

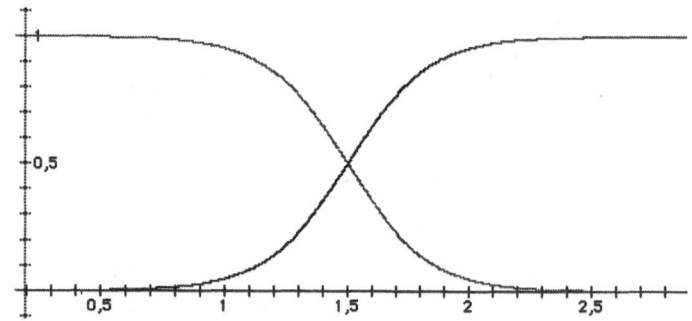

Fig. 4.1. Initial fuzzy sets of "x_1 is low" and "x_1 is high", $b_1 = 6$ and $c_1 = 1.5$.

The fuzzy sets $L_1 =$ "USD/DEM is low" and $H_1 =$ "USD/DEM is high" are given by the following membership functions

$$L_1(t) = \frac{1}{1 + \exp(b_1(t - c_1))},$$

$$H_1(t) = \frac{1}{1 + \exp(-b_1(t - c_1))}.$$

It is easy to check that the equality $L_1(t) + H_1(t) = 1$ holds for all t.

The fuzzy sets $L_2 =$ "USD/SEK is low" and $H_2 =$ "USD/SEK is high" are given by the following membership functions

$$H_2(t) = \frac{1}{1 + \exp(b_2(t - c_2))},$$

$$L_2(t) = \frac{1}{1 + \exp(-b_2(t - c_2))}$$

It is easy to check that the equality $L_2(t) + H_2(t) = 1$ holds for all t. The

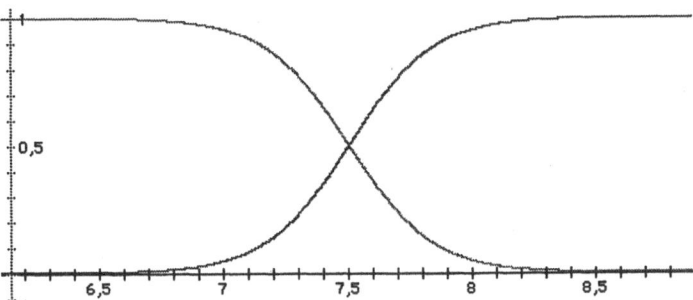

Fig. 4.2. Initial fuzzy sets for "x_2 is low" and "x_2 is high", $b_2 = 6$ and $c_2 = 7.5$.

fuzzy sets $L_3 = $ "USD/FIM is low" and $H_3 = $ "USD/FIM is high" are given by the following membership function

$$L_3(t) = \frac{1}{1 + \exp(b_3(t - c_3))},$$

$$H_3(t) = \frac{1}{1 + \exp(-b_3(t - c_3))}.$$

It is easy to check that the equality $L_3(t) + H_3(t) = 1$ holds for all t. The

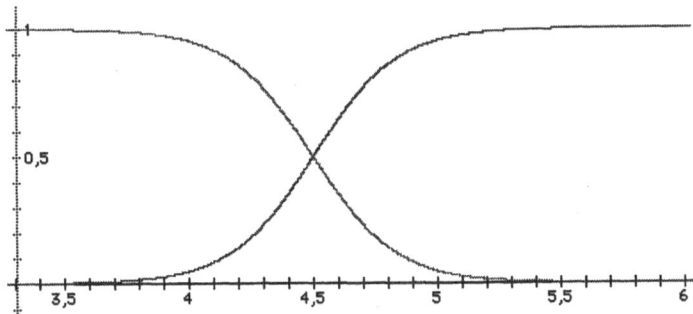

Fig. 4.3. Initial fuzzy sets for "x_3 is low" and "x_3 is high", $b_1 = 6$ and $c_1 = 4.5$.

fuzzy sets $VB = $ "portfolio value is very big" and $VB = $ "portfolio value is very small" are given by the following membership functions

$$VS(t) = \frac{1}{1 + \exp(b_4(t - c_4 - c_5))},$$

$$VB(t) = \frac{1}{1 + \exp(-b_4(t - c_4 + c_5))},$$

The fuzzy sets $B = $ "portfolio value is big" and $S = $ "portfolio value is small" are given by the following membership function

$$B(t) = \frac{1}{1 + \exp(-b_4(t - c_4))},$$

$$S(t) = \frac{1}{1 + \exp(b_4(t - c_4))}.$$

It is easy to check that the equality $B(t) + S(t) = 1$ holds for all t.

We evaluate the daily portfolio value by Tsukamoto's reasoning mechanism, i.e.

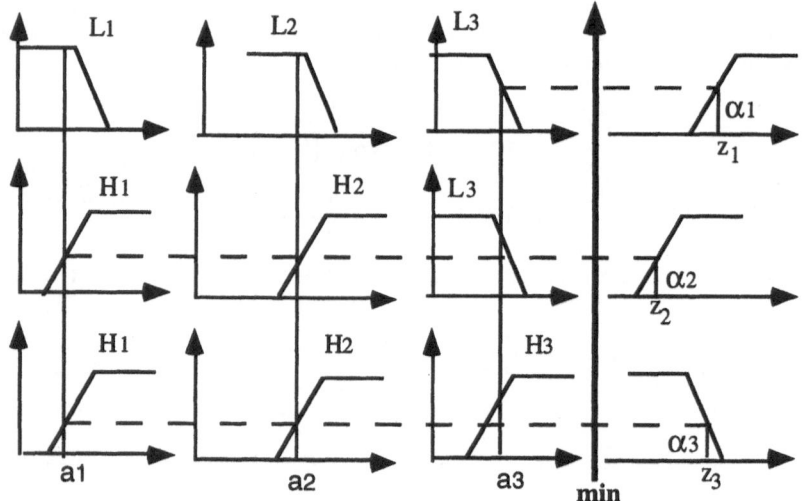

Fig. 4.4. Tsukamoto's reasoning mechanism with three inference rules.

- The firing levels of the rules are computed by

$$\alpha_1 = L_1(a_1) \wedge L_2(a_2) \wedge L_3(a_3),$$

$$\alpha_2 = H_1(a_1) \wedge H_2(a_2) \wedge L_3(a_3),$$

$$\alpha_3 = H_1(a_1) \wedge H_2(a_2) \wedge H_3(a_3),$$

- The individual rule outputs are derived from the relationships

$$z_1 = VB^{-1}(\alpha_1) = c_4 + c_5 + \frac{1}{b_4} \ln \frac{1 - \alpha_1}{\alpha_1}, \tag{4.1}$$

$$z_2 = B^{-1}(\alpha_2) = c_4 + \frac{1}{b_4} \ln \frac{1 - \alpha_2}{\alpha_2} \tag{4.2}$$

$$z_3 = S^{-1}(\alpha_3) = c_4 - \frac{1}{b_4} \ln \frac{1 - \alpha_3}{\alpha_3} \tag{4.3}$$

- The overall system output is expressed as

$$z_0 = \frac{\alpha_1 z_1 + \alpha_2 z_2 + \alpha_3 z_3}{\alpha_1 + \alpha_2 + \alpha_3}$$

where a_1, a_2 and a_3 are the inputs to the system.

4.1.1 Tuning the membership functions

We describe a simple method for learning of membership functions of the antecedent and consequent parts of fuzzy if-then rules. A hybrid neural net [33] computationally identical to our fuzzy system is shown in Fig. 4.5.

- **Layer 1** The output of the node is the degree to which the given input satisfies the linguistic label associated to this node.
- **Layer 2** Each node computes the firing strength of the associated rule. The output of top neuron is

$$\alpha_1 = L_1(a_1) \wedge L_2(a_2) \wedge L_3(a_3),$$

the output of the middle neuron is

$$\alpha_2 = H_1(a_1) \wedge H_2(a_2) \wedge L_3(a_3),$$

and the output of the bottom neuron is

$$\alpha_3 = H_1(a_1) \wedge H_2(a_2) \wedge H_3(a_3).$$

All nodes in this layer is labeled by T, because we can choose other t-norms for modeling the logical *and* operator. The nodes of this layer are called *rule nodes*.

- **Layer 3** Every node in this layer is labeled by N to indicate the normalization of the firing levels.
  The output of the top, middle and bottom neuron is the normalized firing level of the corresponding rule

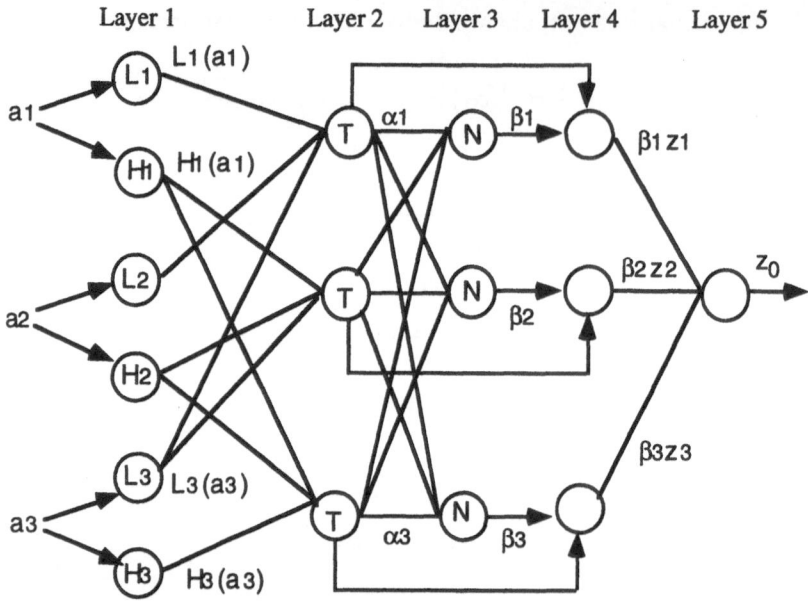

Fig. 4.5. A hybrid neural net ANFIS architecture which is computationally equivalent to Tsukomato's reasoning method.

$$\beta_1 = \frac{\alpha_1}{\alpha_1 + \alpha_2 + \alpha_3},$$

$$\beta_2 = \frac{\alpha_2}{\alpha_1 + \alpha_2 + \alpha_3},$$

$$\beta_3 = \frac{\alpha_3}{\alpha_1 + \alpha_2 + \alpha_3},$$

- **Layer 4** The output of the top, middle and bottom neuron is the product of the normalized firing level and the individual rule output of the corresponding rule

$$\beta_1 z_1 = \beta_1 V B^{-1}(\alpha_1),$$
$$\beta_2 z_2 = \beta_2 B^{-1}(\alpha_2),$$
$$\beta_3 z_3 = \beta_3 S^{-1}(\alpha_3),$$

- **Layer 5** The single node in this layer computes the overall system output as the sum of all incoming signals, i.e.

$$z_0 = \beta_1 z_1 + \beta_2 z_2 + \beta_3 z_3.$$

Suppose we have the following crisp training set

$$\{(x_1, y_1), \ldots, (x_K, y_K)\}$$

where x_k is the vector of the actual exchange rates and y_k is the real value of our portfolio at time k. We define the measure of error for the k-th training pattern as usually

$$E_k = \frac{1}{2}(y_k - o_k)^2$$

where o_k is the computed output from the fuzzy system $\Re$ corresponding to the input pattern x_k, and y_k is the real output, $k = 1, \ldots, K$.

The steepest descent method is used to learn the parameters of the conditional and the consequence parts of the fuzzy rules. We show now how to tune the shape parameters b_4, c_4 and c_5 of the portfolio value. From (4.1), (4.2) and (4.3) we get the following learning rule for the slope, b_4, of the portfolio values

$$b_4(t+1) = b_4(t) - \eta\frac{\partial E_k}{\partial b_4} = b_4(t) - \frac{\eta}{b_4^2}\delta_k\frac{\alpha_1 + \alpha_2 - \alpha_3}{\alpha_1 + \alpha_2 + \alpha_3},$$

In a similar manner we can derive the learning rules for the center c_4

$$c_4(t+1) = c_4(t) - \eta\frac{\partial E_k}{\partial c_4} = c_4(t) + \eta\delta_k\frac{\alpha_1 + \alpha_2 + \alpha_3}{\alpha_1 + \alpha_2 + \alpha_3} = c_4(t) + \eta\delta_k,$$

and for the *shifting value* c_5

$$c_5(t+1) = c_5(t) - \eta\frac{\partial E_k}{\partial c_5} = c_5(t) + \eta\delta_k\frac{\alpha_1}{\alpha_1 + \alpha_2 + \alpha_3}$$

where $\delta_k = (y_k - o_k)$ denotes the error, $\eta > 0$ is the learning rate and t indexes the number of the adjustments. The learning rules for the shape parameters of the antecedent part of the rules can be derived in a similar way.

Table 4.1 shows some mean exchange rates, the computed portfolio values (CPV) with the initial membership functions and real portfolio values (RPV) from 1995.

Date	USD/DEM	USD/SEK	USD/FIM	CPV	RPV
January 11	1.534	7.530	4.779	14.88	19
May 19	1.445	7.393	4.398	17.55	19.4
August 11	1.429	7.146	4.229	19.25	22.6
August 28	1.471	7.325	4.369	17.71	20

Table 4.1. Performance of the fuzzy system before the training.

Table 4.2 shows some mean exchange rates, the computed portfolio values with the final membership functions and real portfolio values from 1995.

Date	USD/DEM	USD/SEK	USD/FIM	CPV	RPV
January 11	1.534	7.530	4.779	18.92	19
May 19	1.445	7.393	4.398	19.37	19.4
August 11	1.429	7.146	4.229	22.64	22.6
August 28	1.471	7.325	4.369	19.9	20

Table 4.2. Performance of the fuzzy system after the training.

4.2 Exercises

Exercise 4.1 *Interpret the following fuzzy set.*

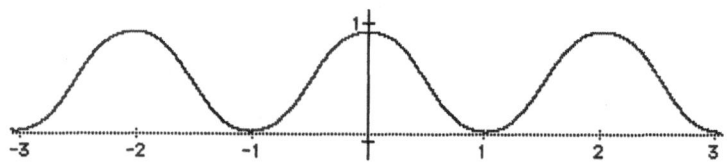

Fig. 4.6. A fuzzy set.

Solution 4.1 *The fuzzy set from the Figure 4.6 can be interpreted as:*

x is close to -2 **or** *x is close to 0* **or** *x is close to 2*

Exercise 4.2 *Suppose we have a fuzzy partition of the universe of discourse* $[-1000, 1000]$ *with three fuzzy terms* $\{N, ZE, P\}$*, where*

$$N(t) = \begin{cases} 1 - \dfrac{t + 1000}{500} & \text{if } -1000 \leq t \leq -500 \\ 1 & \text{if } t \leq -1000 \\ 0 & \text{if } t \geq -500 \end{cases}$$

$$P(t) = \begin{cases} 1 - \dfrac{1000 - t}{500} & \text{if } 500 \leq t \leq 1000 \\ 1 & \text{if } t \geq 1000 \\ 0 & \text{if } t \leq 500 \end{cases}$$

$$ZE(t) = \begin{cases} 1 & \text{if } -500 \le t \le 500 \\ 0 & \text{if } t \ge 1000 \\ 0 & \text{if } t \le -1000 \\ 1 + \dfrac{t + 500}{500} & \text{if } -1000 \le t \le -500 \\ 1 - \dfrac{t - 500}{500} & \text{if } 500 \le t \le 1000 \end{cases}$$

Find the biggest ϵ for which this fuzzy partition satisfies the property ϵ-completeness.

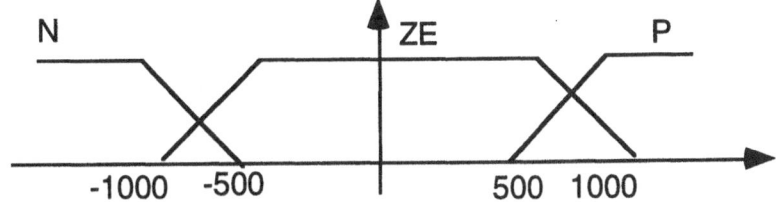

Fig. 4.7. Initial membership functions of $\{N, ZE, P\}$.

Solution 4.2 $\epsilon = 0.5$.

Exercise 4.3 *Show that if $\gamma \le \gamma'$ then the relationship*

$$H_\gamma(a, b) \ge H_{\gamma'}(a, b),$$

holds for all $x, y \in [0, 1]$, i.e. the family of parametized Hamacher's t-norms, $\{H_\gamma\}$, is monoton decreasing.

Solution 4.3 *Let $0 \le \gamma \le \gamma'$. Then from the relationship*

$$\gamma' ab + (1 - \gamma')ab(a + b - ab) \ge \gamma ab + (1 - \gamma)ab(a + b - ab)$$

it follows that

$$\begin{aligned} H_\gamma(a, b) &= \frac{ab}{\gamma + (1 - \gamma)(a + b - ab)} \\ &\ge \frac{ab}{\gamma' + (1 - \gamma')(a + b - ab)} \\ &= H_{\gamma'}(a, b). \end{aligned}$$

Which ends the proof.

Exercise 4.4 *Consider two fuzzy relations R and G, where R is interpreted linguistically as "x is approximately equal to y" and the linguistic interpretation of G is "y is very close to z". Assume R and G have the following membership functions*

$$R = \begin{pmatrix} & y_1 & y_2 & y_3 \\ x_1 & 1 & 0.1 & 0.1 \\ x_2 & 0 & 1 & 0 \\ x_3 & 0.9 & 1 & 1 \end{pmatrix}$$

$$G = \begin{pmatrix} & z_1 & z_2 & z_3 \\ y_1 & 0.4 & 0.9 & 0.3 \\ y_2 & 0 & 0.4 & 0 \\ y_3 & 0.9 & 0.5 & 0.8 \end{pmatrix}$$

What is the membership function of their sup-min composition $R \circ G$? What can be the linguistic interpretation of $R \circ G$?

Solution 4.4

$$R \circ G = \begin{pmatrix} & y_1 & y_2 & y_3 \\ x_1 & 1 & 0.1 & 0.1 \\ x_2 & 0 & 1 & 0 \\ x_3 & 0.9 & 1 & 1 \end{pmatrix} \circ \begin{pmatrix} & z_1 & z_2 & z_3 \\ y_1 & 0.4 & 0.9 & 0.3 \\ y_2 & 0 & 0.4 & 0 \\ y_3 & 0.9 & 0.5 & 0.8 \end{pmatrix}$$

$$= \begin{pmatrix} & z_1 & z_2 & z_3 \\ x_1 & 0.4 & 0.9 & 0.3 \\ x_2 & 0 & 0.4 & 0 \\ x_3 & 0.9 & 0.9 & 0.8 \end{pmatrix}$$

$R \circ G$ can be interpreted as "x is very close to z".

Exercise 4.5 *Assume the membership function of the fuzzy set A, big pressure is*

$$A(u) = \begin{cases} 1 & \text{if } u \geq 5 \\ 1 - \dfrac{5 - u}{4} & \text{if } 1 \leq u \leq 5 \\ 0 & \text{otherwise} \end{cases}$$

Assume the membership function of the fuzzy set B, small volume is

$$B(v) = \begin{cases} 1 & \text{if } v \leq 1 \\ 1 - \dfrac{v - 1}{4} & \text{if } 1 \leq v \leq 5 \\ 0 & \text{otherwise} \end{cases}$$

What is the truth value of the proposition?

"4 is big pressure" → *"3 is small volume"*

where → *is the Łukasiewicz implication.*

Solution 4.5 *Using the definition of Łukasiewicz implication we get*

"4 is big pressure" → *"3 is small volume"*

$$\equiv A(4) \to B(3) = \min\{1 - A(4) + B(3), 1\}$$

and from $A(4) = 0.75$ *and* $B(3) = 0.5$ *we get*

"4 is big pressure" → *"3 is small volume"*

$$= \min\{1 - 0.75 + 0.5, 1\} = 0.75.$$

Exercise 4.6 *Let* A, A', $B \in \mathcal{F}$. *Show that the* Generalized Modus Ponens *inference rule with Gödel implication satisfies*

- **Basic property:** $A \circ (A \to B) = B$
- **Total indeterminance:** $\neg A \circ (A \to B) = \mathbf{1}$ *where* $\mathbf{1}(t) = 1$ *for* $t \in \mathbb{R}$
- **Subset property:** *If* $A' \subset A$ *then* $A' \circ (A \to B) = B$
- **Superset property:** $A' \circ (A \to B) = B' \subset B$ *holds for any* $A' \in \mathcal{F}$.

Solution 4.6 *The Generalized Modus Ponens inference rule says*

premise	if x is A then	y is B
fact	x is A'	
consequence:		y is B'

where the consequence B' *is determined as a composition of the fact and the fuzzy implication operator*

$$B' = A' \circ (A \to B)$$

that is,

$$B'(v) = \sup_{u \in U} \min\{A'(u), (A \to B)(u, v)\}, \quad v \in V.$$

Let us choose Gödel implication operator

$$A(x) \to B(y) := \begin{cases} 1 & \text{if } A(x) \leq B(y) \\ B(y) & \text{otherwise} \end{cases}$$

Proof. **Basic property.**

Let $A' = A$ and let $x, y \in \mathbb{R}$ be arbitrarily fixed. On the one hand from the definition of Gödel implication operator we obtain

$$\min\{A(x), A(x) \to B(y)\} = \begin{cases} A(x) & \text{if } A(x) \le B(y) \\ B(y) & \text{if } A(x) > B(y) \end{cases}$$

That is,

$$B'(y) = \sup_x \min\{A(x), A(x) \to B(y)\} \le B(y)$$

On the other hand from continuity and normality of A it follows that there exists an $x' \in \mathbb{R}$ such that $A(x') = B(y)$. So

$$B'(y) = \sup_x \min\{A(x), A(x) \to B(y)\}$$
$$\ge \min\{A(x'), A(x') \to B(y)\}$$
$$= B(y).$$

Total indeterminance.

Let $x' \notin \mathrm{supp}(A)$ be arbitrarily chosen. Then from $A(x') = 0$ it follows that

$$B'(y) = \sup_x \min\{1 - A(x), A(x) \to B(y)\}$$
$$\ge \min\{1 - A(x'), A(x') \to B(y)\}$$
$$= 1.$$

for any $y \in \mathbb{R}$.

Subset.

Let $A'(x) \le A(x), \forall x \in \mathbb{R}$. Then

$$B'(y) = \sup_x \min\{A'(x), A(x) \to B(y)\}$$
$$\le \sup_x \min\{A(x), A(x) \to B(y)\}$$
$$= B(y).$$

Superset.

From $A' \in \mathcal{F}$ it follows that there exists an $x' \in \mathbb{R}$ such that $A'(x') = 1$. Then

$$B'(y) = \sup_x \min\{A'(x), A(x) \to B(y)\}$$
$$\ge \min\{A'(x'), A(x') \to B(y)\}$$
$$= A(x') \to B(y)$$
$$\ge B(y).$$

Which ends the proof.

Exercise 4.7 *Construct a single-neuron network, which computes the material implication function. The training set is*

	x_1	x_2	$o(x_1, x_2)$
1.	1	1	1
2.	1	0	0
3.	0	1	1
4.	0	0	1

Solution 4.7 *A solution to the material implication function is shown in Fig. 4.8.*

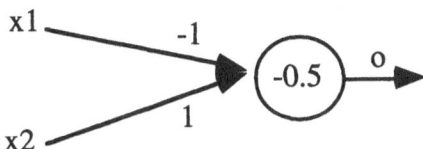

Fig. 4.8. A single-neuron network for the material implication.

Exercise 4.8 *Suppose we have the following fuzzy rule base.*

if	x is SMALL	and	y is BIG	then	$z = x - y$
if	x is BIG	and	y is SMALL	then	$z = x + y$
if	x is BIG	and	y is BIG	then	$z = x + 2y$

where the membership functions SMALL and BIG are defined by

$$\mu_{SMALL}(v) = \begin{cases} 1 & \text{if } v \leq 1 \\ 1 - \dfrac{v-1}{4} & \text{if } 1 \leq v \leq 5 \\ 0 & \text{otherwise} \end{cases}$$

$$\mu_{BIG}(u) = \begin{cases} 1 & \text{if } u \geq 5 \\ 1 - \dfrac{5-u}{4} & \text{if } 1 \leq u \leq 5 \\ 0 & \text{otherwise} \end{cases}$$

Suppose we have the inputs $x_0 = 3$ and $y_0 = 3$. What is the output of the system, z_0, if we use Sugeno's inference mechanism?

Solution 4.8 *The firing level of the first rule is*

$$\alpha_1 = \min\{\mu_{SMALL}(3), \mu_{BIG}(3)\} = \min\{0.5, 0.5\} = 0.5$$

the individual output of the first rule is

$$z_1 = x_0 - y_0 = 3 - 3 = 0$$

The firing level of the second rule is

$$\alpha_1 = \min\{\mu_{BIG}(3), \mu_{SMALL}(3)\} = \min\{0.5, 0.5\} = 0.5$$

the individual output of the second rule is

$$z_2 = x_0 + y_0 = 3 + 3 = 6$$

The firing level of the third rule is

$$\alpha_1 = \min\{\mu_{BIG}(3), \mu_{BIG}(3)\} = \min\{0.5, 0.5\} = 0.5$$

the individual output of the third rule is

$$z_3 = x_0 + 2y_0 = 3 + 6 = 9$$

and the system output, z_0, is computed from the equation

$$z_0 = \frac{0 \times 0.5 + 6 \times 0.5 + 9 \times 0.5}{0.5 + 0.5 + 0.5} = 5.0.$$

Exercise 4.9 *Why do we use differentiable transfer functions in multi-layer feedforward neural networks?*

Solution 4.9 *We use differentiable transfer functions in multi-layer networks, because the derivative of the error function is used in the generalized delta learning rule.*

Exercise 4.10 *What is the meaning of the error correction learning procedure?*

Solution 4.10 *The error correction learning procedure is simple enough in conception. The procedure is as follows: During training an input is put into the network and flows through the network generating a set of values on the output units. Then, the actual output is compared with the desired target, and a match is computed. If the output and target match, no change is made to the net. However, if the output differs from the target a change must be made to some of the connections.*

Exercise 4.11 *Let $A = (a, \alpha, \beta)$ be a triangular fuzzy number. Calculate $[A]^\gamma$ as a function of a, α and β.*

Solution 4.11 *The γ-cut of triangular fuzzy number $A = (a, \alpha, \beta)$ is*

$$[A]^\gamma = [a - (1 - \gamma)\alpha, a + (1 - \gamma)\beta], \quad \forall \gamma \in [0, 1].$$

Especially, $[A]^1 = \{a\}$ and $[A]^0 = [a - \alpha, a + \beta]$.

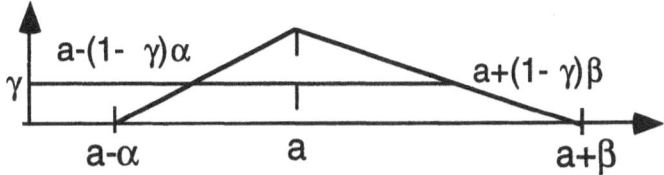

Fig. 4.9. γ-cut of a triangular fuzzy number.

Exercise 4.12 *Consider some alternative with the following scores on five criteria*

Criteria:	C_1	C_2	C_3	C_4	C_5
Importance:	VH	VH	M	L	VL
Score:	M	L	EX	VH	EX

Calculate the unit score of this alternative.

Solution 4.12 *In this case we have*

$$\mathcal{U} = \min\{Neg(VH) \vee M, Neg(VH) \vee L, Neg(M) \vee EX,$$
$$Neg(L) \vee VH, Neg(VL) \vee EX\}$$
$$= \min\{VL \vee M, VL \vee L, M \vee EX, H \vee VH, VH \vee EX\}$$
$$= \min\{M, L, EX, VH, EX\}$$
$$= L.$$

Exercise 4.13 *Let $A = (a, \alpha)$ and $B = (b, \beta)$ be fuzzy numbers of symmetrical triangular form. Calculate their Hausdorff distances, $D(A, B)$, as a function of a, b, α and β.*

Solution 4.13 *The γ-cuts of A and B can be written in the form*

$$[A]^\gamma = [a_1(\gamma), a_2(\gamma)] = [a - (1 - \gamma)\alpha, a + (1 - \gamma)\alpha], \ \forall \gamma \in [0, 1].$$

$$[B]^\gamma = [b_1(\gamma), b_2(\gamma)] = [b - (1 - \gamma)\beta, b + (1 - \gamma)\beta], \ \forall \gamma \in [0, 1].$$

and from the definition of Hausdorff distance

$$D(A, B) = \sup_{\gamma \in [0,1]} \max\{|a_1(\gamma) - b_1(\gamma)|, |a_2(\gamma) - b_2(\gamma)|\}$$

we get

$$D(A, B) = \sup_{\gamma \in [0,1]} \max\{|a - b + (1 - \gamma)(\alpha - \beta)|, |a - b + (1 - \gamma)(\beta - \alpha)|\}$$

That is,

$$D(A, B) = \max\{|a - b + \alpha - \beta|, |a - b + \beta - \alpha|\}.$$

Exercise 4.14 *The error function to be minimized is given by*

$$E(w_1, w_2) = \frac{1}{2}[(w_2 - w_1)^2 + (1 - w_1)^2]$$

Find analytically the gradient vector

$$E'(w) = \begin{bmatrix} \partial_1 E(w) \\ \partial_2 E(w) \end{bmatrix}$$

Find analytically the weight vector w^* that minimizes the error function such that

$$E'(w) = 0.$$

Derive the steepest descent method for the minimization of E.

Solution 4.14 *The gradient vector of E is*

$$E'(w) = \begin{bmatrix} (w_1 - w_2) + (w_1 - 1) \\ (w_2 - w_1) \end{bmatrix} = \begin{bmatrix} 2w_1 - w_2 - 1 \\ w_2 - w_1 \end{bmatrix}$$

and $w^*(1, 1)^T$ is the unique solution to the equation

$$\begin{bmatrix} 2w_1 - w_2 - 1 \\ w_2 - w_1 \end{bmatrix} = \begin{bmatrix} 0 \\ 0 \end{bmatrix}.$$

The steepest descent method for the minimization of E reads

$$\begin{bmatrix} w_1(t+1) \\ w_2(t+1) \end{bmatrix} = \eta \begin{bmatrix} 2w_1(t) - w_2(t) - 1 \\ w_2(t) - w_1(t) \end{bmatrix}.$$

where $\eta > 0$ is the learning constant and t indexes the number of iterations. That is,

$$w_1(t+1) = w_1(t) - \eta(2w_1(t) - w_2(t) - 1)$$

$$w_2(t+1) = w_2(t) - \eta(w_2(t) - w_1(t))$$

Exercise 4.15 *Let f be a bipolar sigmoidal activation function of the form*

$$f(t) = \frac{2}{1 + \exp(-t)} - 1.$$

Show that f satisfies the following differential equality

$$f'(t) = \frac{1}{2}\left(1 - f^2(t)\right).$$

Solution 4.15 *By using the chain rule for derivatives of composed functions we get*

$$f'(t) = \frac{2\exp(-t)}{[1+\exp(-t)]^2}.$$

From the identity

$$\frac{1}{2}\left[1 - \left(\frac{1-\exp(-t)}{1+\exp(-t)}\right)^2\right] = \frac{2\exp(-t)}{[1+\exp(-t)]^2}$$

we get

$$\frac{2\exp(-t)}{[1+\exp(-t)]^2} = \frac{1}{2}(1 - f^2(t)).$$

Which completes the proof.

Exercise 4.16 *Let f be a unipolar sigmoidal activation function of the form*

$$f(t) = \frac{1}{1+\exp(-t)}.$$

Show that f satisfies the following differential equality

$$f'(t) = f(t)(1 - f(t)).$$

Solution 4.16 *By using the chain rule for derivatives of composed functions we get*

$$f'(t) = \frac{\exp(-t)}{[1+\exp(-t)]^2}$$

and the identity

$$\frac{1}{[1+\exp(-t)]^2} = \frac{\exp(-t)}{1+\exp(-t)}\left(1 - \frac{\exp(-t)}{1+\exp(-t)}\right)$$

verifies the statement of the exercise.

Exercise 4.17 *Construct a hybid neural net implementing Tsukumato's reasoning mechanism with two input variables, two linguistiuc values for each input variable and two fuzzy IF-THEN rules.*

Solution 4.17 *Consider the fuzzy rule base*

$$\Re_1 : \text{if } x \text{ is } A_1 \text{ and } y \text{ is } B_1 \text{ then } z \text{ is } C_1$$

$$\Re_2 : \text{if } x \text{ is } A_2 \text{ and } y \text{ is } B_2 \text{ then } z \text{ is } C_2$$

where all linguistic terms are supposed to have monotonic membership functions.

The firing levels of the rules are computed by

$$\alpha_1 = A_1(x_0) \times B_1(y_0)$$

$$\alpha_2 = A_2(x_0) \times B_2(y_0),$$

where the logical *and* can be modelled by any continuous t-norm, e.g

$$\alpha_1 = A_1(x_0) \wedge B_1(y_0)$$

$$\alpha_2 = A_2(x_0) \wedge B_2(y_0),$$

In this mode of reasoning the individual crisp control actions z_1 and z_2 are computed as

$$z_1 = C_1^{-1}(\alpha_1) \text{ and } z_2 = C_2^{-1}(\alpha_2)$$

and the overall control action is expressed as

$$z_0 = \frac{\alpha_1 z_1 + \alpha_2 z_2}{\alpha_1 + \alpha_2} = \beta_1 z_1 + \beta_2 z_2$$

where β_1 and β_1 are the normalized values of α_1 and α_2 with respect to the sum $(\alpha_1 + \alpha_2)$, i.e.

$$\beta_1 = \frac{\alpha_1}{\alpha_1 + \alpha_2}, \quad \beta_2 = \frac{\alpha_2}{\alpha_1 + \alpha_2}.$$

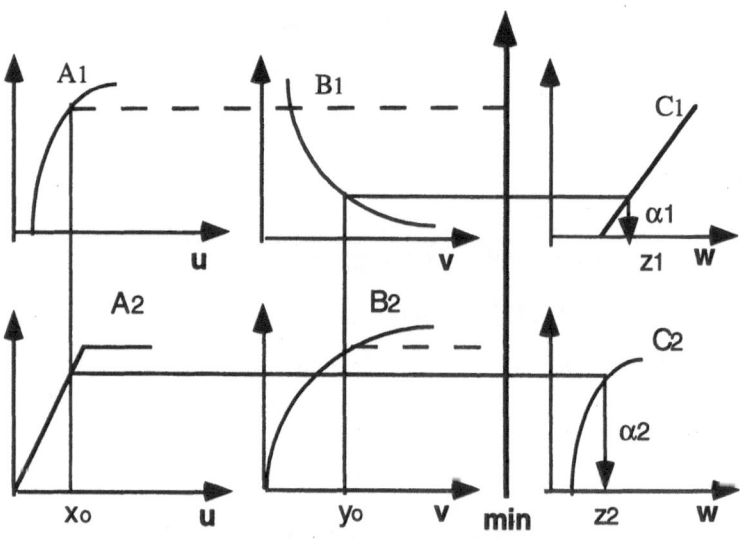

Fig. 4.10. Tsukamoto's inference mechanism.

A hybrid neural net computationally identical to this type of reasoning is shown in the Fig. 4.11.

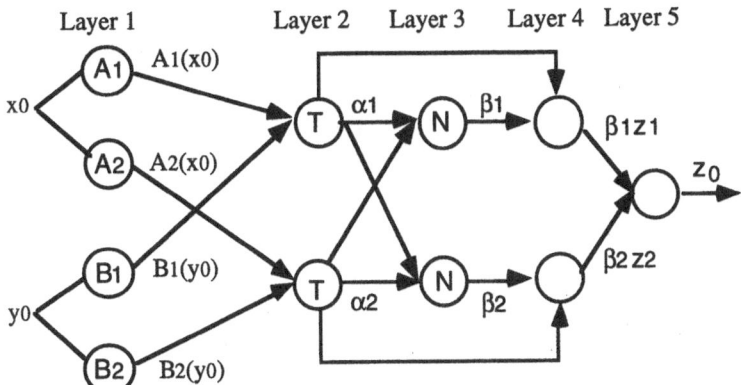

Fig. 4.11. A hybrid neural net (ANFIS architecture) which is computationally equivalent to Tsukomato's reasoning method.

- **Layer 1** The output of the node is the degree to which the given input satisfies the linguistic label associated to this node.
- **Layer 2** Each node computes the firing strength of the associated rule. The output of top neuron is

$$\alpha_1 = A_1(x_0) \times B_1(y_0) = A_1(x_0) \wedge B_1(y_0),$$

and the output of the bottom neuron is

$$\alpha_2 = A_2(x_0) \times B_2(y_0) = A_2(x_0) \wedge B_2(y_0)$$

Both nodes in this layer is labeled by T, because we can choose other t-norms for modeling the logical *and* operator. The nodes of this layer are called *rule nodes*.

- **Layer 3** Every node in this layer is labeled by N to indicate the normalization of the firing levels.
  The output of top neuron is the normalized (with respect to the sum of firing levels) firing level of the first rule

$$\beta_1 = \frac{\alpha_1}{\alpha_1 + \alpha_2},$$

and the output of the bottom neuron is the normalized firing level of the second rule

$$\beta_2 = \frac{\alpha_2}{\alpha_1 + \alpha_2},$$

- **Layer 4** The output of top neuron is the product of the normalized firing level and the individual rule output of the first rule

$$\beta_1 z_1 = \beta_1 C_1^{-1}(\alpha_1).$$

The output of top neuron is the product of the normalized firing level and the individual rule output of the second rule

$$\beta_2 z_2 = \beta_2 C_2^{-1}(\alpha_2),$$

- **Layer 5** The single node in this layer computes the overall system output as the sum of all incoming signals, i.e.

$$z_0 = \beta_1 z_1 + \beta_2 z_2.$$

Exercise 4.18 *Show that fuzzy inference systems with simplified fuzzy IF-THEN rules are universal approximators.*

Solution 4.18 *Consider a fuzzy inference systems with two simplified fuzzy IF-THEN rules*

$$\Re_1: \quad \text{if } x_1 \text{ is } A_{11} \text{ and } x_2 \text{ is } A_{12} \text{ then } y = z_1$$
$$\Re_2: \quad \text{if } x_1 \text{ is } A_{21} \text{ and } x_2 \text{ is } A_{22} \text{ then } y = z_2$$

Suppose that the output of the system $\Re = \{\Re_1, \Re_2\}$ for a given input is computed by

$$z = \frac{\alpha_1 z_1 + \alpha_2 z_2}{\alpha_1 + \alpha_2} \tag{4.4}$$

where α_1 and α_2 denote the firing strengths of the rules with respect to given input vector. Let z' be the output of the system $\Re$ for some other input.

We recall the Stone-Weierstrass theorem:

Theorem 4.2.1 *Let domain K be a compact space of n dimensions, and let $\mathcal{G}$ be a set of continuous real-valued functions on K, satisfying the following criteria:*

1. *The constant function $f(x) = 1$ is in $\mathcal{G}$.*
2. *For any two points $x_1 \neq x_2$ in K, there is an f in $\mathcal{G}$ such that $f(x_1) \neq f(x_2)$.*
3. *If f_1 and f_2 are two functions in $\mathcal{G}$, then fg and $\alpha_1 f_1 + \alpha_2 f_2$ are in $\mathcal{G}$ for any two real numbers α_1 and α_2.*

Then $\mathcal{G}$ is dense in $C(K)$, the set of continuous real-valued functions on K. In other words, for any $\epsilon > 0$ and any function g in $C(K)$, there exists g in $\mathcal{G}$ such that

$$\|f - g\|_\infty = \sup_{x \in K} |f(x) - g(x)| \leq \epsilon.$$

Proof. We show that $az + bz'$, $\forall a, b \in \mathbb{R}$ and zz' can be written in the form (4.4) which means that fuzzy inference systems with simplified fuzzy IF-THEN rules satisfy the conditions of Stone-Weierstrass theorem, i.e. they can approximate all continuous functions on a compact domain.

For $az + bz'$ we get

$$az + bz' = a\frac{\alpha_1 z_1 + \alpha_2 z_2}{\alpha_1 + \alpha_2} + b\frac{\alpha'_1 z'_1 + \alpha'_2 z'_2}{\alpha'_1 + \alpha'_2} =$$

$$\frac{a(\alpha_1 z_1 + \alpha_2 z_2)(\alpha'_1 + \alpha'_2) + b(\alpha'_1 z'_1 + \alpha'_2 z'_2)(\alpha_1 + \alpha_2)}{(\alpha_1 + \alpha_2)(\alpha'_1 + \alpha'_2)} =$$

$$\frac{\alpha_1\alpha'_1(az_1 + bz'_1) + \alpha_1\alpha'_2(az_1 + bz'_2) + \alpha_2\alpha'_1(az_2 + bz'_1) + \alpha_2\alpha'_2(az_2 + bz'_2)}{\alpha_1\alpha'_1 + \alpha_1\alpha'_2 + \alpha_2\alpha'_1 + \alpha_2\alpha'_2}$$

So, zz' is the output of a fuzzy inference system with four simplified fuzzy IF-THEN, where the individual rule outputs are: $az_1 + bz'_1$, $az_1 + bz'_2$, $az_2 + bz'_1$ and $az_2 + bz'_2$, and the firing strengths of the associated rules are $\alpha_1\alpha'_1$, $\alpha_1\alpha'_2$, $\alpha_2\alpha'_1$ and $\alpha_2\alpha'_2$, respectively.

Finally, for zz' we obtain

$$zz' = \frac{\alpha_1 z_1 + \alpha_2 z_2}{\alpha_1 + \alpha_2} \times \frac{\alpha'_1 z'_1 + \alpha'_2 z'_2}{\alpha'_1 + \alpha'_2}$$

$$= \frac{\alpha_1\alpha'_1 z_1 z'_1 + \alpha_1\alpha'_2 z_1 z'_2 + \alpha_2\alpha'_1 z_2 z'_1 + \alpha_2\alpha'_2 z_2 z'_2}{\alpha_1\alpha'_1 + \alpha_1\alpha'_2 + \alpha_2\alpha'_1 + \alpha_2\alpha'_2}$$

So, zz' is the output of a fuzzy inference system with four simplified fuzzy IF-THEN, where the individual rule outputs are: $z_1 z'_1$, $z_1 z'_2$, $z_2 z'_1$ and $z_2 z'_2$, and the firing strengths of the associated rules are $\alpha_1\alpha'_1$, $\alpha_1\alpha'_2$, $\alpha_2\alpha'_1$ and $\alpha_2\alpha'_2$, respectively.

Which completes the proof.

Exercise 4.19 *Let $A_1 = (a_1, \alpha)$ and $A_2 = (a_2, \alpha)$ be fuzzy numbers of symmetric triangular form. Compute analytically the membership function of their product-sum, $A_1 \oplus A_2$, defined by*

$$(A_1 \oplus A_2)(y) = \sup_{x_1 + x_2 = y} T_P(A_1(x_1), A_2(x_2)) = \sup_{x_1 + x_2 = y} A_1(x_1)A_2(x_2).$$

Solution 4.19 *The membership functions of $A_1 = (a_1, \alpha)$ and $A_2 = (a_2, \alpha)$ are defined by*

$$A_1(t) = \begin{cases} 1 - \dfrac{|a_1 - t|}{\alpha} & \text{if } |a_1 - t| \leq \alpha \\ 0 & \text{otherwise} \end{cases}$$

$$A_2(t) = \begin{cases} 1 - \dfrac{|a_2 - t|}{\alpha} & \text{if } |a_2 - t| \le \alpha \\ 0 & \text{otherwise} \end{cases}$$

First we show hat the support of the product-sum, $A_1 \oplus A_2$, is equal to the sum of the supports of A_1 and A_2, i.e.

$$\text{supp}(A_1 \oplus A_2) = \text{supp}(A_1) + \text{supp}(A_2) =$$

$$(a_1 - \alpha, a_1 + \alpha) + (a_2 - \alpha, a_2 + \alpha) = (a_1 + a_2 - 2\alpha, a_1 + a_2 + 2\alpha).$$

Really, the product $A_1(x_1)A_2(x_2)$ is positive, if and only if $A_1(x_1) > 0$ and $A_2(x_2) > 0$, i.e. $x_1 \in (a_1 - \alpha, a_1 + \alpha)$ and $x_2 \in (a_2 - \alpha, a_2 + \alpha)$. This means that $(A_1 \oplus A_2)(y)$ is positive if and only if y is represented as a sum of x_1 from $\text{supp}(A_1)$ and x_2 from $\text{supp}(A_2)$.

From the definition of product-sum it follows that $(A_1 \oplus A_2)(y)$, $y \in [a_1 + a_2 - 2\alpha, a_1 + a_2]$, is equal to the optimal value of the following mathematical programming problem

$$\left(1 - \frac{a_1 - x}{\alpha}\right)\left(1 - \frac{a_2 - y + x}{\alpha}\right) \tag{4.5}$$

$$\text{subject to} \quad a_1 - \alpha \le x \le a_1, \ a_2 - \alpha \le y - x \le a_2.$$

Using Lagrange's multipliers method for the solution of (4.5) we get that its optimal value is

$$\left(1 - \frac{a_1 + a_2 - y}{2\alpha}\right)^2$$

and its unique solution is

$$x = \frac{a_1 - a_2 + y}{2}$$

(where the derivative of the objective function vanishes).

In order to determine $(A_1 \oplus A_2)(y)$, $y \in [a_1 + a_2, a_1 + a_2 + 2\alpha]$, we need to solve the following mathematical programming problem

$$\left(1 - \frac{a_1 - x}{\alpha}\right)\left(1 - \frac{a_2 - y + x}{\alpha}\right) \tag{4.6}$$

$$\text{subject to} \quad a_1 \le x \le a_1 + \alpha, \ a_2 \le y - x \le a_2 + \alpha.$$

Using Lagrange's multipliers method for the solution of (4.6) we get that its optimal value is

$$\left(1 - \frac{y - (a_1 + a_2)}{2\alpha}\right)^2.$$

Summarizing these findings we obtain that

$$(A_1 \oplus A_2)(y) = \begin{cases} \left(1 - \dfrac{|a_1 + a_2 - y|}{2\alpha}\right)^2 & \text{if } |a_1 + a_2 - y| \leq 2\alpha \\ 0 & \text{otherwise} \end{cases} \qquad (4.7)$$

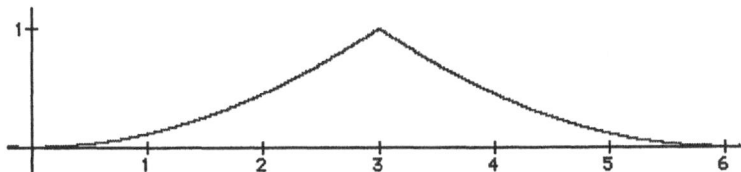

Fig. 4.12. Product-sum of fuzzy numbers $(1, 3/2)$ and $(2, 3/2)$.

Exercise 4.20 Let $A_i = (a_i, \alpha)$, $i \in \mathbb{N}$ be fuzzy numbers of symmetric triangular form. Suppose that

$$a := \sum_{i=1}^{\infty} a_i$$

exists and it is finite. Find the limit distribution of the product-sum

$$\bigoplus_{i=1}^{n} A_i$$

when $n \to \infty$.

Solution 4.20 Let us denote B_n the product sum of A_i, $i = 1, \ldots, n$, i.e.

$$B_n = A_1 \oplus \cdots \oplus A_n$$

Making an induction argument on n we show that

$$B_n(y) = \begin{cases} \left(1 - \dfrac{|a_1 + \cdots + a_n - y|}{n\alpha}\right)^n & \text{if } |a_1 + \cdots + a_n - y| \leq n\alpha \\ 0 & \text{otherwise} \end{cases} \qquad (4.8)$$

From (4.7) it follows that (4.8) holds for $n = 2$. Let us assume that it holds for some $n \in \mathbb{N}$. Then using the definition of product-sum we obtain

$$\begin{aligned}
B_{n+1}(y) &= (B_n + A_{n+1})(y) \\
&= \sup_{x_1 + x_2 = y} B_n(x_1) A_{n+1}(x_2) \\
&= \sup_{x_1 + x_2 = y} \left(1 - \frac{|a_1 + \cdots + a_n - x_1|}{n\alpha}\right)^2 \left(1 - \frac{|a_{n+1} - x_2|}{\alpha}\right) \\
&= \left(1 - \frac{|a_1 + \cdots + a_{n+1} - y|}{(n+1)\alpha}\right)^{n+1}.
\end{aligned}$$

This ends the proof.

From (4.8) we obtain the limit distribution of B_n's as

$$\lim_{n \to \infty} B_n(y) = \lim_{n \to \infty} \left(1 - \frac{|a_1 + \cdots + a_n - y|}{n\alpha} \right)^n$$

$$= \exp \left(- \frac{|a - y|}{\alpha} \right).$$

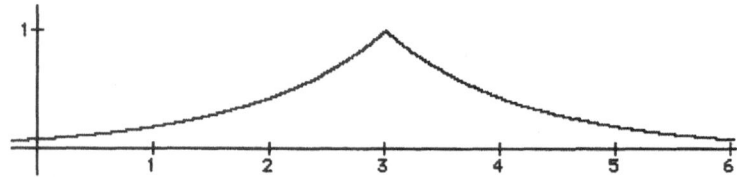

Fig. 4.13. The limit distribution of the product-sum of A_i, $i \in \mathbb{N}$.

Exercise 4.21 *Prove the following theorem:*

Theorem 4.2.2 *[78] Let T be an Archimedean t-norm with additive generator f and let $\tilde{a}_i = (a_i, b_i, \alpha, \beta)_{LR}$, $i = 1, \dots, n$, be fuzzy numbers of LR-type. If L and R are twice differentiable, concave functions, and f is twice differentiable, strictly convex function then the membership function of the T-sum $\tilde{A}_n = \tilde{a}_1 + \cdots + \tilde{a}_n$ is*

$$\tilde{A}_n(z) = \begin{cases} 1 & \text{if } A_n \leq z \leq B_n \\ f^{[-1]}\left(n \times f\left(L\left(\frac{A_n - z}{n\alpha} \right) \right) \right) & \text{if } A_n - n\alpha \leq z \leq A_n \\ f^{[-1]}\left(n \times f\left(R\left(\frac{z - B_n}{n\beta} \right) \right) \right) & \text{if } B_n \leq z \leq B_n + n\beta \\ 0 & \text{otherwise} \end{cases}$$

where $A_n = a_1 + \cdots + a_n$ and $B_n = b_1 + \cdots + b_n$.

Solution 4.21 *Recall that a t-norm T is Archimedean iff T is continuous and $T(x, x) < x$ for all $x \in (0, 1)$. Every Archimedean t-norm T is representable by a continuous and decreasing function $f \colon [0, 1] \to [0, \infty]$ with $f(1) = 0$ and*

$$T(x, y) = f^{[-1]}(f(x) + f(y)),$$

where $f^{[-1]}$ is the pseudo-inverse of f, defined by

$$f^{[-1]}(y) = \begin{cases} f^{-1}(y) & \text{if } y \in [0, f(0)] \\ 0 & \text{otherwise.} \end{cases}$$

The function f is the additive generator of T.

If T is an Archimedean t-norm and $\tilde{a}_1$ and $\tilde{a}_2$ are fuzzy sets of the real line (i.e. fuzzy quantities) then their T-sum $\tilde{A}_2 := \tilde{a}_1 + \tilde{a}_2$ is defined by

$$\tilde{A}_2(z) = \sup_{x_1+x_2=z} T(\tilde{a}_1(x_1), \tilde{a}_2(x_2)), \quad z \in \mathbb{R},$$

which expression can be written in the form

$$\tilde{A}_2(z) = f^{[-1]}(f(\tilde{a}_1(x_1)) + f(\tilde{a}_2(x_2))),$$

where f is the additive generator of T. By the associativity of T, the membership function of the T-sum $\tilde{A}_n := \tilde{a}_1 + \cdots + \tilde{a}_n$ of fuzzy quantities $\tilde{a}_1, \ldots, \tilde{a}_n$ can be written as

$$\tilde{A}_n(z) = \sup_{x_1+\cdots+x_n=z} f^{[-1]}\left(\sum_{i=1}^{n} f(\tilde{a}_i(x_i))\right).$$

Since f is continuous and decreasing, $f^{[-1]}$ is also continuous and nonincreasing, we have

$$\tilde{A}_n(z) = f^{[-1]}\left(\inf_{x_1+\cdots+x_n=z} \sum_{i=1}^{n} f(\tilde{a}_i(x_i))\right).$$

As it was mentioned above, the investigated membership function is

$$\tilde{A}_n(z) = f^{[-1]}\left(\inf_{x_1+\cdots+x_n=z} f(\tilde{a}_1(x_1)) + \cdots + f(\tilde{a}_n(x_n))\right). \tag{4.9}$$

It is easy to see that the support of $\tilde{A}_n$ is included in the interval $[A_n - n\alpha, B_n+n\beta]$. From the decomposition rule of fuzzy numbers into two separate parts it follows that the peak of $\tilde{A}_n$ is $[A_n, B_n]$.

Moreover, if we consider the right hand side of $\tilde{A}_n$ (i.e. $B_n \leq z \leq B_n+n\beta$) then only the right hand sides of terms $\tilde{a}_n$ come into account in (4.9) (i.e. $b_i \leq x_i \leq b_i + \beta$, $i = 1, \ldots, n$). The same thing holds for the left hand side of $\tilde{A}_n$ (i.e. if $A_n - n\alpha \leq z \leq A_n$ then $a_i - \alpha \leq x_i \leq a_i$, $i = 1, \ldots, n$).

Let us now consider the right hand side of $\tilde{A}_n$, so let $B_n \leq z \leq B_n + n\beta$. (A similar method can be used if $A_n - n\alpha \leq z \leq A_n$.) The constraints

$$x_1 + \cdots + x_n = z, \quad b_i \leq x_i \leq b_i + \beta, \quad i = 1, \cdots, n$$

determine a compact and convex domain $\mathcal{K} \subset \mathbb{R}^n$ which can be considered as the section of the brick

$$\mathcal{B} := \{(x_1, \ldots, x_n) \in \mathbb{R}^n \mid b_i \leq x_i \leq b_i + \beta, \quad i = 1, \ldots, n\}$$

by the hyperplane

$$\mathcal{P} := \{(x_1, \ldots, x_n) \in \mathbb{R}^n \mid x_1 + \cdots + x_n = z\}$$

In order to determine $\bar{A}_n(z)$ we need to find the conditional minimum value of the function $\phi : \mathcal{B} \to \mathbb{R}$,

$$\phi(x_1, \ldots, x_n) := f(\tilde{a}_1(x_1)) + \cdots + f(\tilde{a}_n(x_n))$$

subject to condition $(x_1, \ldots, x_n) \in \mathcal{K}$. $\mathcal{K}$ is compact and ϕ is continuous, this is why we could change the infimum with minimum.

Following the Lagrange's multipliers method we are searching the stationary points of the auxiliary function $\Phi : \mathcal{B} \to \mathbb{R}$

$$\Phi(x_1, \ldots, x_n) := \phi(x_1, \ldots, x_n) + \lambda(z - (x_1 + x_2 + \cdots + x_n))$$

i.e. the points $(x_1, \cdots, x_n, \lambda) \in \mathcal{B} \times \mathbb{R}$ where its derivative vanishes.

It is clear, that Φ is twice differentiable, and its partial derivative with respect to x_i is

$$\partial_i \Phi(x_1, \ldots, x_n, \lambda) = f'(R(\sigma_i(x_i)))R'(\sigma_i(x_i))\frac{1}{\beta} - \lambda,$$

where $\sigma_i(x_i) := (x_i - b_i)/\beta$ and with respect to λ is

$$\partial_\lambda \Phi(x_1, \ldots, x_n, \lambda) = z - (x_1 + \cdots + x_n)$$

We get the system of equations:

$$f'(R(\sigma_i(x_i)))R'(\sigma_i(x_i))\frac{1}{\beta} - \lambda = 0, \ i = 1, \ldots, n,$$

$$z - (x_1 + \cdots + x_n) = 0.$$

If we take

$$\hat{x}_i = b_i + \frac{z - B_n}{n}, \ i = 1, \ldots, n,$$

then

$$\sigma_1(x_1) = \cdots = \sigma_n(x_n)$$

and we can define

$$\lambda = f'(R(\sigma_i(x_i)))R'(\sigma_i(x_i))\frac{1}{\beta}.$$

It is easy to verify that $(\hat{x}_1, \ldots, \hat{x}_n, \lambda)$ is a solution of the system of equations above.

This means that ϕ attains its absolute conditional minimum at the point $\hat{x}$. Thus we have obtained a stationary point of the auxiliary function Φ. It is left to show, that $\Phi|\mathcal{K}$ attains its absolute minimum at $\hat{x} := (\hat{x}_1, \ldots, \hat{x}_n) \in \mathcal{K}$. For, then

$$\tilde{A}_n(z) = f^{[-1]}\left(f(\tilde{a}_1(\hat{x}_1)) + \cdots + f(\tilde{a}_n(\hat{x}_n))\right) =$$

$$f^{[-1]}(f(R(\sigma_1(\hat{x}_1))) + \cdots + f(R(\sigma_n(\hat{x}_n)))) =$$

$$f^{[-1]}\left(n \cdot f\left(R\left(\frac{z - B_n}{n\beta}\right)\right)\right).$$

It is easy to see that the partial derivatives of the second order of ϕ at any point $x = (x_1, \ldots, x_n) \in \mathcal{B}$ have the following properties:

$$\partial_{ij}\phi(x) = 0, \ \ if \ i \neq j$$

and for $i = 1, \ldots, n$,

$$\partial_{ii}\phi(x) = \frac{1}{\beta^2}\left[f''(R(\sigma_i(x_i)))(R'(\sigma_i(x_i)))^2 + f'(R(\sigma_i(x_i)))R''(\sigma_i(x_i))\right].$$

Therefore, we have $\partial_{ii}(x) > 0$ for each $i = 1, \ldots, n$. Indeed,

- *$R'(\sigma_i(x_i)) \neq 0$ since R is non-increasing and concave, hence strictly decreasing in a neighbourhood of $\sigma_i(x_i)$);*
- *$f' < 0$, $f'' > 0$ and $R'' < 0$ hold by monotonicity and strict convexity of f and concavity of R.*

The matrix of the derivative of the socond order of ϕ at any point inside $\mathcal{B}$ has only (nonzero) elements in its diagonal, which are positive. Therefore, it is positive definite in $\mathcal{B}$. Now we show that $\phi(\hat{x})$ is the minimum of ϕ in $\mathcal{K}$.

Consider an arbitrary point $x = (x_1, \ldots, x_n) \in \mathcal{K}$. From convexity of $\mathcal{K}$ it follows that the segment $[\hat{x}, x]$ lies within $\mathcal{K}$. By virtue of Lagrange's mean value theorem, there exists $\xi = (\xi_1, \ldots, \xi_n) \in [\hat{x}, x]$ such that

$$\phi(x) = \phi(\hat{x}) + \sum_{i=1}^{n}\partial_i\phi(\hat{x})(x_i - \hat{x}_i) + \sum_{i,j=1}^{n}\partial_{ij}\phi(\xi)(x_i - \hat{x}_i)(x_j - \hat{x}_j)$$

and using the properties $\partial_1\phi(\hat{x}) = \cdots = \partial_n\phi(\hat{x}) = 0$, $\partial_{ij}\phi(\hat{x}) = 0$, if $i \neq j$ and $\partial_{ii}\phi(\hat{x}) > 0$, for each i, we obtain that $\phi(x) > \phi(\hat{x})$. This means that ϕ attains its absolute conditional minimum at the point $\hat{x}$.

Exercise 4.22 *Suppose the unknown mapping to be realized by fuzzy systems can be represented as $y = f(x_1, x_2)$, and we have the following two input/output training pairs*

$$\{(1, 1; 1), (2, 2; 2)\}$$

(i.e. if the input vector is $(1, 1)$ then the desired output is equal to 1, and if the input vector is $(2, 2)$ then the desired output is equal to 2)

For modeling the unknown mapping f, we employ four fuzzy IF-THEN rules

> if x_1 is small   and x_2 is small   then $y = ax_1 - bx_2$
> if x_1 is small   and x_2 is big   then $y = ax_1 + bx_2$
> if x_1 is big   and x_2 is small   then $y = bx_1 + ax_2$
> if x_1 is big   and x_2 is big   then $y = bx_1 - ax_2$

where the membership functions of fuzzy numbers "small" and "big" are given by

$$\text{small}(v) = \begin{cases} 1 - \dfrac{v}{2} & \text{if } 0 \leq v \leq 2 \\ 0 & \text{otherwise} \end{cases}$$

$$\text{big}(v) = \begin{cases} 1 - \dfrac{2 - v}{2} & \text{if } 0 \leq v \leq 2 \\ 0 & \text{otherwise} \end{cases}$$

a and b are the unknown parameters.

The overall system output is computed by Sugeno's reasoning mechanism.

Construct the error functions, $E_1(a, b)$, $E_2(a, b)$ for the first and second training pairs!

Solution 4.22 Let $(1, 1)$ be the input to the fuzzy system. The firing levels of the rules are computed by

$$\alpha_1 = small(1) \wedge small(1) = 0.5,$$
$$\alpha_2 = small(1) \wedge big(1) = 0.5,$$
$$\alpha_3 = big(1) \wedge small(1) = 0.5,$$
$$\alpha_4 = big(1) \wedge big(1) = 0.5,$$

and the output of the system is computed by

$$o_1 = \frac{a + b}{2}$$

We define the measure of error for the first training pattern as

$$E_1 = \frac{1}{2}\left(\frac{a + b}{2} - 1\right)^2$$

and in the case of the second training pattern we get

$$\alpha_1 = small(2) \wedge small(2) = 0,$$
$$\alpha_2 = small(2) \wedge big(2) = 0,$$
$$\alpha_3 = big(2) \wedge small(2) = 0,$$
$$\alpha_4 = big(2) \wedge big(2) = 1,$$

and the output of the system is computed by

$$o_2 = 2b - 2a$$

The measure of error for the second training pattern is as

$$E_2 = \frac{1}{2}\left(2b - 2a - 2\right)^2$$

Exercise 4.23 *Suppose that we have a classification problem shown in Fig. 4.14. Find a fuzzy partition (with 3×3 terms) which correctly classifies all but two patterns !*

Construct the fuzzy rules derived from this fuzzy partition ! Optimize the rule base (minimize the number of the rules)!

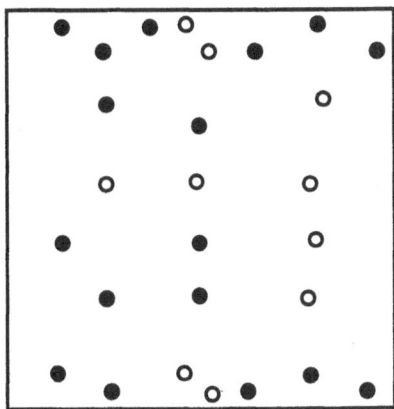

Fig. 4.14. Open and closed circles represent the given pattens from Class 1 and Class 2, respectively.

Solution 4.23 *We can classify the patterns with the following 7 fuzzy IF-THEN rules*

$\Re_1:$ If x_1 is small *then Class* 2

$\Re_2:$ If x_1 is medium *and* x_2 is small *then Class* 1

$\Re_3:$ If x_1 is medium *and* x_2 is medium *then Class* 2

$\Re_4:$ If x_1 is medium *and* x_2 is big *then Class* 1

$\Re_5:$ If x_1 is big *and* x_2 is small *then Class* 2

$\Re_6:$ If x_1 is big *and* x_2 is medium *then Class* 1

$\Re_7:$ If x_1 is big *and* x_2 is big *then Class* 2

Or, equivalently, with the following 5 fuzzy IF-THEN rules

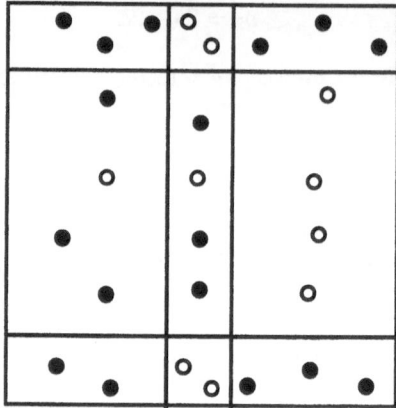

Fig. 4.15. A possible fuzzy partition.

$\Re_1$:   If x_1 is small *then Class* 2

$\Re_2$:   If x_1 is medium *and* x_2 *is* not medium *then Class* 1

$\Re_3$:   If x_1 is medium *and* x_2 *is* medium *then Class* 2

$\Re_4$:   If x_1 is big *and* x_2 *is* not medium *then Class* 2

$\Re_5$:   If x_1 is big *and* x_2 *is* medium *then Class* 1

Exercise 4.24 *Suppose the unknown nonlinear mapping to be realized by fuzzy systems can be represented as*

$$y^k = f(x^k) = f(x_1^k, \ldots, x_n^k) \tag{4.10}$$

for $k = 1, \ldots, K$, i.e. we have the following training set

$$\{(x^1, y^1), \ldots, (x^K, y^K)\}$$

For modeling the unknown mapping in (4.10), we employ three simplified fuzzy IF-THEN rules of the following type

$$\text{if } x \text{ is } small \text{ then} \qquad y = z_1$$
$$\text{if } x \text{ is } medium \text{ then} \quad y = z_2$$
$$\text{if } x \text{ is } big \text{ then} \qquad y = z_3$$

where the linguistic terms $A_1 = "small"$, $A_2 = "medium"$ and $A_3 = "big"$ are of triangular form with membership functions (see Figure 4.16)

$$A_1(v) = \begin{cases} 1 & \text{if } v \leq c_1 \\ \dfrac{c_2 - x}{c_2 - c_1} & \text{if } c_1 \leq v \leq c_2 \\ 0 & \text{otherwise} \end{cases}$$

$$A_2(u) = \begin{cases} \dfrac{x - c_1}{c_2 - c_1} & \text{if } c_1 \le u \le c_2 \\[2mm] \dfrac{c_3 - x}{c_3 - c_2} & \text{if } c_2 \le u \le c_3 \\[2mm] 0 & \text{otherwise} \end{cases}$$

$$A_3(u) = \begin{cases} 1 & \text{if } u \ge c_3 \\[2mm] \dfrac{x - c_2}{c_3 - c_2} & \text{if } c_2 \le u \le c_3 \\[2mm] 0 & \text{otherwise} \end{cases}$$

Derive the steepest descent method for tuning the premise parameters $\{c_1, c_2, c_3\}$ and the consequent parameters $\{y_1, y_2, y_3\}$.

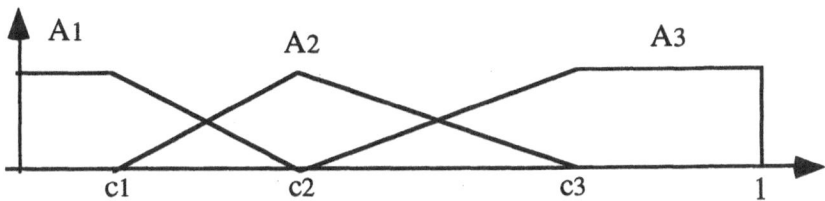

Fig. 4.16. Initial fuzzy partition with three linguistic terms.

Solution 4.24 *Let x be the input to the fuzzy system. The firing levels of the rules are computed by*

$$\alpha_1 = A_1(x), \quad \alpha_2 = A_2(x), \quad \alpha_3 = A_3(x),$$

and the output of the system is computed by

$$o = \frac{\alpha_1 z_1 + \alpha_2 z_2 + \alpha_3 z_3}{\alpha_1 + \alpha_2 + \alpha_3} = \frac{A_1(x) z_1 + A_2(x) z_2 + A_3(x) z_3}{A_1(x) + A_2(x) + A_3(x)}$$

$$= A_1(x) z_1 + A_2(x) z_2 + A_3(x) z_3$$

where we have used the identity

$$A_1(x) + A_2(x) + A_3(x) = 1$$

for all $x \in [0, 1]$. We define the measure of error for the k-th training pattern as usually

$$E_k = E_k(c_1, c_2, c_3, z_1, z_2, z_3) = \frac{1}{2}(o^k(c_1, c_2, c_3, z_1, z_2, z_3) - y^k)^2$$

where o^k is the computed output from the fuzzy system corresponding to the input pattern x^k and y^k is the desired output, $k = 1, \ldots, K$.

The steepest descent method is used to learn z_i in the consequent part of the i-th fuzzy rule. That is,

$$z_1(t+1) = z_1(t) - \eta \frac{\partial E_k}{\partial z_1} = z_1(t) - \eta(o^k - y^k)A_1(x^k)$$

$$z_2(t+1) = z_2(t) - \eta \frac{\partial E_k}{\partial z_2} = z_2(t) - \eta(o^k - y^k)A_2(x^k)$$

$$z_3(t+1) = z_3(t) - \eta \frac{\partial E_k}{\partial z_3} = z_3(t) - \eta(o^k - y^k)A_3(x^k)$$

where x^k is the input to the system, $\eta > 0$ is the learning constant and t indexes the number of the adjustments of z_i.

In a similar manner we can tune the centers of A_1, A_2 and A_3.

$$c_1(t+1) = c_1(t) - \eta \frac{\partial E_k}{\partial c_1},$$

$$c_2(t+1) = c_2(t) - \eta \frac{\partial E_k}{\partial c_2},$$

$$c_3(t+1) = c_3(t) - \eta \frac{\partial E_k}{\partial c_3},$$

where $\eta > 0$ is the learning constant and t indexes the number of the adjustments of the parameters.

The partial derivative of the error function E_k with respect to c_1 can be written as

$$\frac{\partial E_k}{\partial c_1} = (o^k - y^k)\frac{\partial o^k}{\partial c_1} = (o^k - y^k)\frac{(x - c_1)}{(c_2 - c_1)^2}(z_1 - z_2)$$

if $c_1 \leq x^k \leq c_2$, and zero otherwise.

It should be noted that the adjustments of a center can not be done independently of other centers, because the inequality

$$0 \leq c_1(t+1) < c_2(t+1) < c_3(t+1) \leq 1$$

must hold for all t.

Index